Module Theory

Module Theory
An Approach to Linear Algebra

Second Edition

T. S. BLYTH

*Professor of Pure Mathematics,
University of St Andrews*

CLARENDON PRESS · OXFORD
1990

Oxford University Press, Walton Street, Oxford OX2 6DP
Oxford New York Toronto
Delhi Bombay Calcutta Madras Karachi
Petaling Jaya Singapore Hong Kong Tokyo
Nairobi Dar es Salaam Cape Town
Melbourne Auckland
and associated companies in
Berlin Ibadan

Oxford is a trade mark of Oxford University Press

Published in the United States
by Oxford University Press, New York

© T. S. Blyth 1990

All rights reserved. No part of this publication may be reproduced,
stored in a retrieval system, or transmitted, in any form or by any means,
electronic, mechanical, photocopying, recording, or otherwise, without
the prior permission of Oxford University Press

This book is sold subject to the condition that it shall not, by way
of trade or otherwise, be lent, re-sold, hired out or otherwise circulated
without the publisher's prior consent in any form of binding or cover
other than that in which it is published and without a similar condition
including this condition being imposed on the subsequent purchaser

British Library Cataloguing in Publication Data
Blyth, T. S. (Thomas Scott)
Module theory.—2nd ed.
1. Linear algebra. Modules
I. Title
512.4
ISBN 0-19-853293-8
ISBN 0-19-853389-6 (pbk.)

Library of Congress Cataloging in Publication Data
Blyth, T. S. (Thomas Scott)
Module theory: an approach to linear algebra / T. S. Blyth.—2nd
ed.
p. cm.
1. Modules (Algebra) 2. Vector spaces. 3.Algebras, Linear.
I. Title.
QA247.B57 1990 512'.4—dc20 89-72127
ISBN 0-19-853293-8
ISBN 0-19-853389-6 (pbk.)

Printed in Great Britain by
Bookcraft (Bath) Ltd
Midsomer Norton, Avon

Preface

Many branches of algebra are linked by the theory of modules. Since the notion of a module is obtained essentially by a modest generalisation of that of a vector space, it is not surprising that it plays an important role in the theory of linear algebra. Modules are also of great importance in the higher reaches of group theory and ring theory, and are fundamental to the study of advanced topics such as homological algebra, category theory, and algebraic topology. The aim of this text is to develop the basic properties of modules and to show their importance, mainly in the theory of linear algebra.

The first eleven sections can easily be used as a self-contained course for first-year honours students. Here we cover all the basic material on modules and vector spaces required for embarkation on advanced courses. Concerning the prerequisite algebraic background for this, we mention that any standard course on groups, rings, and fields will suffice. Although we have kept the discussion as self-contained as possible, there are places where references to standard results are unavoidable; readers who are unfamiliar with such results should consult a standard text on abstract algebra. The remainder of the text can be used, with a few omissions to suit any particular instructor's objectives, as an advanced course. In this, we develop the foundations of multilinear and exterior algebra. In particular, we show how exterior powers lead to determinants. In this edition we include some results of a ring-theoretic nature that are directly related to modules and linear algebra. In particular, we establish the celebrated Wedderburn–Artin theorem that every simple ring is isomorphic to the ring of endomorphisms of a finite-dimensional module over a division ring. Finally, we discuss in detail the structure of finitely generated modules over a principal ideal domain, and apply the fundamental structure theorems to obtain, on the one hand, the structure of all finitely generated abelian groups and, on the other, important decomposition theorems for vector spaces which lead naturally to various canonical forms for matrices.

At the end of each section we have supplied a number of exercises. These provide ample opportunity to consolidate the results in the body of the text, and we include lots of hints to help the reader gain the satisfaction of solving problems.

Although this second edition is algebraically larger than the first edition, it is geometrically smaller. The reason is simple: the first edition was produced at a time when rampant inflation had caused typesetting to become very expensive and, regrettably, publishers were choosing to produce texts from camera-ready material (the synonym of the day for typescript). Nowadays, texts are still produced from camera-ready material but there is an enormous difference in the quality. The intervening years have seen the march of technology: typesetting by computer has arrived and, more importantly, can be done by the authors themselves. This is the case with the present edition. It was set entirely by the author, without scissors, paste, or any cartographic assistance, using the mathematical typsetting system TeX developed by Professor Donald Knuth, and the document preparation system LaTeX developed by Dr Leslie Lamport. To be more precise, it was set on a Macintosh II computer using the package MacTeX developed by FTL systems Inc. of Toronto. We record here our gratitude to Lian Zerafa, President of FTL, for making this wonderful system available.

St Andrews
August 1989   T.S.B.

Contents

1. Modules, vector spaces, and algebras — 1
2. Submodules; intersections and sums — 8
3. Morphisms; exact sequences — 15
4. Quotient modules; basic isomorphism theorems — 32
5. Chain conditions; Jordan-Hölder towers — 45
6. Products and coproducts — 56
7. Free modules; bases — 77
8. Groups of morphisms; projective modules — 96
9. Duality; transposition — 115
10. Matrices; linear equations — 126
11. Inner product spaces — 147
12. Injective modules — 163
13. Simple and semisimple modules — 172
14. The Jacobson radical — 185
15. Tensor products; flat modules; regular rings — 194
16. Tensor products; tensor algebras — 217
17. Exterior algebras, determinants — 235
18. Modules over a principal ideal domain; finitely generated abelian groups — 263
19. Vector space decomposition theorems; canonical forms under similarity — 289
20. Diagonalisation; normal transformations — 318

Index — 358

1 Modules, vector spaces, and algebras

In this text our objective will be to develop the foundations of the branch of mathematics that is called *linear algebra*. From the various elementary courses that readers have followed, they will recognise this as essentially the study of vector spaces and linear transformations, notions that have applications in several areas of mathematics.

Now in most elementary introductions to linear algebra the notion of a determinant is defined for square matrices, and it is assumed that the elements of the matrices in question lie in some *field* (usually the field $\mathbb{R}$ of real numbers). But, come the consideration of eigenvalues (or latent roots), the matrix whose determinant has to be found is of the form

$$\begin{bmatrix} x_{11} - \lambda & x_{12} & \cdots & x_{1n} \\ x_{21} & x_{22} - \lambda & \cdots & x_{2n} \\ \vdots & \vdots & \ddots & \vdots \\ x_{n1} & x_{n2} & \cdots & x_{nn} - \lambda \end{bmatrix}$$

and therefore has its entries in a polynomial *ring*. This prompts the question of whether the various properties of determinants should not really be developed in a more general setting, and leads to the wider question of whether the scalars in the definition of a vector space should not be restricted to lie in a field but should more generally belong to a ring (which, as in the case of a polynomial ring, may be required at some stage to be commutative).

It turns out that the modest generalisation so suggested is of enormous importance and leads to what is arguably the most important structure in the whole of algebra, namely that of a *module*. The importance of this notion lies in a greatly extended domain of application, including the higher reaches of group theory and ring theory, and such areas as homological algebra, category theory, algebraic topology, etc.

Before giving a formal definition of a module, we ask the reader to recall the following elementary notions. If E is a non-empty set then a *law of composition* on E is a mapping $f : E \times E \to E$. Given $(x,y) \in E \times E$ it is common practice to write $f(x,y)$ as $x+y$, or xy, except when it might cause confusion to use such additive or multiplicative notations, in which case notations such as $x \star y$, $x\top y$, $x \odot y$, etc., are useful. A set on which there is defined a law of composition that is associative is called a *semigroup*. By a *group* we mean a semigroup with an identity element in which every element has an inverse. By an *abelian group* we mean a group in which the law of composition is commutative. By a *ring* we mean a set E endowed with two laws of composition, these being traditionally denoted by $(x,y) \mapsto x+y$ and $(x,y) \mapsto xy$, such that

(1) E is an abelian group under addition;
(2) E is a semigroup under multiplication;
(3) $(\forall x,y,z \in E)$ $x(y+z) = xy + xz$, $(y+z)x = yx + zx$.

A ring R is said to be *unitary* if it has a multiplicative identity element, such an element being written 1_R. A ring is *commutative* if the multiplication is commutative. By a *division ring* we mean a unitary ring in which every non-zero element has a multiplicative inverse (so that the non-zero elements form a group under multiplication). By a *field* we mean a commutative division ring.

In what follows we shall have occasion to consider mappings of the form $f : F \times E \to E$ where F and E are non-empty sets. Such a mapping will be denoted by $(\lambda, x) \mapsto \lambda x$, and called a *left action* on E by elements of F. Although here λx is simply the juxtaposition of $\lambda \in F$ and $x \in E$ with λ on the left, it is often also called *left multiplication of elements of E by elements of F*. In this context the elements of F are often called *scalars*. In a similar way we can define a *right action* on E by elements of F to be a mapping $f : E \times F \to E$ described by $(x, \lambda) \mapsto x\lambda$.

- It should be noted that a particular case of an action is obtained by taking $F = E$, in which case we obtain a mapping $f : E \times E \to E$, i.e. a law of composition on E.

Definition Let R be a unitary ring. By an *R-module*, or a *module over R*, we shall mean an additive abelian group M together with a left action $R \times M \to M$, described by $(\lambda, x) \mapsto \lambda x$, such that

(1) $(\forall \lambda \in R)(\forall x, y \in M)$ $\lambda(x+y) = \lambda x + \lambda y$;
(2) $(\forall \lambda, \mu \in R)(\forall x \in M)$ $(\lambda + \mu)x = \lambda x + \mu x$;
(3) $(\forall \lambda, \mu \in R)(\forall x \in M)$ $\lambda(\mu x) = (\lambda \mu)x$;
(4) $(\forall x \in M)$ $1_R x = x$.

A module over a field is called a *vector space*.

- An R-module, as we have defined it, is often called a *left R-module*. The reason for this is that the scalars are written on the left. By writing $x\lambda$ instead of λx throughout and altering (3) and (4) to

(3') $(\forall x \in M)(\forall \lambda, \mu \in R)$ $(x\lambda)\mu = x(\lambda\mu)$;
(4') $(\forall x \in M)$ $x 1_R = x$,

we obtain what is called a *right R-module*, the external law in this case being a right action on M. In what follows we shall make the convention that the term *R-module* will always mean a *left* R-module, and when we have occasion to talk about a *right* R-module we shall use that adjective.

- Some authors prefer not to include the identity element 1_R in the above definition. What we have called an *R-module* they would call a *unitary R-module*.

If M is an R-module then we shall denote the additive identity element of M by 0_M, and that of R by 0_R. The following elementary properties will be used without reference in what follows.

Theorem 1.1 *Let M be an R-module. Then*

(1) $(\forall \lambda \in R)\ \lambda 0_M = 0_M$;
(2) $(\forall x \in M)\ 0_R x = 0_M$;
(3) $(\forall \lambda \in R)(\forall x \in M)\ \lambda(-x) = -(\lambda x) = (-\lambda) x$.

Moreover, when R is a division ring,

(4) $\lambda x = 0_M$ *implies* $\lambda = 0_R$ *or* $x = 0_M$.

Proof (1) We have $\lambda 0_M = \lambda(0_M + 0_M) = \lambda 0_M + \lambda 0_M$ whence it follows that $\lambda 0_M = 0_M$.

(2) We have $0_R x = (0_R + 0_R) x = 0_R x + 0_R x$, whence $0_R x = 0_M$.

(3) By (1) we have $0_M = \lambda 0_M = \lambda[x + (-x)] = \lambda(x) + \lambda(-x)$, whence $\lambda(-x) = -\lambda x$; and by (2) we have $0_M = 0_R x = [\lambda + (-\lambda)] x = \lambda x + (-\lambda) x$, whence $(-\lambda) x = -\lambda x$.

(4) Suppose now that R is a division ring and that $\lambda x = 0_M$ with $\lambda \neq 0_R$. Then using the fact that λ has a multiplicative inverse we have

$$x = 1_R x = (\lambda^{-1} \lambda) x = \lambda^{-1}(\lambda x) = \lambda^{-1} 0_M = 0_M.\ \Diamond$$

Example 1.1 Every unitary ring R is an R-module; the action $R \times R \to R$ is the multiplication in R. Likewise, any field F is an F-vector space.

Example 1.2 Every additive abelian group M can be regarded as a $\mathbb{Z}$-module; here the action $\mathbb{Z} \times M \to M$ is given by $(m, x) \mapsto mx$ where

$$mx = \begin{cases} \underbrace{x + x + \cdots + x}_{m} & \text{if } m > 0; \\ 0 & \text{if } m = 0; \\ -|m|x & \text{if } m < 0. \end{cases}$$

Example 1.3 The field $\mathbb{C}$ of complex numbers can be regarded as an $\mathbb{R}$-vector space; the action $\mathbb{R} \times \mathbb{C} \to \mathbb{C}$ is described by

$$(\lambda, x+iy) \mapsto \lambda(x+iy) = \lambda x + i\lambda y.$$

More generally, if R is a unitary ring and S is a subring of R that contains 1_R then R can be considered as an S-module; here the action is given by $(s, r) \mapsto sr$.

Example 1.4 If R is a unitary ring and n is a positive integer consider the abelian group R^n of all n-tuples of elements of R under the component-wise addition

$$(x_1, \ldots, x_n) + (y_1, \ldots, y_n) = (x_1 + y_1, \ldots, x_n + y_n).$$

Define a left action $R \times R^n \to R^n$ in the obvious way, namely by

$$r(x_1, \ldots, x_n) = (rx_1, \ldots, rx_n).$$

Then R^n becomes an R-module. Similarly, if F is a field then F^n is an F-vector space.

Example 1.5 Let R be a unitary ring and let $R^{\mathbb{N}}$ denote the set of all mappings $f : \mathbb{N} \to R$ (i.e. the set of all sequences of elements of R). Endow $R^{\mathbb{N}}$ with the obvious addition, namely for $f, g \in R^{\mathbb{N}}$ define $f + g$ by the prescription

$$(f+g)(n) = f(n) + g(n).$$

Clearly, $R^{\mathbb{N}}$ forms an abelian group under this law of composition. Now define an action $R \times R^{\mathbb{N}} \to R^{\mathbb{N}}$ by $(r, f) \mapsto rf$ where $rf \in R^{\mathbb{N}}$ is given by the prescription

$$(rf)(n) = r f(n).$$

This then makes $R^{\mathbb{N}}$ into an R-module.

Each of the above examples can be made into a right module in the obvious way.

Definition Let R be a commutative unitary ring. An *R-algebra* (or *algebra over R*) is an R-module A together with a law of composition $A \times A \to A$, described by $(x, y) \mapsto xy$ and called *multiplication*, which is distributive over addition and is linked to the action of R on A by the identities

$$(\forall \lambda \in R)(\forall x, y \in A) \quad \lambda(xy) = (\lambda x)y = x(\lambda y).$$

Modules, vector spaces, and algebras

By imposing conditions on the multiplication in the above definition we obtain various types of algebra. For example, if the multiplication is associative then A is called an *associative algebra* (note that in this case A is a ring under its laws of addition and multiplication); if the multiplication is commutative then A is called a *commutative algebra*; if there is a multiplicative identity element in A then A is said to be *unitary*. A unitary associative algebra in which every non-zero element has an inverse is called a *division algebra*.

Example 1.6 $\mathbb{C}$ is a division algebra over $\mathbb{R}$.

Example 1.7 Let R be a commutative unitary ring and consider the R-module $R^{\mathbb{N}}$ of Example 1.5. Given $f, g \in R^{\mathbb{N}}$, define the product map $fg : \mathbb{N} \to R$ by the prescription

$$(fg)(n) = \sum_{i=1}^{n} f(i)g(n-i).$$

It is readily verified that the law of composition described by $(f, g) \mapsto fg$ makes $R^{\mathbb{N}}$ into an R-algebra. This R-algebra is called the *algebra of formal power series with coefficients in R*.

The reason for this traditional terminology is as follows. Let $t \in R^{\mathbb{N}}$ be given by the prescription

$$t(n) = \begin{cases} 1 & \text{if } n = 1; \\ 0 & \text{otherwise.} \end{cases}$$

Then for every positive integer m the m-fold composite map

$$t^m = \underbrace{t \circ t \circ \cdots \circ t}_{m}$$

is given by

$$t^m(n) = \begin{cases} 1 & \text{if } n = m; \\ 0 & \text{otherwise.} \end{cases}$$

Consider now (without worrying how to imagine the sum of an infinite number of elements of $R^{\mathbb{N}}$ or even questioning the lack of any notion of convergence) the *formal power series* associated with $f \in R^{\mathbb{N}}$ given by

$$\vartheta = f(0)t^0 + f(1)t^1 + f(2)t^2 + \cdots + f(m)t^m + \cdots,$$

where $t^0 = \text{id}_R$, the identity map on R. Since, as is readily seen,

$$(\forall n \in \mathbb{N}) \quad \vartheta(n) = f(n),$$

it is often said that f can be represented symbolically by the above formal power series.

Example 1.8 If R is a unitary ring then the set $\text{Mat}_{n \times n}(R)$ of $n \times n$ matrices over R is a unitary associative R-algebra.

EXERCISES

1.1 Let M be an abelian group and let $\text{End}\, M$ be the set of all endomorphisms on M, i.e. the set of all group morphisms $f : M \to M$. Show that $\text{End}\, M$ is an abelian group under the law of composition $(f,g) \mapsto f + g$ where

$$(\forall x \in M) \quad (f+g)(x) = f(x) + g(x).$$

Show also that

1. $(\text{End}\, M, +, \circ)$ is a unitary ring;
2. M is an $\text{End}\, M$-module under the action $\text{End}\, M \times M \to M$ given by $(f, m) \mapsto f \cdot m = f(m)$;
3. if R is a unitary ring and $\mu : R \to \text{End}\, M$ is a ring morphism such that $\mu(1_R) = \text{id}_M$, then M is an R-module under the action $R \times M \to M$ given by $(\lambda, m) \mapsto \lambda m = [\mu(\lambda)](m)$.

1.2 Let R be a unitary ring and M an abelian group. Prove that M is an R-module if and only if there is a 1-preserving ring morphism $f : R \to \text{End}\, M$.

[Hint. $\Rightarrow$: For every $r \in R$ define $f_r : M \to M$ by $f_r(m) = rm$. Show that $f_r \in \text{End}\, M$ and let f be given by $r \mapsto f_r$.

$\Leftarrow$: Use Exercise 1.1(3).]

1.3 Let G be a finite abelian group with $|G| = m$. Show that if $n, t \in \mathbb{Z}$ then

$$n \equiv t \pmod{m} \Rightarrow (\forall g \in G) \quad ng = tg.$$

Deduce that G is a $\mathbb{Z}/m\mathbb{Z}$-module under the action $\mathbb{Z}/m\mathbb{Z} \times G \to G$ given by $(n + m\mathbb{Z}, g) \mapsto ng$. Conclude that every finite abelian group whose order is a prime p can be regarded as a vector space over a field of p elements.

1.4 Let S be a non-empty set and R a unitary ring. If F is the set of all mappings $f : S \to R$ such that $f(s) = 0$ for almost all $s \in S$, i.e. all but a finite number of $s \in S$, show that F is an R-module under the addition defined by

$$(\forall s \in S) \quad (f+g)(s) = f(s) + g(s)$$

and the action defined by

$$(\forall \lambda \in R)(\forall s \in S) \quad (\lambda f)(s) = \lambda f(s).$$

1.5 If R is a commutative unitary ring show that the set $P_n(R)$ of all polynomials over R of degree less than or equal to n is an R-module. Show also that the set $P(R)$ of all polynomials over R is a unitary associative R-algebra.

1.6 If A is a unitary ring define its *centre* to be

$$\operatorname{Cen} A = \{x \in A \, ; \, (\forall y \in A) \; xy = yx\}.$$

Show that $\operatorname{Cen} A$ is a unitary ring. If R is a commutative unitary ring, prove that A is a unitary associative R-algebra if and only if there is a 1-preserving ring morphism $\vartheta : R \to \operatorname{Cen} A$.

[*Hint.* $\Rightarrow$: Denoting the action of R on A by $(r, a) \mapsto r \cdot a$, define ϑ by $\vartheta(r) = r \cdot 1_A$.
$\Leftarrow$: Define an action by $(r, a) \mapsto r \cdot a = \vartheta(r)a$.]

1.7 Let S and R be unitary rings and let $f : S \to R$ be a 1-preserving ring morphism. If M is an R-module prove that M can be regarded as an S-module under the action $S \times M \to M$ given by $(s, x) \mapsto f(s)x$.

1.8 Show that if V is a vector space over a field F then the set T of linear transformations $f : V \to V$ is a unitary associative F-algebra. If $F[X]$ denotes the ring of polynomials over F and α is a fixed element of T, show that V can be made into an $F[X]$-module by the action $F[X] \times V \to V$ given by

$$(p, x) \mapsto p \cdot_\alpha x = [p(\alpha)](x).$$

2 Submodules; intersections and sums

If S is a non-empty subset of an additive group G then S is said to be a *stable subset* of G, or to be *closed under the operation of G*, if

$$(\forall x, y \in S) \quad x + y \in S.$$

Equivalently, S is a stable subset of G if the restriction to $S \times S$ of the law of composition on G induces a law of composition on S, these laws being denoted by the same symbol + without confusion. In this case it is clear that S is a semigroup. By a subgroup of G we mean a non-empty subset that is stable and which is also a group with respect to the induced law of composition. The reader will recall that a non-empty subset H of a group G is a subgroup of G if and only if

$$(\forall x, y \in H) \quad x - y \in H.$$

Definition By a *submodule* of an R-module M we mean a subgroup N of M that is stable under the action of R on M, in the sense that if $x \in N$ and $\lambda \in R$ then $\lambda x \in N$.

It is clear that a non-empty subset N of an R-module M is a submodule of M if and only if

$$(\forall x, y \in N)(\forall \lambda \in R) \quad x - y \in N \text{ and } \lambda x \in N. \tag{1}$$

These conditions can be combined into the single condition

$$(\forall x, y \in N)(\forall \lambda, \mu \in R) \quad \lambda x + \mu y \in N. \tag{2}$$

To see this, observe that if (1) holds then $\lambda x \in N$ and $-\mu y = (-\mu)y \in N$, whence $\lambda x + \mu y = \lambda x - (-\mu y) \in N$. Conversely, if (2) holds then taking $\lambda = 1_R$ and $\mu = -1_R$ we obtain $x - y \in N$; and taking $\mu = 0_R$ we obtain $\lambda x \in N$.

The notion of a *subspace* of a vector space is defined similarly. Likewise, we say that a non-empty subset B of an R-algebra A is a *subalgebra* of A if

$$(\forall x, y \in B)(\forall \lambda \in R) \quad x - y \in B, xy \in B, \lambda x \in B.$$

Example 2.1 Let R be a unitary ring considered as an R-module (Example 1.1). The submodules of R are precisely the left ideals of R. Likewise, if we consider R as a right R-module the its submodules are precisely its right ideals.

- Although we agree to omit the adjective 'left' when talking about modules, it is essential (except in the case where R is commutative) to retain this adjective when referring to left ideals as submodules of R.

Example 2.2 Borrowing some notions from analysis, let C be the set of continuous functions $f : [a, b] \to \mathbb{R}$. Clearly, C can be given the structure of an $\mathbb{R}$-vector space (essentially as in Example 1.5). The subset D that consists of the differentiable functions on $[a, b]$ is then a subspace of C; for, if $f, g \in D$ then, as is shown in analysis,

$$(\forall \lambda, \mu \in \mathbb{R}) \qquad \lambda f + \mu g \in D.$$

Example 2.3 If G is an abelian group then the submodules of the $\mathbb{Z}$-module G are simply the subgroups of G.

Example 2.4 The vector space C of Example 2.2 becomes an $\mathbb{R}$-algebra when we define a multiplication on C by $(f, g) \mapsto fg$ where

$$(\forall x \in [a, b]) \qquad (fg)(x) = f(x)g(x).$$

It is readily verified that the subspace D is a subalgebra of C.

Theorem 2.1 *If $(M_i)_{i \in I}$ is a family of submodules of an R-module M then $\bigcap_{i \in I} M_i$ is a submodule of M.*

Proof We observe first that $\bigcap_{i \in I} M_i \neq \emptyset$ since every submodule, being a subgroup, contains the identity element 0_M. Now, since each M_i is a submodule, we have

$$x, y \in \bigcap_{i \in I} M_i \Rightarrow (\forall i \in I)\ x, y \in M_i$$
$$\Rightarrow (\forall i \in I)\ x - y \in M_i$$
$$\Rightarrow x - y \in \bigcap_{i \in I} M_i$$

and

$$x \in \bigcap_{i \in I} M_i, \lambda \in R \Rightarrow (\forall i \in I)\ \lambda x \in M_i \Rightarrow \lambda x \in \bigcap_{i \in I} M_i.$$

Consequently, $\bigcap_{i \in I} M_i$ is a submodule of M. ◊

The above result leads to the following observation. Suppose that S is a subset (possibly empty) of an R-module M and consider the collection of all the submodules of M that contain S. By Theorem 2.1, the intersection of this collection is a submodule of M, and it clearly contains S. It is thus the smallest submodule of M to contain S. We call this the *submodule generated by S* and denote it by $\langle S \rangle$. We shall now give an explicit description of this submodule. For this purpose we require the following notion.

Definition Let M be an R-module and let S be a non-empty subset of M. Then $x \in M$ is a *linear combination of elements of S* if there exist elements $x_1, \ldots, x_n$ in S and scalars $\lambda_1, \ldots, \lambda_n$ in R such that

$$x = \sum_{i=1}^{n} \lambda_i x_i = \lambda_1 x_1 + \ldots + \lambda_n x_n.$$

We denote the set of all linear combinations of elements of S by $\mathrm{LC}(S)$.

Theorem 2.2 *Let S be a subset of the R-module M. Then*

$$\langle S \rangle = \begin{cases} \{0_M\} & \text{if } S = \emptyset; \\ \mathrm{LC}(S) & \text{if } S \neq \emptyset. \end{cases}$$

Proof It is clear that if $S = \emptyset$ then the smallest submodule that contains S is the smallest submodule of M, namely the zero submodule $\{0_M\}$. Suppose then that $S \neq \emptyset$. It is clear that $\mathrm{LC}(S)$ is a submodule of M. Moreover, $S \subseteq \mathrm{LC}(S)$ since for every $x \in S$ we have $x = 1_R x \in \mathrm{LC}(S)$. As $\langle S \rangle$ is, by definition, the smallest submodule to contain S, we therefore have $\langle S \rangle \subseteq \mathrm{LC}(S)$. On the other hand, every linear combination of elements of S clearly belongs to every submodule that contains S and so we have the reverse inclusion $\mathrm{LC}(S) \subseteq \langle S \rangle$, whence the result follows. ◊

Definition We say that an R-module M is *generated* by the subset S, or that S is a *set of generators* of M, when $\langle S \rangle = M$. By a *finitely generated R-module* we mean an R-module which has a finite set of generators.

One of the main theorems that we shall eventually establish concerns the structure of finitely generated R-modules where R is a particularly important type of ring (in fact, a principal ideal domain). As we shall see in due course, this structure theorem has far-reaching consequences.

Suppose now that $(M_i)_{i \in I}$ is a family of submodules of an R-module M and consider the submodule of M that is generated by $\bigcup_{i \in I} M_i$. This is the smallest submodule of M that contains every M_i. By abuse of

Submodules; intersections and sums

language it is often referred to as the *submodule generated by the family* $(M_i)_{i \in I}$. It can be characterised in the following way.

Theorem 2.3 *Let $(M_i)_{i \in I}$ be a family of submodules of an R-module M. If $\mathbb{P}^{\star}(I)$ denotes the set of all non-empty finite subsets of I then the submodule generated by $\bigcup_{i \in I} M_i$ consists of all finite sums of the form $\sum_{j \in J} m_j$ where $J \in \mathbb{P}^{\star}(I)$ and $m_j \in M_j$.*

Proof A linear combination of elements of $\bigcup_{i \in I} M_i$ is precisely a sum of the form $\sum_{j \in J} m_j$ for some $J \in \mathbb{P}^{\star}(I)$. The result is therefore an immediate consequence of Theorem 2.2. $\Diamond$

Because of Theorem 2.3, we call the submodule generated by the family $(M_i)_{i \in I}$ the *sum of the family* and denote it by $\sum_{i \in I} M_i$. In the case where the index set I is finite, say $I = \{1, \ldots, n\}$, we often write $\sum_{i \in I} M_i$ as $\sum_{i=1}^{n} M_i$ or as $M_1 + \ldots + M_n$. With this notation we have the following immediate consequences of the above.

Corollary 1 [Commutativity of $\sum$] *If $\sigma : I \to I$ is a bijection then*

$$\sum_{i \in I} M_i = \sum_{i \in I} M_{\sigma(i)}. \quad \Diamond$$

Corollary 2 [Associativity of $\sum$] *If $(I_k)_{k \in A}$ is a family of non-empty subsets of I with $I = \bigcup_{k \in A} I_k$ then*

$$\sum_{i \in I} M_i = \sum_{k \in A} \left(\sum_{i \in I_k} M_i \right).$$

Proof A typical element of the right-hand side is $\sum_{k \in J} \left(\sum_{i \in J_k} m_i \right)$ with $J_k \in \mathbb{P}^{\star}(I_k)$ and $J \in \mathbb{P}^{\star}(A)$. By associativity of addition in M this can be written as $\sum_{i \in K} m_i$ where $K = \bigcup_{k \in J} J_k \in \mathbb{P}^{\star}(I)$. Thus the right-hand side is contained in the left-hand side. As for the converse inclusion, a typical element of the left-hand side is $\sum_{i \in J} m_i$ where $J \in \mathbb{P}^{\star}(I)$. Now $J = J \cap I = \bigcup_{k \in A} (J \cap I_k)$ so that if we define $J_k = J \cap I_k$ we have $J_k \in \mathbb{P}^{\star}(I_k)$ and, by the associativity of addition in M, $\sum_{i \in J} m_i = \sum_{k \in B} \left(\sum_{i \in J_k} m_i \right)$ where $B \in \mathbb{P}^{\star}(A)$. Thus the left-hand side is contained in the right-hand side. $\Diamond$

Corollary 3 $(\forall i \in I) \quad \sum_{i \in I} M_i = M_i + \sum_{j \neq i} M_j.$

Proof Take $A = \{1,2\}, I_1 = \{i\}$ and $I_2 = I \setminus I_1$ in the above. ◊

- Note that $\bigcup_{i \in I} M_i \neq \sum_{i \in I} M_i$ in general, for $\bigcup_{i \in I} M_i$ need not be a submodule. For example, take $I = \{1,2\}$ and let M_1, M_2 be the subspaces of the vector space $\mathbb{R}^2$ given by $M_1 = \{(x,0) \; ; \; x \in \mathbb{R}\}$ and $M_2 = \{(0,y) \; ; \; y \in \mathbb{R}\}$. We have $M_1 + M_2 = \mathbb{R}^2$ whereas $M_1 \cup M_2 \subset \mathbb{R}^2$.

Suppose now that M is an R-module and that A, B are submodules of M. We know that $A + B$ is the smallest submodule of M that contains both A and B, and that $A \cap B$ is the largest submodule contained in both A and B. The set of submodules of M, ordered by set inclusion, is therefore such that every two-element subset $\{A, B\}$ has a supremum (namely $A + B$) and an infimum (namely $A \cap B$). Put another way, the set of submodules of M, ordered by set inclusion, forms a *lattice*. An important property of this lattice is that it is *modular*, by which we mean the following.

Theorem 2.4 [Modular law] *If M is an R-module and if A, B, C are submodules of M with $C \subseteq A$ then*

$$A \cap (B + C) = (A \cap B) + C.$$

Proof Since $C \subseteq A$ we have $A + C = A$. Now $(A \cap B) + C \subseteq A + C$ and $(A \cap B) + C \subseteq B + C$ and so we have

$$(A \cap B) + C \subseteq (A + C) \cap (B + C) = A \cap (B + C).$$

To obtain the reverse inclusion, let $a \in A \cap (B + C)$. Then $a \in A$ and there exist $b \in B, c \in C$ such that $a = b + c$. Since $C \subseteq A$ we have $c \in A$ and therefore $b = a - c \in A$. Consequently $b \in A \cap B$ and so $a = b + c \in (A \cap B) + C$. ◊

EXERCISES

2.1 Determine all the subspaces of the $\mathbb{R}$-vector space $\mathbb{R}^2$. Give a geometric interpretation of these subspaces. Do the same for $\mathbb{R}^3$.

2.2 Let M be an R-module. If S is a non-empty subset of M, define the *annihilator* of S in R by

$$\text{Ann}_R S = \{\lambda \in R \; ; \; (\forall x \in S) \; \lambda x = 0_M\}.$$

Show that $\text{Ann}_R S$ is a left ideal of R and that it is a two-sided ideal whenever S is a submodule of M.

2.3 Describe the kernel of the ring morphism μ of Exercise 1.1.

2.4 Prove that the ring of endomorphisms of the abelian group $\mathbb{Z}$ is isomorphic to the ring $\mathbb{Z}$, and that the ring of endomorphisms of the abelian group $\mathbb{Q}$ is isomorphic to the field $\mathbb{Q}$.

[*Hint.* Use Exercises 1.1 and 2.3; note that if $f \in \text{End}\,\mathbb{Z}$ then $f = \mu[f(1)]$.]

2.5 Let M be an R-module. If $r, s \in R$ show that

$$r - s \in \text{Ann}_R M \Rightarrow (\forall x \in M)\ rx = sx.$$

Deduce that M can be considered as an $R/\text{Ann}_R M$-module. Show that the annihilator of M in $R/\text{Ann}_R M$ is zero.

2.6 Let R be a commutative unitary ring and let M be an R-module. For every $r \in R$ let $rM = \{rx\ ;\ x \in M\}$ and $M_r = \{x \in M\ ;\ rx = 0_M\}$. Show that rM and M_r are submodules of M. In the case where $R = \mathbb{Z}$ and $M = \mathbb{Z}/n\mathbb{Z}$, suppose that $n = rs$ where r and s are mutually prime. Show that $rM = M_s$.

[*Hint.* Use the fact that there exist $a, b \in \mathbb{Z}$ such that $ra + sb = 1$.]

2.7 Let $(M_i)_{i \in I}$ be a family of submodules of an R-module M. Suppose that, for every finite subset J of I, there exists $k \in I$ such that $(\forall j \in J)\ M_j \subseteq M_k$. Show that $\bigcup_{i \in I} M_i$ and $\sum_{i \in I} M_i$ coincide. Show that in particular this arises when $I = \mathbb{N}$ and the M_i form an ascending chain $M_0 \subseteq M_1 \subseteq M_2 \subseteq \cdots$.

2.8 An R-module M is said to be *simple* if it has no submodules other than M and $\{0_M\}$. Prove that M is simple if and only if M is generated by every non-zero $x \in M$.

2.9 If R is a unitary ring prove that R is a simple R-module if and only if R is a division ring.

[*Hint.* Observe that, for $x \neq 0_M$, the set $Rx = \{rx\ ;\ r \in R\}$ is a non-zero submodule, whence it must coincide with R and so contains 1_R.]

2.10 Find subspaces A, B, C of $\mathbb{R}^2$ such that

$$(A \cap B) + (A \cap C) \subset A \cap (B + C).$$

2.11 If M is an R-module and A, B, C are submodules of M such that

$$A \subseteq B, \quad A + C = B + C, \quad A \cap C = B \cap C,$$

prove that $A = B$.

[*Hint.* $A = A + (A \cap C) = \cdots$; use the modular law.]

2.12 Let V be a vector space over a field F and let $\alpha : V \to V$ be a linear transformation on V. Consider V as an $F[X]$-module under the action defined via α as in Exercise 1.8. Let W be an $F[X]$-submodule of V. Prove that W is a subspace of V that satisfies the property

$$x \in W \Rightarrow \alpha(x) \in W.$$

Conversely, show that every subspace W of V that satisfies this property is an $F[X]$-submodule of V.

3 Morphisms; exact sequences

The reader will recall that in the theory of groups, for example, an important part is played by the structure-preserving mappings or morphisms. Precisely, if G and H are groups whose laws of composition are each denoted by + for convenience then a mapping $f : G \to H$ is called a *morphism* (or *homomorphism*) if

$$(\forall x, y \in G) \quad f(x+y) = f(x) + f(y).$$

Such a mapping sends G onto a subgroup of H, namely the subgroup

$$\operatorname{Im} f = \{f(x) \; ; \; x \in G\}.$$

For such a mapping f we have, with 0_G and 0_H denoting respectively the identity elements of G and H,

(α) $f(0_G) = 0_H$;
(β) $(\forall x \in G) \; f(-x) = -f(x)$.

In fact, $f(0_G) = f(0_G + 0_G) = f(0_G) + f(0_G)$ whence, by cancellation, $f(0_G) = 0_H$; and $f(x) + f(-x) = f[x + (-x)] = f(0_G) = 0_H$ so that $f(-x) = -f(x)$.

We shall now define the notion of a morphism from one R-module to another. This will obviously be an extension of the notion of a group morphism, so that (α) and (β) above will hold.

Definition If M and N are R-modules then a mapping $f : M \to N$ is called an *R-morphism* if

(1) $(\forall x, y \in M) \; f(x+y) = f(x) + f(y)$;
(2) $(\forall x \in M)(\forall \lambda \in R) \; f(\lambda x) = \lambda f(x)$.

When R is a field an R-morphism is traditionally called a *linear transformation*. An R-morphism f is called an *R-monomorphism* if it is injective; an *R-epimorphism* if it is surjective; and an *R-isomorphism* if it is bijective. An R-morphism $f : M \to M$ is often called an *R-endomorphism* on M.

Example 3.1 If M and N are abelian groups regarded as $\mathbb{Z}$-modules then a $\mathbb{Z}$-morphism $f : M \to N$ is simply a group morphism. For, as is readily seen by induction, we have

$$(\forall n \in \mathbb{N}) \quad f(nx) = nf(x)$$

and consequently
$$(\forall n \in \mathbb{Z}) \quad f(nx) = nf(x).$$

Example 3.2 If M is an R-module and n is a positive integer then for $i = 1,\ldots,n$ the mapping $\mathrm{pr}_i : M^n \to M$ described by
$$\mathrm{pr}_i(x_1,\ldots,x_n) = x_i$$
is an R-epimorphism, called the *i-th projection of M^n onto M*.

An important property of an R-morphism $f : M \to N$ is that it induces mappings between the lattices of submodules. In fact if we define, for every submodule X of M,
$$f^{\to}(X) = \{f(x) \; ; \; x \in X\}$$
and, for every submodule Y of N,
$$f^{\leftarrow}(Y) = \{x \in M \; ; \; f(x) \in Y\}$$
then we have the following result.

Theorem 3.1 *Let M and N be R-modules and $f : M \to N$ a morphism. Then for every submodule X of M the set $f^{\to}(X)$ is a submodule of N, and for every submodule Y of N the set $f^{\leftarrow}(Y)$ is a submodule of M.*

Proof We note first that $f^{\to}(X) \neq \emptyset$ since X contains 0_M and so $f^{\to}(X)$ contains $f(0_M) = 0_N$. If now $y, z \in f^{\to}(X)$ then there exist $a, b \in X$ such that $y = f(a), z = f(b)$ whence, since X is a submodule of M,
$$y - z = f(a) - f(b) = f(a - b) \in f^{\to}(X).$$
Also, for every $\lambda \in R$ we have, again since X is a submodule,
$$\lambda y = \lambda f(a) = f(\lambda a) \in f^{\to}(X).$$
Thus $f^{\to}(X)$ is a submodule of N.

Suppose now that Y is a submodule of N. Then $f^{\leftarrow}(Y) \neq \emptyset$ since it clearly contains 0_M. If now $a, b \in f^{\leftarrow}(Y)$ we have $f(a), f(b) \in Y$ whence, since Y is a submodule of N, $f(a-b) = f(a) - f(b) \in Y$ and so $a - b \in f^{\leftarrow}(Y)$. Also, if $\lambda \in R$ then $f(\lambda a) = \lambda f(a) \in Y$ so that $\lambda a \in f^{\leftarrow}(Y)$. Thus $f^{\leftarrow}(Y)$ is a submodule of M. $\diamond$

If $L(M)$ denotes the lattice of submodules of M then the previous result shows that we can define mappings $f^{\to} : L(M) \to L(N)$ and $f^{\leftarrow} : L(N) \to L(M)$, described respectively by $X \mapsto f^{\to}(X)$ and $Y \mapsto f^{\leftarrow}(Y)$. A simple consequence of the definitions is that each of these

Morphisms; exact sequences

induced mappings is *inclusion-preserving* in the sense that if X_1, X_2 are submodules of M such that $X_1 \subseteq X_2$ then $f^{\rightarrow}(X_1) \subseteq f^{\rightarrow}(X_2)$; and if Y_1, Y_2 are submodules of N such that $Y_1 \subseteq Y_2$ then $f^{\leftarrow}(Y_1) \subseteq f^{\leftarrow}(Y_2)$.

For an R-morphism $f : M \to N$ the submodule $f^{\rightarrow}(M)$ of N is called the *image* of f and is written $\operatorname{Im} f$; and the submodule $f^{\leftarrow}\{0_N\}$ of M is called the *kernel* of f and is written $\operatorname{Ker} f$.

- In the case of vector spaces and linear transformations the terms *range* $R(f)$ and *null-space* $N(f)$ are sometimes used instead of image and kernel respectively.

It is clear that a necessary and sufficient condition for a morphism f to be an epimorphism is that $\operatorname{Im} f$ be as large as possible, namely $\operatorname{Im} f = N$. Likewise, a necessary and sufficient condition for f to be a monomorphism is that $\operatorname{Ker} f$ be as small as possible, namely $\operatorname{Ker} f = \{0_M\}$. In fact, if $\operatorname{Ker} f = \{0_M\}$ and $x, y \in M$ are such that $f(x) = f(y)$ then $f(x - y) = f(x) - f(y) = 0_M$ gives $x - y \in \operatorname{Ker} f = \{0_M\}$ and so $x = y$; conversely, if f is injective and $x \in \operatorname{Ker} f$ then $f(x) = 0_N = f(0_M)$ gives $x = 0_M$ so that $\operatorname{Ker} f = \{0_M\}$. Note that no use is made here of the left action; the results are purely group-theoretic.

Theorem 3.2 *Let $f : M \to N$ be an R-morphism. If A is a submodule of M and B is a submodule of N then*

(1) $f^{\rightarrow}[A \cap f^{\leftarrow}(B)] = f^{\rightarrow}(A) \cap B$;

(2) $f^{\leftarrow}[B + f^{\rightarrow}(A)] = f^{\leftarrow}(B) + A$.

Proof (1) Observe first that if $y \in f^{\leftarrow}(B)$ then $f(y) \in B$ and therefore we have that $f^{\rightarrow}[f^{\leftarrow}(B)] \subseteq B$. The fact that $f^{\rightarrow}$ is inclusion-preserving now implies that the left-hand side is contained in the right-hand side. To obtain the reverse inclusion, suppose that $y \in f^{\rightarrow}(A) \cap B$. Then $y = f(a)$ and $y \in B$. Since then $f(a) \in B$ we have $a \in f^{\leftarrow}(B)$ and $y \in f^{\rightarrow}[A \cap f^{\leftarrow}(B)]$.

(2) Since for $a \in A$ we have $f(a) \in f^{\rightarrow}(A)$ we see that

$$A \subseteq f^{\leftarrow}[f^{\rightarrow}(A)].$$

The fact that $f^{\leftarrow}$ is inclusion-preserving now implies that the left-hand side of (2) contains the right-hand side. For the reverse inclusion, let $x \in f^{\leftarrow}[B + f^{\rightarrow}(A)]$. Then $f(x) \in B + f^{\rightarrow}(A)$ and so $f(x) \in B$ and $f(x) = f(a)$ for some $a \in A$. This gives $x - a \in \operatorname{Ker} f = f^{\leftarrow}\{0_N\} \subseteq f^{\leftarrow}(B)$ and therefore $x \in f^{\leftarrow}(B) + A$. ◊

Corollary *If A is a submodule of M and B is a submodule of N then*

(3) $f^{\rightarrow}[f^{\leftarrow}(B)] = B \cap \operatorname{Im} f$;

(4) $f^{\leftarrow}[f^{\rightarrow}(A)] = A + \text{Ker } f$.

Proof To obtain (3), take $A = M$ in (1); and to obtain (4) take $B = \{0_N\}$ in (2). ◊

Just as with group morphisms, we can compose R-morphisms in the appropriate situation to form new R-morphisms. The basic facts concerning this are the following, which we shall use in the sequel without reference:

(a) *if $f : M \to N$ and $g : N \to P$ are R-morphisms then the composite map $g \circ f : M \to P$ is also an R-morphism.*

Since this is true of group morphisms it suffices to note that, for all $x, y \in M$ and all $\lambda \in R$, $(g \circ f)(\lambda x) = g[f(\lambda x)] = g[\lambda f(x)] = \lambda g[f(x)] = \lambda(g \circ f)(x)$.

(b) *if $f : M \to N$ and $g : N \to P$ are R-epimorphisms then so is $g \circ f$.*

This is equally true for group morphisms, as are the following:

(c) *if $f : M \to N$ and $g : N \to P$ are R-monomorphisms then so is $g \circ f$.*

(d) *if $g \circ f$ is an epimorphism then so is g.*

(e) *if $g \circ f$ is a monomorphism then so is f.*

Concerning composite morphisms we shall now consider the following 'diagram-completing' problems. Suppose that we are given a diagram of modules and morphisms of the form

We pose the question : under what conditions does there exist a morphism $h : B \to C$ such that $h \circ f = g$? We can also formulate the *dual* problem, obtained essentially by reversing all the arrows. Specifically, given a diagram of modules and morphisms of the form

under what conditions does there exist a morphism $h : C \to B$ such that $f \circ h = g$?

Morphisms; exact sequences

Let us first solve these problems when A, B, C are simply *sets* and f, g are simply *mappings*.

Theorem 3.3 (a) *If A, B, C are non-empty sets and $f : A \to B, g : A \to C$ are mappings then the following conditions are equivalent:*

(1) *there exists a mapping $h : B \to C$ such that $h \circ f = g$;*
(2) $(\forall x, y \in A) \quad f(x) = f(y) \Rightarrow g(x) = g(y).$

(b) *If A, B, C are non-empty sets and $f : B \to A, g : C \to A$ are mappings then the following conditions are equivalent:*

(3) *there exists a mapping $h : C \to B$ such that $f \circ h = g$;*
(4) $\operatorname{Im} g \subseteq \operatorname{Im} f.$

Proof (1) $\Rightarrow$ (2) : If $h : B \to C$ exists such that $h \circ f = g$ and if $x, y \in A$ are such that $f(x) = f(y)$ then clearly we have $g(x) = h[f(x)] = h[f(y)] = g(y)$.

(2) $\Rightarrow$ (1) : Consider the subset G of $\operatorname{Im} f \times C$ given by

$$G = \{(y, z) \; ; \; (\exists x \in A) \; y = f(x), \; z = g(x)\}.$$

We note that $G \neq \emptyset$; for, given any $x \in A$ we have $(f(x), g(x)) \in G$. Now given any $y \in \operatorname{Im} f$ there is a unique $z \in C$ such that $(y, z) \in G$. In fact, if $y = f(x)$ choose $z = g(x)$ to see that such an element z exists. To see that such an element z is unique, suppose that $(y, z) \in G$ and $(y, z') \in G$; then by the definition of G we have

$$(\exists x, x' \in A) \qquad y = f(x) = f(x'), \quad z = g(x), \quad z' = g(x')$$

whence, by (2), $g(x) = g(x')$ and consequently $z = z'$. We can therefore define a mapping $t : \operatorname{Im} f \to C$ by the prescription

$$(\forall x \in A) \qquad t[f(x)] = g(x).$$

We now construct a mapping $h : B \to C$ by the prescription

$$h(y) = \begin{cases} t(y) & \text{if } y \in \operatorname{Im} f; \\ \text{any } c \in C & \text{otherwise.} \end{cases}$$

Then for every $x \in A$ we have $h[f(x)] = t[f(x)] = g(x)$ so that $h \circ f = g$.

As for the dual problem, we now establish the equivalence of (3) and (4).

(3) $\Rightarrow$ (4) : If $h : C \to B$ exists such that $f \circ h = g$ then for every $x \in C$ we have $g(x) = f[h(x)] \in \operatorname{Im} f$ and so $\operatorname{Im} g \subseteq \operatorname{Im} f$.

(4) $\Rightarrow$ (3) : If (4) holds then for every $x \in C$ there exists $y \in B$ such that $g(x) = f(y)$. Given any $x \in C$, label (courtesy of the axiom of choice)

as y_x any element of B such that $g(x) = f(y_x)$. We can thus define a mapping $h : C \to B$ by the prescription $h(x) = y_x$. Then $f[h(x)] = f(y_x) = g(x)$ gives $f \circ h = g$. ◊

Corollary (a) *If A, B are non-empty sets and $f : A \to B$ is a mapping then the following statements are equivalent :*

(α) *f is injective;*
(β) *there exists $g : B \to A$ such that $g \circ f = \text{id}_A$;*
(γ) *f is left cancellable, in the sense that for every non-empty set C and all mappings $h, k : C \to A$,*

$$f \circ h = f \circ k \Rightarrow h = k.$$

(b) *If A, B are non-empty sets and $f : A \to B$ is a mapping then the following statements are equivalent :*

(α') *f is surjective;*
(β') *there exists $g : B \to A$ such that $f \circ g = \text{id}_B$;*
(γ') *f is right cancellable, in the sense that for every non-empty set C and all mappings $h, k : B \to C$,*

$$h \circ f = k \circ f \Rightarrow h = k.$$

Proof (α) ⇔ (β) : This follows immediately from (1) ⇔ (2) on taking $C = A$ and $g = \text{id}_A$.

(β) ⇒ (γ) : If $f \circ h = f \circ k$ then composing each side on the left with g and using the fact that $g \circ f = \text{id}_A$ we obtain $h = k$.

(γ) ⇒ (α) : Suppose that f is not injective. Then for some $x, y \in A$ with $x \neq y$ we have $f(x) = f(y)$. Let C be any non-empty set and let $h, k : C \to A$ be the constant mappings given by

$$(\forall c \in C) \qquad h(c) = x, \ k(c) = y.$$

Then clearly $h \neq k$ and

$$(\forall c \in C) \qquad f[h(c)] = f(x) = f(y) = f[k(c)]$$

so that $f \circ h = f \circ k$. Thus if (α) does not hold then neither does (γ), and consequently (γ) ⇒ (α).

We now establish the equivalence of (α'), (β'), (γ').

(α') ⇔ (β') : This is immediate from (3) ⇔ (4).

(β') ⇒ (γ') : If $h \circ f = k \circ f$ then composing each side on the right with g and using the fact that $f \circ g = \text{id}_B$ we obtain $h = k$.

Morphisms; exact sequences

$(\gamma') \Rightarrow (\alpha')$: If B is a singleton then f is automatically surjective and there is nothing to prove. Suppose then that B contains at least two distinct elements p, q. Let $h, k : B \to B$ be given by

$$h(x) = \begin{cases} x & \text{if } x \in \operatorname{Im} f; \\ p & \text{otherwise,} \end{cases} \qquad k(x) = \begin{cases} x & \text{if } x \in \operatorname{Im} f; \\ q & \text{otherwise.} \end{cases}$$

Then for every $y \in A$ we have $h[f(y)] = f(y) = k[f(y)]$ and so $h \circ f = k \circ f$. Applying (γ') we deduce that $h = k$. Now if $\operatorname{Im} f \neq B$ we must have $\operatorname{Im} f \subset B$ whence there exists $x \in B$ with $x \notin \operatorname{Im} f$. For such an element x we have $h(x) = p$ and $k(x) = q$ whence, since $h = k$, we obtain the contradiction $p = q$. We conclude therefore that $\operatorname{Im} f = B$ so that f is surjective. ◇

One is tempted to conjecture that Theorem 3.3 and its Corollary can be made into module-theoretic results by replacing 'non-empty set' by 'R-module' and 'mapping' by 'R-morphism' throughout. However, as the following examples show, such a conjecture is in general false.

Example 3.1 Consider the diagram of $\mathbb{Z}$-modules and $\mathbb{Z}$-morphisms

$$\begin{array}{ccc} \mathbb{Z} & \xrightarrow{\mathrm{id}_{\mathbb{Z}}} & \mathbb{Z} \\ {\scriptstyle \times 2} \downarrow & & \\ \mathbb{Z} & & \end{array}$$

in which $\mathrm{id}_{\mathbb{Z}}$ is the identity morphism and $\times 2$ is the $\mathbb{Z}$-morphism described by $n \mapsto 2n$. Although, by Theorem 3.3(a), there is a *mapping* $h : \mathbb{Z} \to \mathbb{Z}$ such that $h \circ (\times 2) = \mathrm{id}_{\mathbb{Z}}$, no such $\mathbb{Z}$-*morphism* can exist. For, suppose that h were such a $\mathbb{Z}$-morphism. Then for every $n \in \mathbb{Z}$ we would have $2h(n) = h(2n) = n$. In particular, we would have $2h(1) = 1$; and this is impossible since the equation $2x = 1$ has no solution in $\mathbb{Z}$.

Example 3.2 For a given prime p, consider the subgroup $\mathbb{Q}_p$ of $\mathbb{Q}$ that is given by

$$\mathbb{Q}_p = \{x \in \mathbb{Q} \,;\, (\exists k \in \mathbb{Z})(\exists n \in \mathbb{N}) \; x = k/p^n\}.$$

Observe that $\mathbb{Z}$ is a subgroup of $\mathbb{Q}_p$ so we can form the quotient group $\mathbb{Q}_p/\mathbb{Z}$. Consider now the diagram of $\mathbb{Z}$-modules and $\mathbb{Z}$-morphisms

$$\begin{array}{ccc} & & \mathbb{Q}_p/\mathbb{Z} \\ & & \downarrow \mathrm{id} \\ \mathbb{Q}_p/\mathbb{Z} & \xrightarrow{f} & \mathbb{Q}_p/\mathbb{Z} \end{array}$$

where id is the identity morphism and f is the $\mathbb{Z}$-morphism described by $x \mapsto px$. Since for all k and n we have

$$\frac{k}{p^n} + \mathbb{Z} = p\left(\frac{k}{p^{n+1}} + \mathbb{Z}\right)$$

we see that $\operatorname{Im} f = \mathbb{Q}_p/\mathbb{Z} = \operatorname{Im} \operatorname{id}$. By Theorem 3.3($b$) there is therefore a *mapping* $h : \mathbb{Q}_p/\mathbb{Z} \to \mathbb{Q}_p/\mathbb{Z}$ such that $f \circ h = \operatorname{id}$. However, no such $\mathbb{Z}$-morphism can exist. For, suppose that h were such a $\mathbb{Z}$-morphism. Then we would have

$$\frac{1}{p} + \mathbb{Z} = f\left[h\left(\frac{1}{p} + \mathbb{Z}\right)\right] = p\left[h\left(\frac{1}{p} + \mathbb{Z}\right)\right]$$
$$= h\left[p\left(\frac{1}{p} + \mathbb{Z}\right)\right] = h(1 + \mathbb{Z}) = 0 + \mathbb{Z}$$

which is nonsense since $x + \mathbb{Z} = 0 + \mathbb{Z} \iff x \in \mathbb{Z}$.

Despite the above examples, there are certain situations in which, given some extra conditions, we do have module-theoretic analogues of Theorem 3.3. The following two results indicate such situations; we shall see others later.

Theorem 3.4 *Consider the diagram*

$$\begin{array}{ccc} A & \xrightarrow{g} & C \\ {\scriptstyle f}\downarrow & & \\ B & & \end{array}$$

of R-modules and R-morphisms in which f is an R-epimorphism. The following conditions are equivalent :
  (1) *there is a unique R-morphism $h : B \to C$ such that $h \circ f = g$;*
  (2) $\operatorname{Ker} f \subseteq \operatorname{Ker} g$.

Moreover, such an R-morphism h is a monomorphism if and only if $\operatorname{Ker} f = \operatorname{Ker} g$.

Proof (1) $\Rightarrow$ (2) : Suppose that (1) holds and that $x \in \operatorname{Ker} f$. Then $g(x) = h[f(x)] = h(0) = 0$ and (2) follows.

(2) $\Rightarrow$ (1) : Suppose now that $\operatorname{Ker} f \subseteq \operatorname{Ker} g$. Given $x, y \in A$ we have

$$\begin{aligned} f(x) = f(y) &\Rightarrow f(x - y) = f(x) - f(y) = 0_B \\ &\Rightarrow x - y \in \operatorname{Ker} f \subseteq \operatorname{Ker} g \\ &\Rightarrow g(x) - g(y) = g(x - y) = 0_C \\ &\Rightarrow g(x) = g(y). \end{aligned}$$

By Theorem 3.3(a) we can therefore define a mapping $h : B \to C$ such that $h \circ f = g$. Since f is surjective by hypothesis, it follows by the Corollary to Theorem 3.3 that f is right cancellable and so h is unique. It remains to show that h is in fact an R-morphism. Since f is surjective, this follows from the equalities

$$h[f(x)+f(y)] = h[f(x+y)] = g(x+y) = g(x)+g(y) = h[f(x)]+h[f(y)],$$
$$h[\lambda f(x)] = h[f(\lambda x)] = g(\lambda x) = \lambda g(x) = \lambda h[f(x)].$$

Finally we observe that if h is injective then since $g(x) = h[f(x)]$ we have

$$x \in \operatorname{Ker} g \Rightarrow f(x) \in \operatorname{Ker} h = \{0_B\} \Rightarrow x \in \operatorname{Ker} f,$$

and so $\operatorname{Ker} g \subseteq \operatorname{Ker} f$ whence we have equality by (2). Conversely, suppose that $\operatorname{Ker} g = \operatorname{Ker} f$ and let $x \in \operatorname{Ker} h$. Since f is surjective we have $x = f(y)$ for some $y \in A$ and so $0_B = h(x) = h[f(y)] = g(y)$ and consequently $y \in \operatorname{Ker} g = \operatorname{Ker} f$ whence $x = f(y) = 0_B$ and h is injective. ◊

Theorem 3.5 *Consider the diagram*

of R-modules and R-morphisms in which f is an R-monomorphism. The following conditions are equivalent :

(1) *there is a unique R-morphism $h : C \to B$ such that $f \circ h = g$;*
(2) $\operatorname{Im} g \subseteq \operatorname{Im} f.$

Moreover, such an R-morphism h is an epimorphism if and only if $\operatorname{Im} g = \operatorname{Im} f$.

Proof (1) $\Rightarrow$ (2) : If (1) holds then, for every $c \in C$, $g(x) = f[h(x)] \in \operatorname{Im} f$, whence (2) holds.

(2) $\Rightarrow$ (1) : If (2) holds then by Theorem 3.3(b) there is a mapping $h : C \to B$ such that $f \circ h = g$. Since f is injective by hypothesis, it follows by the Corollary to Theorem 3.3 that that f is left cancellable and so h is unique. Now for all $c, d \in C$ and $\lambda \in R$ we have the equalities

$$f[h(c+d)] = g(c+d) = g(c)+g(d) = f[h(c)]+f[h(d)] = f[h(c)+h(d)],$$
$$f[h(\lambda c)] = g(\lambda c) = \lambda g(c) = \lambda f[h(c)] = f[\lambda h(c)].$$

Since f is injective we deduce that $h(c+d) = h(c) + h(d)$ and $h(\lambda c) = \lambda h(c)$, so that h is indeed an R-morphism.

Finally, we observe that if h is surjective then for every $b \in B$ there exists $c \in C$ such that $b = h(c)$, whence $f(b) = f[h(c)] = g(c)$ and consequently $\operatorname{Im} f \subseteq \operatorname{Im} g$, whence we have equality by (2). Conversely, if $\operatorname{Im} f = \operatorname{Im} g$ then for every $b \in B$ there exists $c \in C$ such that $f(b) = g(c) = f[h(c)]$ whence $b = h(c)$, since f is injective. Consequently, h is surjective. ◊

In the discussion to follow we shall on several occasions be faced with the problem of finding a morphism that will «complete» a given diagram in an agreeable way, just as we were able in Theorems 3.4 and 3.5 to find morphisms that 'completed' the triangles there in such a way that, loosely speaking, following the arrows, all paths with the same departure set and same arrival set are equal. To be somewhat more precise, we introduce the following concept.

Definition Given a diagram of sets and mappings, we say that the diagram is *commutative* if all composite mappings from any given departure set to any given arrival set are equal.

By way of illustration, we note that the triangle

is commutative if and only if $h \circ f = g$. Also, the diagram

$$\begin{array}{ccccc}
A & \xrightarrow{f} & B & \xrightarrow{g} & C \\
\downarrow \alpha & & \downarrow \beta & & \downarrow \gamma \\
A' & \xrightarrow[f']{} & B' & \xrightarrow[g']{} & C'
\end{array}$$

is commutative if and only if $f' \circ \alpha = \beta \circ f$ and $g' \circ \beta = \gamma \circ g$; i.e. if and only if each of its squares is commutative.

The notion of a commutative diagram will appear many times in the discussion to follow. Linked with this is another important concept which we now introduce.

Definition By a *sequence of modules and morphisms* we shall mean a diagram of the form

$$\cdots \longrightarrow M_{i-1} \xrightarrow{f_{i-1}} M_i \xrightarrow{f_i} M_{i+1} \longrightarrow \cdots$$

Morphisms; exact sequences

Such a sequence is said to be *exact at* M_i if $\text{Im } f_{i-1} = \text{Ker } f_i$, and to be *exact* if it is exact at each M_i.

The above sequence is therefore exact if, at each stage, the image of the input morphism coincides with the kernel of the output morphism.

Simple examples of exact sequences are given in the following result, in which all zero modules are written 0.

Theorem 3.6 *If* $f : M \to N$ *is an R-morphism and if* $0 \to M, N \to 0$ *denote the inclusion map and the zero map respectively then* f *is*

(1) *a monomorphism if and only if* $0 \longrightarrow M \xrightarrow{f} N$ *is exact*:

(2) *an epimorphism if and only if* $M \xrightarrow{f} N \longrightarrow 0$ *is exact*:

(3) *an isomorphism if and only if* $0 \longrightarrow M \xrightarrow{f} N \longrightarrow 0$ *is exact*.

Proof This is immediate from the definitions. ◊

Example 3.3 If $f : A \to B$ is a morphism of abelian groups then we have the exact sequence

$$0 \longrightarrow \text{Ker } f \xrightarrow{\iota} A \xrightarrow{\natural} A/\text{Ker } f \longrightarrow 0$$

in which ι is the inclusion map and $\natural$ is the natural epimorphism. Likewise, we have the exact sequence

$$0 \longrightarrow \text{Im } f \longrightarrow B \longrightarrow B/\text{Im } f \longrightarrow 0.$$

As we shall see in due course, exact sequences of the form

$$0 \longrightarrow M' \xrightarrow{f} M \xrightarrow{g} M'' \longrightarrow 0$$

are of especial importance. They are called *short exact sequences*.

- Note that in an exact sequence the composite of two successive morphisms is the zero morphism. The converse of this is not true in general, for $f \circ g = 0$ is equivalent to $\text{Im } g \subseteq \text{Ker } f$. Sequences in which $f_i \circ f_{i-1} = 0$ for every index i are called *semi-exact*.

By way of illustrating the foregoing notions we shall derive a useful property of the kernel of an R-morphism. This follows from the following result.

Theorem 3.7 *Given the diagram of R-modules and R-morphisms*

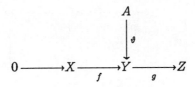

in which the row is exact and $g \circ \vartheta = 0$, there is a unique R-morphism $h : A \to X$ such that the completed diagram is commutative.

Proof Since $g \circ \vartheta = 0$ and since the row is exact we have

$$\text{Im}\, \vartheta \subseteq \text{Ker}\, g = \text{Im}\, f.$$

Since, by Theorem 3.6(1), f is a monomorphism, the result is an immediate consequence of Theorem 3.5. ◊

Theorem 3.8 Let $f : M \to N$ be an R-morphism. If $\iota : \text{Ker}\, f \to M$ is the inclusion map then

(1) $f \circ \iota = 0$;

(2) if P is an R-module and if $g : P \to M$ is an R-morphism such that $f \circ g = 0$ then there is a unique R-morphism $\vartheta : P \to \text{Ker}\, f$ such that the following diagram is commutative :

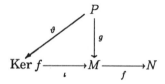

Proof (1) is obvious, and (2) is an immediate consequence of Theorem 3.7. ◊

- It can be shown (see Exercise 3.5 for the details) that the pair $(\text{Ker}\, f, \iota)$ is characterised by the properties of Theorem 3.8. Note that this characterisation involves morphisms and not elements.

In order to give the reader a deeper appreciation of commutative diagrams, we end the present section by illustrating the technique known as 'diagram chasing' .

Theorem 3.9 [The four lemma] Suppose that the diagram of modules and morphisms

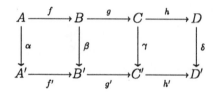

Morphisms; exact sequences 27

is commutative and has exact rows. Then the following hold:

(1) *if α, γ are epimorphisms and δ is a monomorphism then β is an epimorphism;*

(2) *if α is an epimorphism and β, δ are monomorphisms then γ is a monomorphism.*

Proof (1): Let $b' \in B'$. Since γ is surjective there exists $c \in C$ such that $g'(b') = \gamma(c)$. By the commutativity of the right-hand square we then have
$$\delta[h(c)] = h'[\gamma(c)] = h'[g'(b')] = 0,$$
since $h' \circ g' = 0$. Thus $h(c) \in \text{Ker}\,\delta = 0$ and so $h(c) = 0$, giving $c \in \text{Ker}\,h = \text{Im}\,g$ so that $c = g(b)$ for some $b \in B$. Then, by the commutativity of the middle square,
$$g'(b') = \gamma(c) = \gamma[g(b)] = g'[\beta(b)].$$
Consequently $b' - \beta(b) \in \text{Ker}\,g' = \text{Im}\,f'$ so that $b' - \beta(b) = f'(a')$ for some $a' \in A'$. Since α is surjective there exists $a \in A$ such that $a' = \alpha(a)$ and so, by the commutativity of the left-hand square, $b' - \beta(b) = f'[\alpha(a)] = \beta[f(a)]$. We thus have
$$b' = \beta(b) + \beta[f(a)] = \beta[b + f(a)] \in \text{Im}\,\beta.$$
Consequently, β is surjective.

(2): Let $c \in \text{Ker}\,\gamma$. Then $\delta[h(c)] = h'[\gamma(c)] = h'(0) = 0$ and so $h(c) \in \text{Ker}\,\delta = 0$. Thus $c \in \text{Ker}\,h = \text{Im}\,g$ so that $c = g(b)$ for some $b \in B$. Now $0 = \gamma(c) = \gamma[g(b)] = g'[\beta(b)]$ so $\beta(b) \in \text{Ker}\,g' = \text{Im}\,f'$ whence $\beta(b) = f'(a')$ for some $a' \in A'$. Now $a' = \alpha(a)$ for some $a \in A$, so $\beta(b) = f'[\alpha(a)] = \beta[f(a)]$. Since β is a monomorphism, we deduce that $b = f(a)$ whence $c = g(b) = g[f(a)] = 0$ since $f \circ g = 0$. $\diamond$

Theorem 3.10 [The five lemma] *Suppose that the diagram of modules and morphisms*

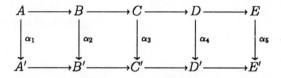

is commutative and has exact rows. If $\alpha_1, \alpha_2, \alpha_4, \alpha_5$ are isomorphisms then so is α_3.

Proof Applying Theorem 3.9(1) to the right-hand three squares we see that α_3 is an epimorphism; and applying Theorem 3.9(2) to the left-hand three squares we see that α_3 is a monomorphism. Thus α_3 is an isomorphism. ◊

Corollary *Suppose that the diagram of modules and morphisms*

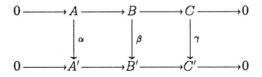

is commutative and has exact rows. If α and γ are isomorphisms then so is β.

Proof Take $A = A' = E = E' = 0$ in the above. ◊

EXERCISES

3.1 Let R be a commutative unitary ring. Prove that a mapping $f : R \times R \to R$ is an R-morphism if and only if there exist $\alpha, \beta \in R$ such that
$$(\forall x, y \in R) \quad f(x,y) = \alpha x + \beta y.$$

3.2 Let M and N be R-modules. Prove that if M is simple (Exercise 2.8) then every non-zero R-morphism $f : M \to N$ is a monomorphism; and that if N is simple then every non-zero R-morphism $f : M \to N$ is an epimorphism. Deduce that if M is a simple R-module then the ring $(\mathrm{End}_R M, +, \circ)$ of R-morphisms $g : M \to M$ is a division ring.

3.3 If $f : M \to N$ is an R-morphism prove that $f^\to \circ f^\leftarrow \circ f^\to = f^\to$ and that $f^\leftarrow \circ f^\to \circ f^\leftarrow = f^\leftarrow$.

3.4 If A and B are submodules of an R-module M, establish a short exact sequence
$$0 \longrightarrow A \cap B \stackrel{\vartheta}{\longrightarrow} A \times B \stackrel{\pi}{\longrightarrow} A + B \longrightarrow 0.$$
[*Hint.* Observe that the 'obvious' definitions of ϑ and π, namely $\vartheta(x) = (x,x)$ and $\pi(x,y) = x + y$, do not work; try $\pi(x,y) = x - y$.]

Morphisms; exact sequences

3.5 Let $f : M \to N$ be an R-morphism and suppose that there is given an R-module X together with an R-monomorphism $j : X \to M$ such that

(1) $f \circ j = 0$;

(2) for every R-module P and every R-morphism $g : P \to M$ such that $f \circ g = 0$ there is a unique R-morphism $\vartheta : P \to X$ such that

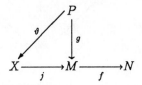

is commutative. Prove that there is a unique R-isomorphism $\xi :$ Ker $f \to X$ such that

is commutative, ι being the inclusion map.

[*Hint.* Take $P =$ Ker f and $g = \iota$ to obtain the existence of an R-morphism ξ. Now take $P = X$ and $g = j$ in Theorem 3.8 to obtain ξ' say. Show, using the Corollary to Theorem 3.3, that $\xi \circ \xi'$ and $\xi' \circ \xi$ are identity morphisms.]

3.6 Given the diagram of R-modules and R-morphisms

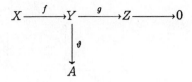

in which the row is exact and $\vartheta \circ f = 0$, prove that there is a unique R-morphism $h : Z \to A$ such that $h \circ g = \vartheta$.

3.7 Consider the diagram of R-modules and R-morphisms

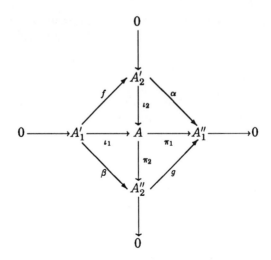

If this diagram is commutative with the row and column exact, prove that

(1) α and β are zero morphisms;

(2) f and g are isomorphisms.

3.8 The diagram of R-modules and R-morphisms

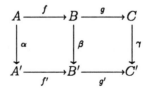

is given to be commutative with α, β, γ isomorphisms. Prove that the top row is exact if and only if the bottom row is exact.

3.9 Suppose that the diagram of R-modules and R-morphisms

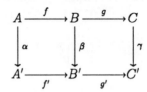

is commutative and has exact rows. Prove that

(1) if α, γ, f' are monomorphisms then so is β;

(2) if α, γ, g are epimorphisms then so is β.

Morphisms; exact sequences

3.10 [The 3 × 3 lemma] Consider the following diagram of R-modules and R-morphisms:

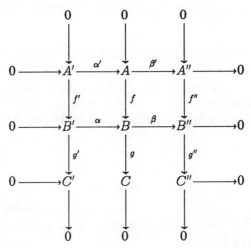

Given that the diagram is commutative, that all three columns are exact, and that the top two rows are exact, prove that there exist unique R-morphisms $\alpha'' : C' \to C$ and $\beta'' : C \to C''$ such that the resulting bottom row is exact and the completed diagram is commutative.

[*Hint.* Observe that $g \circ \alpha \circ f' = 0$ so that $\operatorname{Ker} g' = \operatorname{Im} f' \subseteq \operatorname{Ker} g \circ \alpha$. Use Theorem 3.4 to produce α''. Argue similarly to produce β''. Now chase.]

3.11 A short exact sequence of the form

$$(f, E, g) \equiv 0 \longrightarrow A \xrightarrow{f} E \xrightarrow{g} B \longrightarrow 0$$

is called an *extension* of A by B. Given any R-modules A and B, show that at least one extension of A by B exists.

Two extensions (f_1, E_1, g_1) and (f_2, E_2, g_2) of A by B are said to be *equivalent* if there is an R-morphism $h : E_1 \to E_2$ such that $h \circ f_1 = f_2$ and $g_2 \circ h = g_1$. Prove that such an R-morphism h is an isomorphism.

Show that there are extensions

$$0 \longrightarrow \mathbb{Z}_2 \longrightarrow \mathbb{Z}_2 \times \mathbb{Z}_4 \longrightarrow \mathbb{Z}_4 \longrightarrow 0,$$
$$0 \longrightarrow \mathbb{Z}_2 \longrightarrow \mathbb{Z}_8 \longrightarrow \mathbb{Z}_4 \longrightarrow 0$$

of $\mathbb{Z}_2$ by $\mathbb{Z}_4$ that are not equivalent.

4 Quotient modules; basic isomorphism theorems

We shall now consider an important way of constructing new modules from old ones. This arises from the following problem. Suppose that M is an R-module and that E is an equivalence relation on M. Precisely when can we define laws of composition on the set M/E of equivalence classes in such a way that M/E becomes an R-module with the natural surjection $\natural : M \to M/E$ an epimorphism? This important question is settled in the following result, in which we denote the class of x modulo E by $[x]_E$.

Theorem 4.1 *Let M be an R-module and let E be an equivalence relation on M. Then the following statements are equivalent :*

(1) *there is a unique addition $([x]_E, [y]_E) \mapsto [x]_E + [y]_E$ and a unique R-action $(\lambda, [x]_E) \mapsto \lambda[x]_E$ such that M/E is an R-module and the natural surjection is an R-epimorphism, i.e. the following identities hold :*

$$(\forall x, y \in M)(\forall \lambda \in R) \quad [x]_E + [y]_E = [x+y]_E, \quad \lambda[x]_E = [\lambda x]_E;$$

(2) *E is compatible with the structure of M, in the sense that*

$$x \equiv y(E), \ y \equiv b(E) \Rightarrow x + y \equiv a + b(E),$$
$$x \equiv a(E), \ \lambda \in R \Rightarrow \lambda x \equiv \lambda a(E);$$

(3) *there is a submodule M_E of M such that*

$$x \equiv y(E) \iff x - y \in M_E.$$

Proof (1) $\Leftrightarrow$ (2) : This is immediate on applying Theorem 3.3 to the diagram

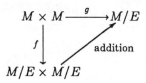

where f is given by $(x, y) \mapsto ([x]_E, [y]_E)$ and g is given by $(x, y) \mapsto [x+y]_E$, and to the diagram

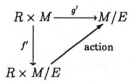

Quotient modules; basic isomorphism theorems

where f' is given by $(\lambda, x) \mapsto (\lambda, x/E)$ and g' is given by $(\lambda, x) \mapsto [\lambda x]_E$. The uniqueness of the laws of composition so obtained follows from the fact that both vertical maps are surjective and so are right cancellable.

(2) $\Rightarrow$ (3) : Suppose that E is compatible with the structure of M. Then $[0]_E$, the class of 0 modulo E, is a submodule of M. In fact, if $x \equiv 0(E)$ and $y \equiv 0(E)$ then, by the compatibility, $x - y \equiv 0 - 0 = 0(E)$ and if $x \equiv 0(E)$ and $\lambda \in R$ then $\lambda x \equiv \lambda 0 = 0(E)$. Moreover, we have

$$x \equiv y(E) \;\Rightarrow\; x - y \equiv y - y = 0(E);$$
$$x - y \equiv 0(E) \;\Rightarrow\; x = (x - y) + y \equiv 0 + y = y(E),$$

so that $x \equiv y(E) \Leftrightarrow x - y \in [0]_E$.

(3) $\Rightarrow$ (2) : Suppose that M_E is a submodule of M such that $x \equiv y(E)$ is equivalent to $x - y \in M_E$. Then from $x \equiv a(E)$ and $y \equiv b(E)$ we have $x - a \in M_E$ and $y - b \in M_E$ so that, M_E being a submodule, $x + y - (a + b) \in M_E$ whence $x + y \equiv a + b(E)$. Similarly, from $x \equiv a(E)$ we have, for every $\lambda \in R$, $\lambda x - \lambda a = \lambda(x - a) \in M_E$ so that $\lambda x \equiv \lambda a(E)$. Thus E is compatible with the structure of M. $\Diamond$

Definition When the situation described in Theorem 4.1 holds we call R/E the *quotient module of M by the compatible equivalence relation E.*

Identifying equivalence relations on M that yield the same quotient set, we now observe that *there is a bijection from the set of compatible equivalences on M to the set of submodules of M*. This is given as follows : for every compatible equivalence relation E on M define $\vartheta(E)$ to be the submodule $[0]_E$. That ϑ is surjective follows from the fact that if N is a submodule of M then the relation F given by

$$x \equiv y(F) \iff x - y \in N$$

is (as is readily seen) a compatible equivalence relation on M with

$$x \equiv 0(F) \iff x \in N,$$

so that $\vartheta(F) = [0]_F = N$. That ϑ is also injective results from the fact that if E, F are compatible equivalence relations on M such that $\vartheta(E) = \vartheta(F)$ then $[0]_E = [0]_F$ and so, by Theorem 4.1(3), $x \equiv y(E)$ is equivalent to $x \equiv y(F)$, whence $E = F$ by the agreed identification.

Because of this bijection, it is standard practice to write M/N for the quotient module M/E where N is the submodule that corresponds to E (namely $N = [0]_E$). This abuse of notation yields a corresponding abuse of language : we call M/N the *quotient module of M by the submodule N*. In this case the equivalence class of x will be written $[x]_N$. Note that, as in the case of quotient groups, $[x]_N$ coincides with the coset $x + N = \{x + n\,;\, n \in N\}$; for we have

$$\begin{aligned}
y \in [x]_N &\iff [y]_N = [x]_N \\
&\iff y - x \in N \\
&\iff (\exists n \in N)\; y = x + n.
\end{aligned}$$

We now consider the question of how to identify the submodules of a quotient module.

Theorem 4.2 [Correspondence theorem] *If N is a submodule of an R-module M then there is an inclusion-preserving bijection from the set of submodules of M/N to the set of submodules of M that contain N.*

Proof Suppose that A is a submodule of M that contains N. Then the set
$$A/N = \{[a]_N \; ; \; a \in A\}$$
is clearly a submodule of M/N. Consider the mapping ϑ from the set of all such submodules A to the set of submodules of M/N described by $\vartheta(A) = A/N$. Since ϑ so defined is the restriction (to the set of submodules that contain N) of $\natural_N^{\rightarrow}$, it is clear that ϑ is inclusion preserving.

We observe from the Corollary to Theorem 3.2 that if $N \subseteq A$ then
$$\natural_N^{\leftarrow}[\vartheta(A)] = \natural_N^{\leftarrow}[\natural_N^{\rightarrow}(A)] = A + \mathrm{Ker}\,\natural_N = A + N = A.$$
Consequently, if $\vartheta(A) = \vartheta(B)$ then $A = B$ and so ϑ is injective.

We now observe that if P is any submodule of M/N then, again by the Corollary to Theorem 3.2,
$$\vartheta[\natural_N^{\leftarrow}(P)] = \natural_N^{\rightarrow}[\natural_N^{\leftarrow}(P)] = P \cap \mathrm{Im}\,\natural_N = P.$$
Consequently ϑ is also surjective. ◊

Corollary *Every submodule of M/N is of the form A/N where A is a submodule of M that contains N.* ◊

Our aim now is to consider certain induced morphisms from one quotient module to another, and to establish some fundamental isomorphisms.

Theorem 4.3 *Let $f : M \to N$ be an R-morphism. If A and B are submodules of M and N respectively then the following statements are equivalent:*

(1) $f^{\rightarrow}(A) \subseteq B$;

(2) *there is a unique R-morphism $f_* : M/A \to N/B$ such that the diagram*

$$\begin{array}{ccc} M & \xrightarrow{f} & N \\ \natural_A \downarrow & & \downarrow \natural_B \\ M/A & \xrightarrow{f_*} & N/B \end{array}$$

is commutative.

Moreover, when such an R-morphism f_* exists, it is

(a) *a monomorphism if and only if* $A = f^{\leftarrow}(B)$;
(b) *an epimorphism if and only if* $B + \operatorname{Im} f = N$.

Proof Applying Theorem 3.4 to the diagram

$$\begin{array}{ccc} M & \xrightarrow{\natural_B \circ f} & N/B \\ {\scriptstyle \natural_A}\downarrow & & \\ M/A & & \end{array}$$

we see that (2) holds if and only if

$$\operatorname{Ker} \natural_A \subseteq \operatorname{Ker}(\natural_B \circ f).$$

Now clearly

$$x \in \operatorname{Ker} \natural_A \iff [x]_A = [0]_A \iff x \in A,$$

and similarly

$$x \in \operatorname{Ker}(\natural_B \circ f) \iff [f(x)]_B = [0]_B \iff f(x) \in B.$$

Thus we see that (2) holds if and only if $x \in A$ implies $f(x) \in B$, which is (1).

As for the last statements, we observe that $f^{\rightarrow}(A) \subseteq B$ is equivalent to $A \subseteq f^{\leftarrow}(B)$ and that therefore

$$\begin{aligned} \operatorname{Ker} f_\star &= \{[x]_A \,;\, f(x) \in B\} \\ &= \{[x]_A \,;\, x \in f^{\leftarrow}(B)\} \\ &= f^{\leftarrow}(B)/A, \end{aligned}$$

so that $f_\star$ is injective if and only if $A = f^{\leftarrow}(B)$.

Finally,

$$\operatorname{Im} f_\star = \{[f(x)]_B \,;\, x \in M\}$$

and so $f_\star$ is surjective if and only if

$$(\forall n \in N)(\exists m \in M) \ [n]_B = [f(x)]_B,$$

which is equivalent to the condition

$$(\forall n \in N)(\exists x \in M) \ n - f(x) \in B,$$

which is clearly equivalent to $N = B + \operatorname{Im} f$. ◊

If $f : M \to N$ is an R-morphism then we shall denote by $f^+ : M \to \operatorname{Im} f$ the R-morphism given by the same prescription as f, namely $f^+(x) = f(x)$. Note that although f and f^+ have the same effect on

$x \in M$ we distinguish between them since they have different arrival sets; f^+ is surjective whereas f need not be.

Theorem 4.4 [First isomorphism theorem] *If $f : M \to N$ is an R-morphism then there is a unique R-isomorphism $\zeta : M/\operatorname{Ker} f \to \operatorname{Im} f$ such that the diagram*

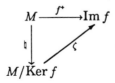

is commutative.

Proof Applying Theorem 4.3 in the case where $N = \operatorname{Im} f, B = \{0_N\}$ and $A = \operatorname{Ker} f$ we obtain the existence of a unique R-morphism

$$\zeta : M/\operatorname{Ker} f \to \operatorname{Im} f$$

such that $\zeta \circ \natural = f^+$. Since f^+ is surjective, so is ζ. Moreover,

$$\operatorname{Ker} f = f^\leftarrow\{0\} = f^\leftarrow(B)$$

and so ζ is also injective. Thus ζ is an isomorphism. ◊

Corollary 1 *If $f : M \to N$ is an R-morphism then there is an inclusion-preserving bijection from the set of submodules of $\operatorname{Im} f$ to the set of submodules of M that contain $\operatorname{Ker} f$.*

Proof This is immediate by Theorem 4.2. ◊

Corollary 2 [Canonical decomposition of a morphism] *Every morphism can be expressed as the composite of an epimorphism, an isomorphism, and a monomorphism.*

Proof With the above notation, the diagram

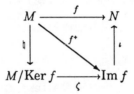

is commutative, ι being the natural inclusion. It follows from this that $f = \iota \circ \zeta \circ \natural$. ◊

Quotient modules; basic isomorphism theorems

- Although the above decomposition is called *canonical* (or *natural*), it is by no means unique; but if

$$M \xrightarrow{\alpha} A \xrightarrow{\beta} B \xrightarrow{\gamma} N$$

is another such decomposition of f then necessarily $A \simeq M/\mathrm{Ker}\, f$ and $B \simeq \mathrm{Im}\, f$ (see Exercise 4.6).

Theorem 4.5 [Second isomorphism theorem] *If M is an R-module and if N, P are submodules of M such that $P \subseteq N$ then N/P is a submodule of M/P and there is a unique R-isomorphism*

$$h : M/N \to (M/P)/(N/P)$$

such that the following diagram is commutative:

$$\begin{array}{ccc} M & \xrightarrow{\natural_P} & M/P \\ \natural_N \downarrow & & \downarrow \natural \\ M/N & \xrightarrow{h} & (M/P)/(N/P) \end{array}$$

Proof We know by the Corollary to Theorem 4.2 that N/P is a submodule of M/P. Since $\natural_P^{\to}(N) = \{[n]_P \,;\, n \in N\} = N/P$, we can apply Theorem 4.3 to the above diagram to obtain the existence of a unique R-morphism $h : M/N \to (M/P)/(N/P)$ making the diagram commutative. Now since, by the commutativity, $h \circ \natural_N$ is an epimorphism, so is h. To show that h is also a monomorphism, it suffices to note that $\natural_P^{\leftarrow}(N/P) = N$ and appeal to Theorem 4.3 again. ◊

The third isomorphism theorem that we shall establish is a consequence of the following.

Given an R-module M and a submodule A of M, it is clear that we have an exact sequence

$$0 \longrightarrow A \xrightarrow{\iota_A} M \xrightarrow{\natural_A} M/A \longrightarrow 0$$

in which ι_A is the natural inclusion and $\natural_A$ is the natural surjection. This therefore generalises to arbitrary R-modules the situation in Example 3.3, in which the abelian groups are considered as $\mathbb{Z}$-modules.

Theorem 4.6 *If A and B are submodules of an R-module M then there*

is a commutative diagram of the form

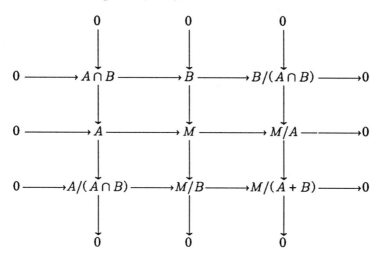

Proof Let $\iota_A : A \to M$ be the natural inclusion. Then we have

$$\vec{\iota_A}(A \cap B) \subseteq B$$

and so we can apply Theorem 4.3 to obtain the commutative diagram

$$\begin{array}{ccc} A & \xrightarrow{\iota_A} & M \\ \natural \downarrow & & \downarrow \natural_B \\ A/(A \cap B) & \xrightarrow[(\iota_A)_\star]{} & M/B \end{array}$$

Considering likewise the natural inclusion $\iota_B : B \to M$, we obtain a similar commutative diagram. These diagrams can be joined together and extended to form all but the bottom right-hand corner of the big diagram, namely

We can now complete the bottom right-hand corner by defining mappings

$$\zeta_B : M/A \to M/(A+B), \quad \zeta_A : M/B \to M/(A+B)$$

Quotient modules; basic isomorphism theorems

by the prescriptions

$$\zeta_B([x]_A) = [x]_{A+B}, \quad \zeta_A([x]_B) = [x]_{A+B}.$$

It is clear that ζ_B, ζ_A are R-morphisms which make the completed diagram commutative. We now show that the bottom row

$$0 \longrightarrow A/(A \cap B) \xrightarrow{(\iota_A)_\star} M/B \xrightarrow{\zeta_A} M/(A+B) \longrightarrow 0$$

is exact. By symmetry, the right-hand column will also be exact. Now since ζ_A is clearly surjective and $(\iota_A)_\star$ is injective (Theorem 4.3) it suffices to show that $\mathrm{Im}(\iota_A)_\star = \mathrm{Ker}\,\zeta_A$. For this purpose, we note that $\mathrm{Im}(\iota_A)_\star = \{[x]_B \,;\, x \in A\}$ and $\mathrm{Ker}\,\zeta_A = \{[x]_B \,;\, x \in A+B\}$. Observing that

$$x \in A + B \Rightarrow (\exists a \in A)(\exists b \in B)\; x = a+b \Rightarrow [x]_B = [a+b]_B = [a]_B,$$

we obtain $\mathrm{Ker}\,\zeta_A \subseteq \mathrm{Im}(\iota_A)_\star$; and observing that

$$x \in A \Rightarrow (\exists a \in A)\; x = a \Rightarrow (\forall b \in B)\; [x]_B = [a]_B = [a+b]_B,$$

we obtain the reverse inclusion. ◊

Corollary [Third isomorphism theorem] *If A and B are submodules of an R-module M then*

$$A/(A \cap B) \simeq (A+B)/B.$$

Proof Since A and B are submodules of $A+B$ we can apply the above in the case where $M = A+B$. The bottom row of the diagram becomes

$$0 \longrightarrow A/(A \cap B) \longrightarrow (A+B)/B \longrightarrow (A+B)/(A+B) \longrightarrow 0.$$

Since $(A+B)/(A+B)$ is a zero module, the exactness of this row together with Theorem 3.6(3) gives the required isomorphism. ◊

The last of the isomorphism theorems that we shall require is the following, in which the diagram is a *Hasse diagram*. The interpretation of this is that an ascending line segment from A to B

indicates that A is a submodule of B.

Theorem 4.7 [The butterfly of Zazzenhaus] *Let M be an R-module and suppose that N, P, N', P' are submodules of M such that $N \subseteq P$ and $N' \subseteq P'$. Then relative to the Hasse diagram*

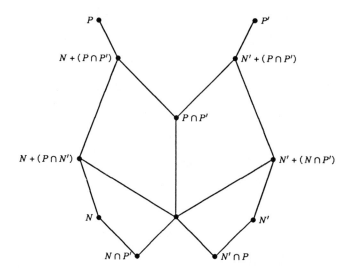

in which the unlabelled submodule is $(N \cap P') + (N' \cap P)$, the following quotient modules are isomorphic :

$$\frac{N+(P\cap P')}{N+(P\cap N')} \simeq \frac{P\cap P'}{(N\cap P')+(N'\cap P)} \simeq \frac{N'+(P\cap P')}{N'+(N\cap P')}.$$

Proof Since $P \cap N' \subseteq P \cap P'$ we have

$$(P \cap P') + N + (P \cap N') = (P \cap P') + N$$

and, by the modular law (Theorem 2.4),

$$(P\cap P') + [N + (P\cap N')] = (P\cap P' \cap N) + (P\cap N') = (P'\cap N) + (P\cap N').$$

Applying the third isomorphism theorem with $A = P \cap P'$ and $B = N + (P \cap N')$, we obtain an isomorphism

$$\frac{P\cap P'}{(N\cap P')+(N'\cap P)} \simeq \frac{N+(P\cap P')}{N+(P\cap N')}.$$

The second isomorphism shown follows by symmetry. ◊

We end this section with some remarks concerning R-algebras. We have defined a subalgebra of an R-algebra A to be a submodule that is also an R-algebra with respect to the multiplication in A.

Definition By an *ideal* of an R-algebra A we mean a subalgebra X of A such that $AX \subseteq X$ and $XA \subseteq X$, where $AX = \{ax \;;\; a \in A, x \in X\}$

Quotient modules; basic isomorphism theorems

and similarly for XA. By an *R-algebra morphism* from an R-algebra A to an R-algebra B we mean an R-morphism $f : A \to B$ that is also a morphism with respect to the semigroup structure; in other words if, for all $x, y \in A$ and all $\lambda \in R$,

$$f(x+y) = f(x) + f(y), \quad f(\lambda x) = \lambda f(x), \quad f(xy) = f(x)f(y).$$

Note that if $f : A \to B$ is an R-algebra isomorphism then so is $f^{-1} : B \to A$. This is readily seen on replacing x, y in the above equalities by $f^{-1}(x), f^{-1}(y)$.

We leave the reader the task of showing that Theorem 4.1 has an analogue in terms of R-algebras in which the role of the associated submodule is assumed by an associated ideal. This analogue leads to the notion of the *quotient algebra of an R-algebra A by an ideal X*. Somewhat later, we shall require the following result concerning R-algebras.

Theorem 4.8 *Let A be an R-algebra and M and R-module. Suppose that there is an R-isomorphism $f : A \to M$. Then there is a unique multiplication on M such that M is an R-algebra with f an R-algebra isomorphism.*

Proof Define a multiplication on M by

$$(x, y) \mapsto x \cdot y = f[f^{-1}(x) f^{-1}(y)].$$

Then since A is an R-algebra and f, f^{-1} are R-morphisms it is readily seen that

$$(\forall \lambda \in R)(\forall x, y \in M) \quad \lambda(x \cdot y) = (\lambda x) \cdot y = x \cdot (\lambda y).$$

Thus M is an R-algebra. Since, from the above definition,

$$f(x) \cdot f(y) = f\left[f^{-1}[f(x)] f^{-1}[f(y)]\right] = f(xy)$$

we see that f is an R-algebra isomorphism.

That the multiplication is unique follows from the fact that if $(x, y) \mapsto x \star y$ is a law of composition on M such that f is an R-algebra isomorphism then, since f^{-1} is also an R-algebra isomorphism, we deduce from $f^{-1}(x \star y) = f^{-1}(x) f^{-1}(y)$ that $x \star y = f[f^{-1}(x) f^{-1}(y)]$. ◊

EXERCISES

4.1 An R-module is said to be *cyclic* if it is generated by a singleton subset. Let $M = Rx$ be a cyclic R-module. Recalling that the annihilator of x is the submodule $\text{Ann}_R\{x\} = \{\lambda \in R \,;\, \lambda x = 0\}$, prove that $M \simeq R/\text{Ann}_R\{x\}$.

Deduce that if R is a principal ideal domain, i.e. a commutative integral domain in which every ideal is generated by a singleton subset, and if $x \in R$ is such that $\mathrm{Ann}_R(x) = p^k R$ for some $p \in R$ (see Exercise 2.2) then the only submodules of M are those in the chain

$$0 = p^k M \subset p^{k-1} M \subset \ldots \subset pM \subset p^0 M = M.$$

[*Hint.* Use the correspondence theorem.]

4.2 Let A, B be submodules of an R-module M. Establish an exact sequence of the form

$$0 \longrightarrow M/(A \cap B) \longrightarrow M/A \times M/B \longrightarrow M/(A+B) \longrightarrow 0.$$

Deduce that

$$(A+B)/(A \cap B) \simeq (A+B)/A \times (A+B)/B \simeq B/(A \cap B) \times A/(A \cap B).$$

4.3 Let R be a commutative unitary ring. Show that if I and J are ideals of R then there is an exact sequence

$$0 \longrightarrow I \cap J \longrightarrow R \longrightarrow R/I \times R/J \longrightarrow R/(I+J) \longrightarrow 0.$$

4.4 Let $f : M \to N$ be an R-morphism. By a *cokernel* of f we mean a pair (P, π) consisting of an R-module P together with an R-epimorphism $\pi : N \to P$ such that

(1) $\pi \circ f = 0$;

(2) for every R-module X and every R-morphism $g : N \to X$ such that $g \circ f = 0$ there is a unique R-morphism $\vartheta : X \to P$ such that the diagram

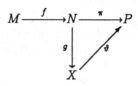

is commutative.

Prove that $(N/\mathrm{Im}\, f, \natural)$ is a cokernel of f. Show also, in a manner dual to that of Exercise 3.5, that cokernels are unique to within R-isomorphism.

Quotient modules; basic isomorphism theorems

4.5 [The snake diagram] Suppose that the diagram of modules and morphisms

$$\begin{array}{ccccccc} A & \xrightarrow{u} & B & \xrightarrow{v} & C & \longrightarrow & 0 \\ & & & & & & \\ \downarrow{\alpha} & & \downarrow{\beta} & & \downarrow{\gamma} & & \\ & & & & & & \\ 0 & \longrightarrow & A' & \xrightarrow{u'} & B' & \xrightarrow{v'} & C' \end{array}$$

is commutative and has exact rows. Show that this can be extended to a diagram

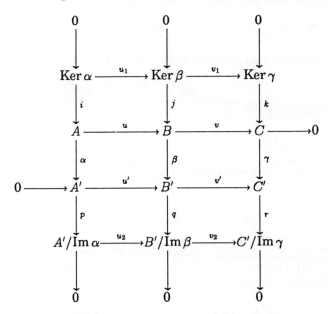

which is also commutative and has exact rows and columns. Show also that there is a 'connecting morphism' $\vartheta : \operatorname{Ker} \gamma \to A'/\operatorname{Im} \alpha$ such that

$$\operatorname{Ker} \alpha \xrightarrow{u_1} \operatorname{Ker} \beta \xrightarrow{v_1} \operatorname{Ker} \gamma \xrightarrow{\vartheta} A'/\operatorname{Im} \alpha \xrightarrow{u_2} B'/\operatorname{Im} \beta \xrightarrow{v_2} C'/\operatorname{Im} \gamma$$

is exact.

[*Hint.* To construct ϑ: given $x \in \operatorname{Ker} \gamma$ let $y \in B$ be such that $v(y) = k(x)$. Show that $\beta(y) \in \operatorname{Ker} v'$ so that there exists a unique $a' \in A'$

such that $u'(a') = \beta(y)$. Show that the prescription $\vartheta(x) = p(a')$ is well defined (i.e. independent of y) and does the trick.]

4.6 Let $f : M \to N$ be an R-morphism and suppose that f can be expressed as the composite map

$$M \xrightarrow{\alpha} A \xrightarrow{\beta} B \xrightarrow{\gamma} N$$

where α is an epimorphism, β is an isomorphism, and γ is a monomorphism. Prove that $A \simeq M/\operatorname{Ker} f$ and $B \simeq \operatorname{Im} f$.

[*Hint*. Use Theorems 3.4 and 3.5.]

4.7 Let R be a commutative unitary ring and let $R_n[X]$ be the R-module of all polynomials of degree less than or equal to n with coefficients in R. Show that, for $n \geq 1$,

$$R_{n-1}[X] \simeq R_n[X]/R.$$

[*Hint*. Consider the differentiation map.]

4.8 If R is a commutative integral domain and x is a non-zero element of R let $Rx^n = \{yx^n \; ; \; y \in R\}$. Show that Rx^n is a submodule of R for every positive integer n and that there is a descending chain of R-modules

$$R \supseteq Rx \supseteq Rx^2 \supseteq \cdots \supseteq Rx^{n-1} \supseteq Rx^n \supseteq \cdots$$

in which $Rx^{n-1}/Rx^n \simeq R/Rx$ for every n.

[*Hint*. Consider $\vartheta : R \to Rx^n/Rx^{n+1}$ given by $\vartheta(r) = rx^n + Rx^{n+1}$. Show that $\operatorname{Ker} \vartheta = Rx$ and use the first isomorphism theorem.]

4.9 Given the diagram $A \xrightarrow{\alpha} B \xrightarrow{\beta} C$ of modules and morphisms, show that there is an exact sequence

$$\operatorname{Ker} \alpha \to \operatorname{Ker}(\beta \circ \alpha) \to \operatorname{Ker} \beta \to B/\operatorname{Im} \alpha \to C/\operatorname{Im}(\beta \circ \alpha) \to C/\operatorname{Im} \beta.$$

5 Chain conditions; Jordan-Hölder towers

Definition An R-module M is said to be *noetherian*, or to *satisfy the ascending chain condition* on submodules, if for every ascending chain

$$M_0 \subseteq M_1 \subseteq M_2 \subseteq \ldots \subseteq M_i \subseteq M_{i+1} \subseteq \cdots$$

of submodules there is a natural number n such that $(\forall k \geq n)\ M_k = M_n$.

- Roughly speaking, this says that after a certain point the increasing sequence $(M_i)_{i \in \mathbb{N}}$ becomes stationary.

Definition We say that an R-module M satisfies the *maximum condition* if every non-empty collection of submodules of M has a maximal member relative to the ordering of set inclusion.

Our immediate task now is to show that the above two definitions are equivalent.

Theorem 5.1 *For an R-module M the following statements are equivalent*:

(1) *M is noetherian*;
(2) *M satisfies the maximum condition*;
(3) *every submodule of M is finitely generated*.

Proof (1) $\Rightarrow$ (2) : Let C be a non-empty collection of submodules of M and choose $M_0 \in C$. If M_0 is not maximal in C then there exists $M_1 \in C$ with $M_0 \subset M_1$. If M_1 is not maximal in C then there exists $M_2 \in C$ such that $M_0 \subset M_1 \subset M_2$. This argument yields an ascending chain of submodules of M. By (1) there exists a natural number n such that $(\forall k \geq n)\ M_k = M_n$. The chain therefore becomes stationary at M_n which is then maximal in C.

(2) $\Rightarrow$ (3) : Let N be a submodule of M. The collection F of all submodules of N that are finitely generated is not empty since it clearly contains the zero submodule of M which is generated by $\emptyset$ (Theorem 2.2). By (2) there is therefore a maximal element, $N^\star$ say, in F. Now for any $x \in N$ the submodule $N^\star + Rx$ of N generated by $N^\star \cup \{x\}$ is finitely generated and so belongs to F. But $N^\star \subseteq N^\star + Rx$ and so, since $N^\star$ is maximal in F, we have $N^\star = N^\star + Rx$, whence $Rx \subseteq N^\star$ and so $x \in N^\star$. Thus we see that $N \subseteq N^\star$ whence we have $N = N^\star$ (since $N^\star$ is a submodule of N), and consequently N is finitely generated.

(3) $\Rightarrow$ (1) : Let $M_0 \subseteq M_1 \subseteq M_2 \subseteq \cdots$ be an ascending chain of submodules of M. We note first that

$$\sum_{i \in \mathbb{N}} M_i = \bigcup_{i \in \mathbb{N}} M_i.$$

In fact, if $x \in \sum_{I \in \mathbb{N}} M_i$ then we have $x = \sum_{I \in I} m_i$ where I is a finite subset of $\mathbb{N}$ and $m_i \in M_i$ for every $i \in I$. If now j denotes the greatest element of I then since $M_i \subseteq M_j$ for every $i \in I$ we clearly have $x \in M_j \subseteq \bigcup_{i \in \mathbb{N}} M_i$. Thus we see that $\sum_{i \in \mathbb{N}} M_i \subseteq \bigcup_{i \in \mathbb{N}} M_i$, with the reverse inclusion obvious.

Now by the hypothesis $\sum_{i \in \mathbb{N}} M_i$ is finitely generated, say by $\{x_1, \ldots, x_n\}$; and since each $x_i \in \sum_{i \in \mathbb{N}} M_i = \bigcup_{i \in \mathbb{N}} M_i$ we have $x_i \in M_j$ for some j. There being only finitely many x_i, there is therefore a natural number n (namely, the largest such j encountered) such that $x_i \in M_n$ for $i = 1, \ldots, r$. Since the set $\{x_1, \ldots, x_r\}$ generates $\sum_{i \in \mathbb{N}} M_i$, it follows that $\sum_{i \in \mathbb{N}} M_i \subseteq M_n$. Again since $\sum_{i \in \mathbb{N}} M_i = \bigcup_{i \in \mathbb{N}} M_i$ we deduce that $\sum_{i \in \mathbb{N}} M_i = M_n$ whence it follows that the given chain terminates and we have (1). $\diamond$

We can of course define the dual concepts of *descending chain condition* on submodules and *minimum condition* in the obvious way. We say that M is *artinian* if it satisfies the descending chain condition on submodules. The analogue of Theorem 5.1 is the following.

Theorem 5.2 *For every R-module M the following statements are equivalent :*

(1) *M is artinian ;*
(2) *M satisfies the minimum condition.*

Proof (1) $\Rightarrow$ (2) : This is similar to (1) $\Rightarrow$ (2) in Theorem 5.1.

(2) $\Rightarrow$ (1) : If M does not satisfy the descending chain condition on submodules then M must have an infinite descending chain of submodules

$$M_0 \supset M_1 \supset M_2 \supset \cdots \supset M_i \supset M_{i+1} \supset \cdots .$$

Clearly, the collection C of all the M_i in this chain has no minimal element and so M cannot satisfy the minimum condition. $\diamond$

Chain conditions have hereditary properties, as we shall now see.

Theorem 5.3 *If an R-module M satisfies either chain condition then every submodule and every quotient submodule of M satisfies the same chain condition.*

Proof The statement concerning submodules is obvious since every submodule of a submodule of M is also a submodule of M. As for quotient modules, the result is an immediate consequence of the correspondence theorem (Theorem 4.2). ◊

The converse of Theorem 5.3 also holds. In fact, there is a much stronger result:

Theorem 5.4 *If M is an R-module and if N is a submodule of M such that N and M/N satisfy the same chain condition then M also satisfies that chain condition.*

Proof We give a proof for the case of the ascending chain condition; that for the descending chain condition is similar.

Suppose that
$$M_0 \subseteq M_1 \subseteq \ldots \subseteq M_i \subseteq M_{i+1} \subseteq \cdots$$
is an ascending chain of submodules of M. Then
$$M_0 \cap N \subseteq M_1 \cap N \subseteq \ldots \subseteq M_i \cap N \subseteq M_{i+1} \cap N \subseteq \cdots$$
is an ascending chain of submodules of N and, by Theorem 3.1,
$$\natural_N^{\to}(M_0) \subseteq \natural_N^{\to}(M_1) \subseteq \ldots \subseteq \natural_N^{\to}(M_i) \subseteq \natural_N^{\to}(M_{i+1}) \subseteq \cdots$$
is an ascending chain of submodules of M/N. Since, by hypothesis, N is noetherian there is a positive integer n such that
$$(\forall k \geq n) \quad M_k \cap N = M_n \cap N;$$
and since M/N is also noetherian there is a positive integer m such that
$$(\forall k \geq m) \quad \natural_N^{\to}(M_k) = \natural_N^{\to}(M_m).$$
Now let $p = \max\{n, m\}$; then we have
$$(\forall k \geq p) \quad M_k \supseteq M_p; \quad M_k \cap N = M_p \cap N; \quad \natural_N^{\to}(M_k) = \natural_N^{\to}(M_p).$$
Suppose that t is any integer greater than or equal to p. Since $\natural_N^{\to}(M_t) = \natural_N^{\to}(M_p)$, given any $y \in M_t$ there exists $x \in M_p$ such that $y + N = x + N$, so that $y - x \in N$. But since $M_p \subseteq M_t$ we have $x \in M_t$ and so $y - x \in M_t$. Thus
$$y - x \in M_t \cap N = M_p \cap N \subseteq M_p$$
and so $y - x = z \in M_p$ whence $y = x + z \in M_p$. It follows that $M_t \subseteq M_p$, whence $M_t = M_p$. Thus M is noetherian. ◊

A natural question arises at this stage, namely whether we can find a characterisation of R-modules which satisfy both chain conditions. This we now proceed to do. For this purpose we require some additional terminology.

Definition An R-module M is said to be *simple* if the only submodules of M are $\{0\}$ and M.

We note by the following result that every simple R-module is finitely generated.

Theorem 5.5 *If M is an R-module then M is simple if and only if*

$$M = Rx = \{rx \; ; \; r \in R\}$$

for every non-zero $x \in M$.

Proof $\Rightarrow$: If M is simple then $x = 1x \in Rx \neq \{0\}$ for every $x \neq 0$; and since Rx is a submodule of M we have $Rx = M$.

$\Leftarrow$: If $Rx = M$ for every non-zero $x \in R$, let $N \neq \{0\}$ be be a submodule of M. Given $n \in N$ with $n \neq 0$ we have $M = Rx \subseteq N$ whence $M = N$ and M is simple. $\Diamond$

Example 5.1 If V is a vector space over a field F then for every non-zero $x \in F$ the subspace $F_x = \{\lambda x \; ; \; \lambda \in F\}$ is simple. In particular, the F-vector space F is simple.

The following result shows how morphisms are affected when there are no proper submodules.

Theorem 5.6 *Let M, N be R-modules and let $f : M \to N$ be a non-zero R-morphism. Then*

   (1) if M is simple, f is a monomorphism;
   (2) if N is simple, f is an epimorphism.

Proof (1) $\operatorname{Ker} f$ is a submodule of M so, since f is not the zero morphism, we must have $\operatorname{Ker} f = \{0\}$ whence f is a monomorphism.

(2) $\operatorname{Im} f$ is a submodule of N so, since f is not the zero morphism, we must have $\operatorname{Im} f = N$ whence f is an epimorphism. $\Diamond$

Corollary [Schur] *If M is a simple R-module then the ring $\operatorname{End}_R M$ of R-morphisms $f : M \to M$ is a division ring.*

Proof By (1) and (2) above, every non-zero $f \in \operatorname{End}_R M$ is an isomorphism and so is an invertible element in the ring. $\Diamond$

Chain conditions; Jordan-Hölder towers

Definition If M is an R-module then by a *tower of submodules* of M we shall mean a finite decreasing chain of submodules

$$M = M_0 \supset M_1 \supset M_2 \supset \ldots \supset M_r = \{0\}.$$

If we have two towers of submodules of M, say

$$T_1 : M = M_0 \supset M_1 \supset M_2 \supset \ldots \supset M_r = \{0\},$$
$$T_2 : M = N_0 \supset N_1 \supset N_2 \supset \ldots \supset N_t = \{0\},$$

then we say that T_2 is a *refinement* of T_1 if for $j = 1, \ldots, r$ there exists $i \in \{1, \ldots, t\}$ such that $N_i = M_j$; in other words, if every module in the chain T_1 appears in the chain T_2. We say that the towers T_1 and T_2 are *equivalent* if $t = r$ and there is a permutation σ on $\{1, \ldots, r\}$ such that

$$N_i/N_{i+1} \simeq M_{\sigma(i)}/M_{\sigma(i)+1}.$$

Theorem 5.7 [Schreier's refinement theorem] *If M is an R-module and if T_1, T_2 are towers of submodules of M then there are refinements S_1 of T_1 and S_2 of T_2 that are equivalent.*

Proof Given the towers

$$T_1 : M = M_0 \supset M_1 \supset M_2 \supset \ldots \supset M_r = \{0\},$$
$$T_2 : M = N_0 \supset N_1 \supset N_2 \supset \ldots \supset N_t = \{0\},$$

define (for $i = 1, \ldots r$ and $j = 1, \ldots, t$)

$$M_{i,j} = M_i + (M_{i-1} \cap N_j);$$
$$N_{j,i} = N_j + (N_{j-1} \cap M_i).$$

Suppose, without loss of generality, that $t \leq r$. Then, defining $M_{i,k} = M_{i,t}$ for $k = t+1, \ldots, r$ we have the descending chains

$$\cdots M_{i-1} = M_{i,0} \supseteq M_{i,1} \supseteq \ldots \supseteq M_{i,t} = \ldots = M_{i,r} = M_i = M_{i+1,0} \cdots$$
$$\cdots N_{j-1} = N_{j,0} \supseteq N_{j,1} \supseteq \ldots \ldots \ldots \ldots \ldots \ldots \supseteq N_{j,r} = N_j = N_{j+1,0} \cdots$$

which are refinements of T_1, T_2 respectively. Consider the quotient modules formed by consecutive entries $M_{i,j-1} \supseteq M_{i,j}$ and $N_{j,i-1} \supseteq N_{j,i}$ in these chains. It is immediate from the Zassenhaus butterfly (Theorem 4.7 with $N = M_i \subset M_{i-1} = P$ and $N' = N_j \subset N_{j-1} = P'$) that for $i, j = 1, \ldots, r$ we have

$$M_{i,j-1}/M_{i,j} \simeq N_{j,i-1}/N_{j,i}.$$

Consequently we see that $M_{i,j-1} = M_{i,j}$ if and only if $N_{j,i-1} = N_{j,i}$. We conclude, therefore, that on deleting from the above chains all entries that are equal to their predecessor we obtain refinements S_1 of T_1 and S_2 of T_2 that are equivalent. ◊

Definition By a *Jordan-Hölder tower of submodules* of an R-module M we shall mean a tower

$$M = M_0 \supset M_1 \supset M_2 \supset \ldots \supset M_r = \{0\}.$$

in which every quotient module M_i/M_{i+1} is simple.

The importance of Jordan-Hölder towers lies in the following observation. The inclusion-preserving bijection from the set of submodules P of M such that $M_i \supseteq P \supseteq M_{i+1}$ to the set of submodules of M_i/M_{i+1} shows immediately that *if T is a Jordan-Hölder tower then T has no proper refinement*; in other words, if T' is a refinement of T then necessarily the entries of T' are precisely those of T. This leads to the following result.

Theorem 5.8 *If M is an R-module and*

$$\begin{aligned} T_1 &: M = M_0 \supset M_1 \supset M_2 \supset \ldots \supset M_r = \{0\}, \\ T_2 &: M = N_0 \supset N_1 \supset N_2 \supset \ldots \supset N_t = \{0\}, \end{aligned}$$

are Jordan-Hölder towers of submodules of M then $t = r$ and T_1, T_2 are equivalent.

Proof By the Schreier refinement theorem, T_1 and T_2 admit equivalent refinements S_1 and S_2 respectively. But since T_1 and T_2 are Jordan-Hölder towers their only refinements are themselves. ◊

This result shows in particular that *the number of non-zero submodules appearing in any Jordan-Hölder tower is independent of the choice of the tower*. This number is called the *height* of the tower. By abuse of language we also call it the height of the module M and denote it by $h(M)$.

- Most authors use the term *composition series* instead of Jordan-Hölder tower, in which case they use the term *length* instead of height.

Theorem 5.9 *An R-module M has a Jordan-Hölder tower of submodules if and only if it is both artinian and noetherian. In this case every tower of submodules has a refinement which is a Jordan-Hölder tower.*

Proof Suppose first that M has a Jordan-Hölder tower and let h be its height. We prove that M satisfies both chain conditions by induction on h. Clearly, if $h = 0$ then $M = \{0\}$ and there is nothing to prove. Assume, therefore, that the result is true for all R-modules having Jordan-Hölder towers of height less than n (where $n > 1$). Let M be an R-module having a Jordan-Hölder tower of height n, say

$$M = M_0 \supset M_1 \supset M_2 \supset \ldots \supset M_{n-1} \supset M_n = \{0\}.$$

Then we observe that

$$M/M_{n-1} = M_0/M_{n-1} \supset M_1/M_{n-1} \supset \cdots \supset M_{n-1}/M_{n-1} = \{0\}$$

is a Jordan-Hölder tower for M/M_{n-1} of height $n-1$. In fact, it is clear that the inclusions in this second tower are strict; and by the second isomorphism theorem (Theorem 4.5) we have, for $i = 1, \ldots, n-1$,

$$(M_i/M_{n-1})/(M_{i+1}/M_{n-1}) \simeq M_i/M_{i+1}$$

amd so each quotient module is simple since M_i/M_{i+1} is simple. By the induction hypothesis, therefore, M/M_{n-1} satisfies both chain conditions. But since $M_{n-1} = M_{n-1}/\{0\} = M_{n-1}/M_n$ we see that M_{n-1} is simple and hence trivially satisfies both chain conditions. It is now immediate by Theorem 5.4 that M satisfies both chain conditions. This then shows that the result holds for all modules of height n and completes the induction.

Conversely, suppose that M satisfies both chain conditions. Let C be the collection of all the submodules of M that have Jordan-Hölder towers. Then $C \neq \emptyset$ since every descending chain of non-zero submodules $M_0 \supset M_1 \supset M_2 \supset \ldots$ terminates at M_p say which must be simple and so has a Jordan-Hölder tower of height 1. We now note that C has a maximal element, M^* say; for otherwise the descending chain condition would be violated. We now show that $M^* = M$, whence M will have a Jordan-Hölder tower. Suppose, by way of obtaining a contradiction, that $M^* \neq M$. Then M/M^* is not a zero module and so, since M/M^* inherits the descending chain condition from M, it follows that M/M^* has simple submodules. There therefore exists M^{**} such that $M^* \subset M^{**} \subseteq M$ with M^{**}/M^* a simple module. Now M^{**} has a Jordan-Hölder tower (since M^* does, and M^{**}/M^* is simple) and so we have $M^{**} \in C$, which contradicts the maximality of M^* in C. This contradiction shows that we must have $M = M^*$, whence M has a Jordan-Hölder tower.

As for the final statement, let T be a tower of submodules of M and let J be a Jordan-Hölder tower of submodules of M. By Schreier's refinement theorem (Theorem 5.7) there are refinements T_0 of T and J_0 of J that are equivalent. But since J is a Jordan-Hölder tower J_0 coincides with J. Thus T_0 is also a Jordan-Hölder tower. Consequently, T has the Jordan-Hölder refinement T_0. ◊

Definition We say that an R-module is of *finite height* if it is both artinian and noetherian.

EXERCISES

5.1 If A is a commutative integral domain prove that the following statements are equivalent

1. A is a principal ideal domain;
2. as an A-module, A is noetherian and the sum of two principal ideals of A is a principal ideal.

[*Hint.* (2) $\Rightarrow$ (1) : If I is an ideal of A then, as an A-module, I is finitely generated, say by $\{x_1, \ldots, x_n\}$; observe that $I = \sum_{i=1}^{n} Ax_i$.]

5.2 Let M be an R-module of finite height. If N is a submodule of M prove that there is a Jordan-Hölder tower

$$M = M_0 \supset M_1 \supset M_2 \supset \ldots \supset M_{r-1} \supset M_r = \{0\}.$$

with, for some index k, $M_k = N$.

[*Hint.* Use Theorem 5.9; consider a tower of submodules of N.]

5.3 Let M and N be R-modules of finite height. Prove that if there is a non-zero R-morphism $f : M \to N$ then there are Jordan-Hölder towers

$$M = M_0 \supset M_1 \supset M_2 \supset \ldots \supset M_{r-1} \supset M_r = \{0\},$$
$$N = N_0 \supset N_1 \supset N_2 \supset \ldots \supset N_{t-1} \supset N_t = \{0\},$$

with the property that, for some i and j,

$$M_i/M_{i+1} \simeq N_j/N_{j+1}.$$

[*Hint.* Use Exercise 5.2; consider a Jordan-Hölder tower through Ker f.]

5.4 If M is an R-module of finite height and if N is a submodule of M prove that
$$h(N) + h(M/N) = h(M).$$

[*Hint.* Use Exercise 5.2.]

Deduce that $N = M$ if and only if $h(N) = h(M)$.

5.5 If M and N are R-modules with M of finite height and if $f : M \to N$ is an R-morphism, prove that $\operatorname{Im} f$ and $\operatorname{Ker} f$ are of finite height with
$$h(\operatorname{Im} f) + h(\operatorname{Ker} f) = h(M).$$

5.6 Let $M_1, \ldots, M_n$ be R-modules of finite height. Prove that if there is an exact sequence
$$0 \longrightarrow M_1 \xrightarrow{f_1} M_2 \xrightarrow{f_2} \cdots \xrightarrow{f_{n-1}} M_n \longrightarrow 0$$
then $\sum_{k=1}^{n}(-1)^k h(M_k) = 0$.

[*Hint.* Use Theorem 3.6 (the case $n = 2$) and induction. For the inductive step use the sequences
$$0 \longrightarrow M_1 \xrightarrow{f_1} M_2 \xrightarrow{f_2} \cdots \longrightarrow M_{n-2} \xrightarrow{\pi} K \longrightarrow 0,$$
$$0 \longrightarrow K \xrightarrow{i} M_{n-1} \xrightarrow{f_{n-1}} M_n \longrightarrow 0$$
in which $K = \operatorname{Ker} f_{n-1} = \operatorname{Im} f_{n-2}$, $\pi = f_{n-2}^+$ and i is the natural inclusion.]

Deduce that if M and N are submodules of finite height of an R-module P then $M + N$ is of finite height with
$$h(M + N) + h(M \cap N) = h(M \times N) = h(M) + h(N).$$

[*Hint.* Apply the above to the exact sequence
$$0 \longrightarrow M \cap N \longrightarrow M \times N \longrightarrow M + N \longrightarrow 0$$
of Exercise 3.4.]

Show also that $P/(M \cap N)$ is of finite height with
$$h[P/(M \cap N)] + h[P/(M + N)] = h(P/M) + h(P/N).$$

[*Hint.* Use the exact sequence of Exercise 4.3.]

5.7 Let M be an R-module of finite height. If $f : M \to N$ is an R-morphism, prove that f is injective if and only if f is surjective.

[*Hint.* Observe that for every positive integer n the R-morphism f^n is injective/surjective whenever f is injective/surjective. Consider the chains
$$0 \subseteq \operatorname{Ker} f \subseteq \operatorname{Ker} f^2 \subseteq \ldots \subseteq \operatorname{Ker} f^n \subseteq \operatorname{Ker} f^{n+1} \subseteq \cdots ;$$
$$M \supseteq \operatorname{Im} f \supseteq \operatorname{Im} f^2 \supseteq \ldots \supseteq \operatorname{Im} f^n \supseteq \operatorname{Im} f^{n+1} \supseteq \cdots .]$$

5.8 Let M and N be noetherian R-modules and let P be an R-module such that there is a short exact sequence

$$0 \longrightarrow M \xrightarrow{f} P \xrightarrow{g} N \longrightarrow 0.$$

If A is a submodule of P show that $A \cap \operatorname{Im} f$ is finitely generated, say by $\{x_1, \ldots, x_r\}$. Now show that there exist $y_1, \ldots, y_n \in A$ such that

$$\{y_1, \ldots, y_n, x_1, \ldots, x_r\}$$

generates A. Deduce that P is also noetherian.

5.9 Determine which of the chain conditions, if any, are satisfied in each of the following modules:

(1) $\mathbb{Z}$ as a $\mathbb{Z}$-module;

(2) $\mathbb{Z}_m$ as a $\mathbb{Z}$-module;

(3) $\mathbb{Z}_m$ as a $\mathbb{Z}_m$-module;

(4) $\mathbb{Q}$ as a $\mathbb{Q}$-module;

(5) $\mathbb{Q}$ as a $\mathbb{Z}$-module;

(6) $\mathbb{Q}[X]$ as a $\mathbb{Q}$-module;

(7) $\mathbb{Q}[X]$ as a $\mathbb{Q}[X]$-module;

(8) $\mathbb{Q}[X]/M$ as a $\mathbb{Q}[X]$-module, where M is a proper submodule.

5.10 Let F be a field and let M_n be the ring of $n \times n$ matrices over F. For $i = 1, \ldots, n$ let $E_i \in M_n$ be the matrix whose (i,i)-th entry is 1 and all other entries are 0. For $i = 1, \ldots, n$ define

$$B_i = M_n(E_1 + \ldots + E_n).$$

Prove that

$$M_n = B_n \supset B_{n-1} \supset \ldots \supset B_1 \supset B_0 = \{0\}$$

is a Jordan-Hölder tower for the M_n-module M_n.

5.11 For a given prime p let

$$\mathbb{Q}_p = \left\{ x \in \mathbb{Q} \; ; \; (\exists k \in \mathbb{Z})(\exists n \in \mathbb{N}) \; x = \frac{k}{p^n} \right\}.$$

Show that the $\mathbb{Z}$-module $\mathbb{Q}_p/\mathbb{Z}$ is artinian but not noetherian.

[*Hint.* Let H be a non-zero subgroup of $\mathbb{Q}_p/\mathbb{Z}$. Let t be the smallest positive integer such that, for some k relatively prime to p, $k/p^t + \mathbb{Z} \notin H$. Show that H coincides with $S_{t-1}/\mathbb{Z}$ where

$$S_{t-1} = \left\{0, \frac{1}{p^{t-1}}, \frac{2}{p^{t-1}}, \ldots, \frac{p^{t-1}-1}{p^{t-1}}\right\}.]$$

6 Products and coproducts

We turn our attention now to another important way of constructing new modules from old ones. For this purpose, we begin with a very simple case of what we shall consider in general.

If A and B are R-modules then the cartesian product set $A \times B$ can be made into an R-module in the obvious component-wise manner, namely by defining

$$(a_1, b_1) + (a_2, b_2) = (a_1 + a_2, b_1 + b_2), \quad \lambda(a_1, a_2) = (\lambda a_1, \lambda a_2).$$

Associated with this *cartesian product module* $A \times B$ are the natural epimorphisms $\mathrm{pr}_A : A \times B \to A$ and $\mathrm{pr}_B : A \times B \to B$ given by $\mathrm{pr}_A(a, b) = a$ and $\mathrm{pr}_B(a, b) = b$. Now there is an interesting 'element-free' characterisation of such a cartesian product module, namely that if X is any R-module and $f_1 : X \to A$ and $f_2 : X \to B$ are R-morphisms then there is a unique R-morphism $\vartheta : X \to A \times B$ such that the diagrams

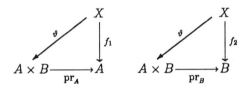

are commutative. In fact, ϑ is given by $\vartheta(x) = \big(f_1(x), f_2(x)\big)$.

Our immediate objective is to generalise this to an arbitrary collection of R-modules. For this purpose, we introduce the following notion.

Definition Let $(M_i)_{i \in I}$ be a family of R-modules. By a *product* of this family we shall mean an R-module P together with a family $(f_i)_{i \in I}$ of R-morphisms $f_i : P \to M_i$ such that, for every R-module M and every family $(g_i)_{i \in I}$ of R-morphisms $g_i : M \to M_i$, there is a unique R-morphism $h : M \to P$ such that every diagram

is commutative. We denote such a product module by $(P, (f_i)_{i \in I})$.

Products and coproducts

Theorem 6.1 *If $(P, (f_i)_{i \in I})$ is a product of a family $(M_i)_{i \in I}$ of R-modules then each f_i is an epimorphism.*

Proof Given $i \in I$, take $M = M_i, g_i = \mathrm{id}_{M_i}$ and g_j the zero map for $j \neq i$ in the above definition. Then from $f_i \circ h = \mathrm{id}_{M_i}$ we deduce that f_i is an epimorphism. ◊

Theorem 6.2 [Uniqueness] *Let $(P, (f_i)_{i \in I})$ be a product of the family $(M_i)_{i \in I}$ of R-modules. Then $(P', (f'_i)_{i \in I})$ is also a product of this family if and only if there is a unique R-isomorphism $h : P' \to P$ such that $(\forall i \in I)\ f_i \circ h = f'_i$.*

Proof Suppose that $(P', (f'_i)_{i \in I})$ is also a product. Then there exist unique morphisms $h : P' \to P$ and $k : P \to P'$ such that, for every $i \in I$, the diagrams

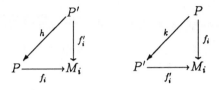

are commutative. Since then $f_i \circ h \circ k = f'_i \circ k = f_i$, the diagram

is commutative for every $i \in I$. But, from the definition of product, only one R-morphism exists making this last diagram commutative for every $i \in I$; and clearly id_P does just that. We deduce, therefore, that $h \circ k = \mathrm{id}_P$. In a similar way we can show that $k \circ h = \mathrm{id}_{P'}$, whence we see that h is an isomorphism with $h^{-1} = k$.

Suppose, conversely, that that the condition holds. Then since $f_i = f'_i \circ h^{-1}$ for every $i \in I$, we can use the fact that $(P, (f_i)_{i \in I})$ is a product to produce a unique R-morphism $\vartheta : M \to P$ such that the diagram

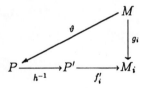

is commutative. Consider now $h^{-1} \circ \vartheta : M \to P'$. We have $f'_i \circ (h^{-1} \circ \vartheta) = g_i$, and if $t : M \to P'$ is any R-morphism such that $f'_i \circ t = g_i$ then the equalities

$$g_i = f'_i \circ t = f'_i \circ h^{-1} \circ h \circ t$$

together with the uniqueness property of ϑ give $h \circ t = \vartheta$ whence $t = h^{-1} \circ \vartheta$. This then shows that $(P', (f'_i)_{i \in I})$ is also a product of $(M_i)_{i \in I}$. ◊

We shall now settle the question concerning the existence of products. For this purpose, we ask the reader to recall that if $(E_i)_{i \in I}$ is a family of sets indexed by I then the cartesian product set $\underset{i \in I}{\times} E_i$ is defined to be the set of all mappings $f : I \to \bigcup_{i \in I} E_i$ such that $f(i) \in E_i$ for each $i \in I$. Following standard practice, we write $f(i)$ as x_i and denote f by $(x_i)_{i \in I}$, so that $\underset{i \in I}{\times} E_i$ consists of those families $(x_i)_{i \in I}$ of elements of $\bigcup_{i \in I} E_i$ such that $x_i \in E_i$ for every $i \in I$.

Given a family $(M_i)_{i \in I}$ of R-modules, the cartesian product set $\underset{i \in I}{\times} M_i$ can be given the structure of an R-module in an obvious way, namely by defining laws of composition by

$$(m_i)_{i \in I} + (n_i)_{i \in I} = (m_i + n_i)_{i \in I}, \quad \lambda(m_i)_{i \in I} = (\lambda m_i)_{i \in I}.$$

We shall also denote the R-module so formed by $\underset{i \in I}{\times} M_i$ without confusion and shall call it the *cartesian product module* of the family $(M_i)_{i \in I}$. For every $j \in I$ the j-th *canonical projection* $\text{pr}_j : \underset{i \in I}{\times} M_i \to M_j$ is defined by the prescription $\text{pr}_j((m_i)_{i \in I}) = m_j$. It is clear that each pr_j is an R-epimorphism.

Theorem 6.3 [Existence] $\left(\underset{i \in I}{\times} M_i, (\text{pr}_i)_{i \in I}\right)$ *is a product of the family* $(M_i)_{i \in I}$.

Proof Let M be an R-module and let $(g_i)_{i \in I}$ be a family of R-morphisms with $g_i : M \to M_i$ for each $i \in I$. Define a mapping $h : M \to \underset{i \in I}{\times} M_i$ as follows : for every $x \in M$ let the i-th component of $h(x)$ be given by $(h(x))_i = g_i(x)$; in other words, $h(x) = (g_i(x))_{i \in I}$. It is then readily verified that h is an R-morphism and is such that $\text{pr}_i \circ h = g_i$ for every $i \in I$. That h is unique follows from the fact that if $k : M \to \underset{i \in I}{\times} M_i$ is also an R-morphism such that $\text{pr}_i \circ k = g_i$ for every $i \in I$ then clearly

$$(\forall i \in I)(\forall x \in M) \qquad (k(x))_i = g_i(x) = (h(x))_i$$

Products and coproducts

whence we see that $k = h$. ◊

Corollary 1 [Commutativity of $\bigtimes$] *If $\sigma : I \to I$ is a bijection then there is an R-isomorphism*
$$\bigtimes_{i \in I} M_i \simeq \bigtimes_{i \in I} M_{\sigma(i)}.$$

Proof It is clear that $\left(\bigtimes_{i \in I} M_{\sigma(i)}, (\mathrm{pr}_{\sigma(i)})_{i \in I}\right)$ is also a product of $(M_i)_{i \in I}$ whence the result follows by Theorem 6.2. ◊

Corollary 2 [Associativity of $\bigtimes$] *If $\{I_k \ ; \ k \in A\}$ is a partition of I then there is an R-isomorphism*
$$\bigtimes_{i \in I} M_i \simeq \bigtimes_{k \in A}\left(\bigtimes_{i \in I_k} M_i\right).$$

Proof Let M be an arbitrary R-module. Given any M_i let $k \in A$ be such that $i \in I_k$. Then, referring to the diagram

$$\begin{array}{c}
 & & & & M \\
 & & & \swarrow f \ \downarrow h_k \ \downarrow g_i & \\
\bigtimes_{k \in A}\left(\bigtimes_{i \in I_k} M_i\right) \xrightarrow{\mathrm{pr}_k} & \bigtimes_{i \in I_k} M_i \xrightarrow{\mathrm{pr}_i} & M_i
\end{array}$$

there is a unique R-morphism $h_k : M \to \bigtimes_{i \in I_k} M_i$ such that $\mathrm{pr}_i \circ h_k = g_i$ for every $i \in I_k$, and a unique R-morphism $f : M \to \bigtimes_{k \in A}\left(\bigtimes_{i \in I_k} M_i\right)$ such that $\mathrm{pr}_k \circ f = h_k$ for every $k \in A$. It follows that f satisfies the property

$$(\forall i \in I_k)(\forall k \in A) \qquad \mathrm{pr}_i \circ \mathrm{pr}_k \circ f = g_i.$$

Suppose now that $f' : M \to \bigtimes_{k \in A}\left(\bigtimes_{i \in I_k} M_i\right)$ is also an R-morphism satisfying this property. Given $x \in M$, let $f(x) = (x_k)_{k \in A}$ where $x_k = (m_i)_{i \in I_k}$ for every $k \in A$, and let $f'(x) = (x'_k)_{k \in A}$ where $x'_k = (m'_i)_{i \in I_k}$ for every $k \in A$. Then we have

$$\begin{cases} g_i(x) = (\mathrm{pr}_i \circ \mathrm{pr}_k)[f(x)] = \mathrm{pr}_i(x_k) = m_i, \\ g_i(x) = (\mathrm{pr}_i \circ \mathrm{pr}_k)[f'(x)] = \mathrm{pr}_i(x'_k) = m'_i, \end{cases}$$

whence it follows that $f'(x) = f(x)$ and consequently that $f' = f$. This then shows that $\bigtimes_{k \in A}\left(\bigtimes_{i \in I_k} M_i\right)$ together with the family $(\mathrm{pr}_i \circ \mathrm{pr}_k)_{i \in I_k, k \in A}$ is a product of the family $(M_i)_{i \in I}$. The required isomorphism now follows by Theorem 6.2. ◊

The above results show that, to within R-isomorphism, there is a unique product of the family $(M_i)_{i \in I}$. As a model of this we can choose the cartesian product module $\underset{i \in I}{\bigtimes} M_i$ together with the canonical projections. In the case where I is finite, say $I = \{1, \ldots, n\}$, we often write $\underset{i \in I}{\bigtimes} M_i$ as $\overset{n}{\underset{i=1}{\bigtimes}} M_i$ or as $M_1 \times \ldots \times M_n$, the elements of which are the n-tuples $(m_1, \ldots, m_n)$ with $m_i \in M_i$ for $i = 1, \ldots, n$. Note that Corollary 2 above implies in particular that

$$\underset{i \in I}{\bigtimes} M_i \simeq M_i \times \underset{j \neq i}{\bigtimes} M_j,$$

and that Corollary 1 implies in particular that $M \times N \simeq N \times M$. Thus, for example, we have

$$M_1 \times (M_2 \times M_3) \simeq \overset{3}{\underset{i=1}{\bigtimes}} M_i \simeq (M_1 \times M_2) \times M_3.$$

We shall now consider the question that is dual to the above, namely that which is obtained by reversing all the arrows.

Definition Let $(M_i)_{i \in I}$ be a family of R-modules. By a *coproduct* of this family we shall mean an R-module C together with a family $(f_i)_{i \in I}$ of R-morphisms $f_i : M_i \to C$ such that, for every R-module M and every family $(g_i)_{i \in I}$ of R-morphisms $g_i : M_i \to M$, there is a unique R-morphism $h : C \to M$ such that every diagram

is commutative. We denote such a coproduct by $(C, (f_i)_{i \in I})$.

Theorem 6.4 *If $(C, (f_i)_{i \in I})$ is a coproduct of the family $(M_i)_{i \in I}$ of R-modules then each f_i is a monomorphism.*

Proof Given $i \in I$, take $M = M_i$, $g_i = \mathrm{id}_{M_i}$ and g_j the zero map for $j \neq i$ in the above definition. Then from $h \circ f_i = \mathrm{id}_{M_i}$ we see that f_i is a monomorphism. ◊

Products and coproducts

Theorem 6.5 [Uniqueness] Let $(C,(f_i)_{i\in I})$ be a coproduct of the family $(M_i)_{i\in I}$ of R-modules. Then $(C',(f'_i)_{i\in I})$ is also a coproduct of $(M_i)_{i\in I}$ if and only if there is an R-isomorphism $h : C \to C'$ such that $(\forall i \in I)\ h \circ f_i = f'_i$.

Proof This is exactly the dual of the proof of Theorem 6.2; we leave the details to the reader. ◇

As to the existence of coproducts, consider the subset of the cartesian product module $\underset{i\in I}{\times} M_i$ that consists of those families $(m_i)_{i\in I}$ of elements of $\underset{i\in I}{\bigcup} M_i$ which are such that $m_i = 0$ for 'almost all' $i \in I$; i.e. $m_i = 0$ for all but finitely many m_i. It is clear that this subset of $\underset{i\in I}{\times} M_i$ is a submodule of $\underset{i\in I}{\times} M_i$. We call it the *external direct sum* of the family $(M_i)_{i\in I}$ and denote it by $\underset{i\in I}{\bigoplus} M_i$. For every $j \in I$ we denote by $\mathrm{in}_j : M_j \to \underset{i\in I}{\bigoplus} M_i$ the mapping given by the prescription $\mathrm{in}_j(x) = (x_i)_{i\in I}$, where $x_i = 0$ for $i \ne j$ and $x_j = x$. It is readily seen that in_j is an R-monomorphism; we call it the j-th *canonical injection* of M_j into $\underset{i\in I}{\bigoplus} M_i$.

Theorem 6.6 [Existence] $\left(\underset{i\in I}{\bigoplus} M_i, (\mathrm{in}_i)_{i\in I}\right)$ is a coproduct of the family $(M_i)_{i\in I}$.

Proof Let M be an R-module and let $(g_i)_{i\in I}$ be a family of R-morphisms with $g_i : M_i \to M$ for every $i \in I$. Define a mapping $h : \underset{i\in I}{\bigoplus} M_i \to M$ by the prescription

$$h((m_i)_{i\in I}) = \sum_{i\in I} g_i(m_i).$$

Note immediately that h is well defined since for every family $(m_i)_{i\in I}$ all but a finite number of the m_i are zero. It is readily verified that h is an R-morphism. Moreover, for every $x \in M_i$ we have

$$h[\mathrm{in}_i(x)] = g_i(x)$$

and so $h \circ \mathrm{in}_i = g_i$ for every $i \in I$.

To show that h is unique, suppose that $k : \underset{i\in I}{\bigoplus} M_i \to M$ is also an R-morphism such that $k \circ \mathrm{in}_i = g_i$ for all $i \in I$. Then for all $(m_i)_{i\in I} \in \underset{i\in I}{\bigoplus} M_i$ we have, recalling that all sums involved are well defined,

$$k((m_i)_{i\in I}) = \sum_{i\in I}(k \circ \mathrm{in}_i)(m_i) = \sum_{i\in I} g_i(m_i) = h((m_i)_{i\in I}),$$

and consequently $k = h$. ◊

Corollary 1 [Commutativity of $\bigoplus$] *If $\sigma : I \to I$ is a bijection then there is an R-isomorphism*

$$\bigoplus_{i \in I} M_i \simeq \bigoplus_{i \in I} M_{\sigma(i)}.$$

Proof Clearly, $\left(\bigoplus_{i \in I} M_{\sigma(i)}, (\operatorname{in}_\sigma(i))_{i \in I}\right)$ is also a coproduct of $(M_i)_{i \in I}$, so the result follows from Theorem 6.5. ◊

Corollary 2 [Associativity of $\bigoplus$] *If $\{I_k\ ;\ k \in A\}$ is a partition of I then there is an R-isomorphism*

$$\bigoplus_{i \in I} M_i \simeq \bigoplus_{k \in A}\left(\bigoplus_{i \in I_k} M_i\right).$$

Proof Let M be an arbitrary R-module. Given any M_i let $k \in A$ be such that $i \in I_k$. Then in the diagram

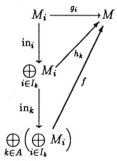

there is a unique R-morphism $h_k : \bigoplus_{i \in I_k} M_i \to M$ such that $h_k \circ \operatorname{in}_i = g_i$ for every $i \in I_k$, and a unique R-morphism f such that $f \circ \operatorname{in}_k = h_k$ for every $k \in A$. It follows that f satisfies the property

$$(\forall i \in I_k)(\forall k \in A) \quad f \circ \operatorname{in}_k \circ \operatorname{in}_i = g_i.$$

Suppose now that $f' : \bigoplus_{k \in A}\left(\bigoplus_{i \in I_k} M_i\right) \to M$ is also an R-morphism that satisfies this property. Let $\operatorname{pr}_j^\oplus : \bigoplus_{i \in I} M_i \to M_j$ be the restriction to $\bigoplus_{i \in I} M_i$ of $\operatorname{pr}_j : \bigtimes_{i \in I} M_i \to M_j$. Then, observing that $\sum_{j \in I}(\operatorname{in}_j \circ \operatorname{pr}_j^\oplus)$ is the identity

Products and coproducts   63

map on $\bigoplus_{j \in I} M_j$, we have

$$\begin{aligned}
f' &= \sum_{k \in A} \sum_{i \in I_k} (f' \circ \mathrm{in}_k \circ \mathrm{in}_i \circ \mathrm{pr}_i^\oplus \circ \mathrm{pr}_k^\oplus) \\
&= \sum_{k \in A} \sum_{i \in I_k} (g_i \circ \mathrm{pr}_i^\oplus \circ \mathrm{pr}_k^\oplus) \\
&= \sum_{k \in A} \sum_{i \in I_k} (f \circ \mathrm{in}_k \circ \mathrm{in}_i \circ \mathrm{pr}_i^\oplus \circ \mathrm{pr}_k^\oplus) \\
&= f.
\end{aligned}$$

This shows that $\bigoplus_{k \in A} \left(\bigoplus_{i \in I_k} M_i \right)$ together with the family $(\mathrm{in}_k \circ \mathrm{in}_i)_{i \in I_k, k \in A}$ is a coproduct of the family $(M_i)_{i \in I}$. The required isomorphism now follows by Theorem 6.5. ◇

The above results show that, to within R-isomorphism, there is a unique coproduct of the family $(M_i)_{i \in I}$ of R-modules. As a model of this we can choose the external direct sum $\bigoplus_{i \in I} M_i$ together with the canonical injections. In the case where I is finite, say $I = \{1, \ldots, n\}$, we often write $\bigoplus_{i \in I} M_i$ as $\bigoplus_{i=1}^{n} M_i$ or as $M_1 \oplus \ldots \oplus M_n$, the elements of which are the n-tuples $(m_1, \ldots, m_n)$ with $m_i \in M_i$ for $i = 1, \ldots, n$.

- Thus we see that *when the index set I is finite, the modules $\bigoplus_{i \in I} M_i$ and $\underset{i \in I}{\times} M_i$ coincide.*

Note also that Corollary 2 above implies in particular that

$$\bigoplus_{i \in I} M_i \simeq M_i \oplus \bigoplus_{j \neq i} M_j,$$

and that Corollary 1 implies in particular that $M_1 \oplus M_2 \simeq M_2 \oplus M_1$. Thus, for example, we have

$$M_1 \oplus (M_2 \oplus M_3) \simeq \bigoplus_{i=1}^{3} M_i \simeq (M_1 \oplus M_2) \oplus M_3.$$

- The reader will find it instructive to verify Theorem 6.6 in the case where $I = \{1, 2\}$.

The following is a useful characterisation of coproducts.

Theorem 6.7 *An R-module M and a family $(i_j)_{j \in I}$ of R-morphisms $i_j : M_j \to M$ form a coproduct of $(M_j)_{j \in I}$ if and only if there is a family $(\pi_j)_{j \in I}$ of R-morphisms $\pi_j : M \to M_j$ such that*

(1) $\pi_k \circ i_j = \begin{cases} \mathrm{id}_{M_j} & \text{if } k = j, \\ 0 & \text{if } k \neq j; \end{cases}$

(2) $(\forall m \in M)\ \pi_j(m) = 0$ *for all but finitely many* $j \in I$ *and*
$$\sum_{j \in I} (i_j \circ \pi_j)(m) = m.$$

Proof Suppose first that $(M, (i_j)_{j \in I})$ is a coproduct of $(M_j)_{j \in I}$. Then by Theorems 6.5 and 6.6 there is an R-isomorphism $f : \bigoplus_{j \in I} M_j \to M$ such that, for every $j \in I$, the diagram

is commutative. For every $j \in I$ define $\pi_j : M \to M_j$ by
$$\pi_j = \mathrm{pr}_j^\oplus \circ f^{-1}$$
where $\mathrm{pr}_j^\oplus$ denotes the restriction to $\bigoplus_{j \in I} M_j$ of the j-th projection $\mathrm{pr}_j : \bigtimes_{j \in I} M_j \to M_j$. Then for all $k, j \in I$ we have
$$\pi_k \circ i_j = \mathrm{pr}_j^\oplus \circ f^{-1} \circ i_j = \mathrm{pr}_j^\oplus \circ \mathrm{in}_j = \begin{cases} \mathrm{id}_{M_j} & \text{if } k = j; \\ 0 & \text{if } k \neq j. \end{cases}$$

Moreover, for every $m \in M$, the equality $\pi_j(m) = \mathrm{pr}_j^\oplus[f^{-1}(m)]$ shows that $\pi_j(m)$ is zero for all but a finite number of $j \in I$. Finally,
$$\sum_{j \in I} (i_j \circ \pi_j)(m) = \sum_{j \in I} (f \circ \mathrm{in}_j \circ \mathrm{pr}_j^\oplus \circ f^{-1})(m)$$
$$= \left(f \circ \sum_{j \in I} (\mathrm{in}_j \circ \mathrm{pr}_j^\oplus) \circ f^{-1} \right)(m)$$
$$= m,$$
since $\sum_{j \in I} (\mathrm{in}_j \circ \mathrm{pr}_j^\oplus)$ is the identity map on $\bigoplus_{j \in I} M_j$.

Suppose, conversely, that (1) and (2) hold. Then, by (2), the prescription
$$(\forall m \in M) \qquad g(m) = \sum_{j \in I} (\mathrm{in}_j \circ \pi_j)(m)$$

Products and coproducts   65

yields an R-morphism $g : M \to \bigoplus_{j \in I} M_j$. By Theorem 6.6, there is an R-morphism $h : \bigoplus_{j \in I} M_j \to M$ such that $h \circ \text{in}_j = i_j$ for every $j \in I$. The diagram

$$\begin{array}{c} M_j \xrightleftharpoons[\pi_j]{i_j} M \\ \text{in}_j \downarrow \; \swarrow g \; \nearrow h \\ \bigoplus_{j \in I} M_j \end{array}$$

now gives, for every $m \in M$,

$$h[g(m)] = \sum_{j \in I}(h \circ \text{in}_j \circ \pi_j)(m) = \sum_{j \in I}(i_j \circ \pi_j)(m) = m$$

and so $h \circ g$ is the identity map on M. Now given $x \in \bigoplus_{j \in I} M_j$ we also have, using the fact that $\sum_{j \in I}(\text{in}_j \circ \text{pr}_j^\oplus)(x) = x$,

$$\begin{aligned} g[h(x)] &= \sum_{j \in I}(\text{in}_j \circ \pi_j)[h(x)] \\ &= \sum_{j \in I}\sum_{k \in I}(\text{in}_j \circ \pi_j \circ h \circ \text{in}_k \circ \text{pr}_k^\oplus)(x) \\ &= \sum_{j \in I}\sum_{k \in I}(\text{in}_j \circ \pi_j \circ i_k \circ \text{pr}_k^\oplus)(x) \\ &= \sum_{j \in I}(\text{in}_j \circ \text{pr}_j^\oplus)(x) \quad \text{by (1)} \\ &= x. \end{aligned}$$

Thus $g \circ h$ is the identity map on $\bigoplus_{j \in I} M_j$. It follows that g, h are mutually inverse R-isomorphisms. Appealing now to Theorem 6.5 we see that $(M,(i_j)_{j \in I})$ is a coproduct of $(M_j)_{j \in I}$. $\Diamond$

- Note that when the index set I is finite, say $I = \{1, \ldots, n\}$, condition (2) of Theorem 6.7 simplifies to $\sum_{j=1}^{n}(i_j \circ \pi_j) = \text{id}_M$.

- There is, of course, a characterisation of products that is dual to that in Theorem 6.7; see Exercise 6.9.

We now turn our attention to a family $(M_i)_{i \in I}$ where each M_i is a submodule of a given R-module M. For every $j \in I$ let $\iota_j : M_j \to M$ be the natural inclusion and let $h : \bigoplus_{i \in I} M_i \to M$ be the unique R-morphism such that the diagram

is commutative for every $j \in I$. From the proof of Theorem 6.6, we know that h is given by

$$h((m_j)_{j \in I}) = \sum_{j \in I} \iota_j(m_j) = \sum_{j \in I} m_j.$$

Thus we see that $\operatorname{Im} h$ is the submodule $\sum_{i \in I} M_i$. Put another way, we have the exact sequence

$$\bigoplus_{i \in I} M_i \xrightarrow{h} M \xrightarrow{\natural} M/\sum_{i \in I} M_i \longrightarrow 0.$$

Definition If M is an R-module and $(M_i)_{i \in I}$ is a family of submodules of M then we shall say that M is the *internal direct sum* of the family $(M_i)_{i \in I}$ if the mapping $h : \bigoplus_{i \in I} M_i \to M$ described above is an isomorphism.

- Since, by the above definition, internal direct sums (when they exist) are isomorphic to external direct sums, we shall adopt the practice of denoting internal direct sums also by the symbol $\bigoplus$. We shall also follow the common practice of dropping the adjectives 'internal' and 'external' as applied to direct sums since the context will always make it clear which one is involved.

Theorem 6.8 *An R-module M is the direct sum of a family $(M_i)_{i \in I}$ of submodules if and only if every $x \in M$ can be written in a unique way as $x = \sum_{i \in I} m_i$ where $m_i \in M_i$ for every $i \in I$ with almost all $m_i = 0$.*

Proof The mapping h described above is surjective if and only if $M = \operatorname{Im} h = \sum_{i \in I} M_i$, which is equivalent to saying that every $x \in M$ can be expressed in the form $x = \sum_{i \in I} m_i$ where $m_i \in M_i$ for every $i \in I$ and almost all $m_i = 0$. Also, since we have $\sum_{i \in I} m_i = h((m_i)_{i \in I})$, we see that h is injective if and only if such expressions are unique. $\Diamond$

Products and coproducts

Theorem 6.9 *Let $(M_i)_{i \in I}$ be a family of submodules of an R-module M. Then the following statements are equivalent:*

(1) $\sum_{i \in I} M_i$ *is the direct sum of* $(M_i)_{i \in I}$;

(2) *if* $\sum_{i \in I} m_i = 0$ *with* $m_i \in M_i$ *for every* $i \in I$ *then* $m_i = 0$ *for every* $i \in I$;

(3) $(\forall i \in I)\ M_i \cap \sum_{j \neq i} M_j = \{0\}$.

Proof $(1) \Rightarrow (2)$: By Theorem 6.8, the only way 0 can be expressed as a sum is the trivial way.

$(2) \Rightarrow (3)$: Let $x \in M_i \cap \sum_{j \neq i} M_j$, say $x = m_i = \sum_{j \neq i} m_j$. Then $m_i + \sum_{j \neq i}(-m_j) = 0$ whence, by (2), $m_i = 0$ and so $x = 0$.

$(3) \Rightarrow (1)$: Suppose that $\sum_{i \in I} m_i = \sum_{i \in I} n_i$ with $m_i, n_i \in M_i$ for each i. Then we have

$$m_i - n_i = \sum_{j \neq i}(n_j - m_j)$$

where the left-hand side belongs to M_i and the right-hand side belongs to $\sum_{j \neq i} M_j$. By (3) we deduce that $m_i - n_i = 0$, whence (1) follows by Theorem 6.8. $\Diamond$

Corollary *If A, B are submodules of M then $M = A \oplus B$ if and only if $M = A + B$ and $A \cap B = \{0\}$.* $\Diamond$

Definition We shall say that submodules A, B of an R-module M are *supplementary* if $M = A \oplus B$. A submodule N of M is called a *direct summand* of M if there is a submodule P of M such that N and P are supplementary.

- It should be noted that an arbitrary submodule of an R-module M need not have a supplement. For example, let $p, q \in \mathbb{Z} \setminus \{0, 1\}$ and consider the submodules $p\mathbb{Z}, q\mathbb{Z}$ of $\mathbb{Z}$. We have $p\mathbb{Z} \cap q\mathbb{Z} \neq \{0\}$. Since all the submodules of $\mathbb{Z}$ are of the form $n\mathbb{Z}$ for some $n \in \mathbb{Z}$, it follows that $p\mathbb{Z}$ (with $p \neq 0, 1$) has no supplement in $\mathbb{Z}$.

- Note also that supplements need not be unique. For example, consider the $\mathbb{R}$-vector space $\mathbb{R}^2$ and the subspaces

$$X = \{(x, 0)\ ;\ x \in \mathbb{R}\},\ Y = \{(0, y)\ ;\ y \in \mathbb{R}\},\ D = \{(r, r)\ ;\ r \in \mathbb{R}\}.$$

It is clear that every $(x, y) \in \mathbb{R}^2$ can be expressed as

$$(x, y) = (x, x) + (0, y - x) = (y, y) + (x - y, 0)$$

whence we see by Corollary 1 of Theorem 6.9 that $\mathbb{R}^2 = D \oplus Y = D \oplus X$. However, as our next result shows, any two supplements of a submodule are isomorphic.

Theorem 6.10 *If M_1, M_2 are supplementary submodules of M. Then $M_2 \simeq M/M_1$.*

Proof Since $M = M_1 \oplus M_2$ the canonical projection $\mathrm{pr}_2 : M \to M_2$, described by $\mathrm{pr}_2(m_1 + m_2) = m_2$, is clearly surjective with $\mathrm{Ker}\,\mathrm{pr}_2 = M_1$. Thus, by the first isomorphism theorem, we have

$$M_2 = \mathrm{Im}\,\mathrm{pr}_2 \simeq M/\mathrm{Ker}\,\mathrm{pr}_2 = M/M_1. \quad \diamond$$

The notion of a direct summand is intimately related to a paticular type of short exact sequence. We shall now describe this connection.

Definition An exact sequence of the form $N \xrightarrow{g} P \longrightarrow 0$ is said to *split* if there is an R-morphism $\pi : P \to N$ such that $g \circ \pi = \mathrm{id}_P$. Similarly, an exact sequence $0 \longrightarrow M \xrightarrow{f} N$ is said to *split* if there is an R-morphism $\rho : N \to M$ such that $\rho \circ f = \mathrm{id}_M$. Such R-morphisms π, ρ will be called *splitting morphisms*.

- Note that in the above definition g is an epimorphism so there always exists a *mapping* $\pi : P \to N$ such that $g \circ \pi = \mathrm{id}_P$; and since f is a monomorphism there always exists a *mapping* $\rho : N \to M$ such that $\rho \circ f = \mathrm{id}_M$.

Definition We shall say that a short exact sequence

$$0 \longrightarrow M \xrightarrow{f} N \xrightarrow{g} P \longrightarrow 0$$

splits on the right whenever $N \xrightarrow{g} P \longrightarrow 0$ splits; and that it *splits on the left* whenever $0 \longrightarrow M \xrightarrow{f} N$ splits.

Theorem 6.11 *For a short exact sequence*

$$0 \longrightarrow M \xrightarrow{f} N \xrightarrow{g} P \longrightarrow 0$$

the following statements are equivalent :
  (1) *the sequence splits on the right;*
  (2) *the sequence splits on the left;*
  (3) $\mathrm{Im}\,f = \mathrm{Ker}\,g$ *is a direct summand of N.*

Proof (1) $\Rightarrow$ (3) : Suppose that $\pi : P \to N$ is a right-hand splitting morphism and consider $\mathrm{Ker}\,g \cap \mathrm{Im}\,\pi$. If $x \in \mathrm{Ker}\,g \cap \mathrm{Im}\,\pi$ then $g(x) = 0_P$ and $x = \pi(p)$ for some $p \in P$ whence

$$0_P = g(x) = g[\pi(p)] = p$$

Products and coproducts 69

and consequently $x = \pi(p) = \pi(0_P) = 0_N$. Thus we see that $\operatorname{Ker} g \cap \operatorname{Im} \pi = \{0\}$. Moreover, for every $n \in N$ we have

$$g\bigl(n - (\pi \circ g)(n)\bigr) = g(n) - (g \circ \pi \circ g)(n) = g(n) - g(n) = 0_P$$

and so $n - (\pi \circ g)(n) \in \operatorname{Ker} g$. Observing that every $n \in N$ can be written

$$n = (\pi \circ g)(n) + n - (\pi \circ g)(n),$$

we see that $N = \operatorname{Im} \pi + \operatorname{Ker} g$. It follows by Corollary 1 of Theorem 6.9 that $N = \operatorname{Im} \pi \oplus \operatorname{Ker} g$.

(3) $\Rightarrow$ (1) : Suppose that $N = \operatorname{Ker} g \oplus A$ for some submodule A of N. Then every $n \in N$ can be written uniquely as $n = y + a$ where $y \in \operatorname{Ker} g$ and $a \in A$. Consider the restriction g_A of g to A. Since $g(n) = g(a)$ and g is surjective, we see that g_A is surjective. It is also injective since if $a \in \operatorname{Ker} g_A$ then $0 = g_A(a) = g(a)$ and so $a \in \operatorname{Ker} g \cap A = \{0\}$. Thus $g_A : A \to P$ is an R-isomorphism. Since then $g_A \circ g_A^{-1} = \operatorname{id}_P$ we see that g_A^{-1} induces a right-hand splitting morphism.

(2) $\Rightarrow$ (3) : Suppose that $\rho : N \to M$ is a left-hand splitting morphism. If $x \in \operatorname{Im} f \cap \operatorname{Ker} \rho$ then $x = f(m)$ for some $m \in M$ and $\rho(x) = 0_M$. Thus $0_M = \rho(x) = \rho[f(m)] = m$ and so $x = f(m) = f(0_M) = 0_N$ whence $\operatorname{Im} f \cap \operatorname{Ker} \rho = \{0\}$. Since every $n \in N$ can be written $n = (f \circ \rho)(n) + n - (f \circ \rho)(n)$ where $n - (f \circ \rho)(n) \in \operatorname{Ker} \rho$, we deduce that $N = \operatorname{Im} f \oplus \operatorname{Ker} \rho$.

(3) $\Rightarrow$ (2) : If $N = \operatorname{Im} f \oplus B$ then every $n \in N$ can be written uniquely as $n = x + b$ where $x \in \operatorname{Im} f$ and $b \in B$. Since f is a monomorphism there is precisely one $m \in M$ such that $f(m) = x$ so we can write this element as $f^{-1}(x)$ without confusion. Now define $\rho : N \to M$ by setting $\rho(n) = \rho(x + a) = f^{-1}(x)$. Then for all $m \in M$ we have $\rho[f(m)] = f^{-1}[f(m)] = m$, so that $\rho \circ f = \operatorname{id}_M$ and ρ is a left-hand splitting morphism. $\Diamond$

- Note from the above result that if the short exact sequence

$$0 \longrightarrow M \xrightarrow{f} N \xrightarrow{g} P \longrightarrow 0$$

splits and if ρ, π are splitting morphisms then we have

$$\operatorname{Ker} g \oplus \operatorname{Im} \pi = N = \operatorname{Im} f \oplus \operatorname{Ker} \rho.$$

Since π and f are monomorphisms we have $\operatorname{Im} \pi \simeq P$ and $\operatorname{Im} f \simeq M$, so

$$N \simeq \operatorname{Ker} g \oplus P = \operatorname{Im} f \oplus P \simeq M \oplus P.$$

Note also, conversely, that

$$0 \longrightarrow M \xrightarrow{i} M \oplus P \xrightarrow{\vartheta} P \longrightarrow 0$$

is split exact, where $i : M \to M \oplus P$ is the canonical injection $m \mapsto (m,0)$ and $\vartheta : M \oplus P \to P$ is the projection $(m,p) \mapsto p$.

Definition Let N and P be supplementary submodules of an R-module M, so that every $x \in M$ can be written uniquely in the form $x = n + p$ where $n \in N$ and $p \in P$. By the *projection on N parallel to P* we shall mean the mapping $p : M \to M$ described by $p(x) = n$. An R-morphism $f : M \to N$ is called a *projection* if there exist supplementary submodules N, P of M such that f is the projection on N parallel to P.

Theorem 6.12 *If M_1 and M_2 are supplementary submodules of an R-module M and if f is the projection on M_1 parallel to M_2 then*

(1) $M_1 = \operatorname{Im} f = \{x \in M \ ; \ f(x) = x\}$;
(2) $M_2 = \operatorname{Ker} f$;
(3) $f \circ f = f$.

Proof (1) It is clear that $M_1 = \operatorname{Im} f \supseteq \{x \in M \ ; \ f(x) = x\}$. If now $m_1 \in M_1$ then its unique representation as a sum is $m_1 = m_1 + 0$ whence $f(m_1) = m_1$ and consequently $m_1 \in \{x \in M \ ; \ f(x) = x\}$.

(2) Let $x \in M$ have the unique representation $x = m_1 + m_2$ where $m_1 \in M_1$ and $m_2 \in M_2$. Then since $f(x) = m_1$ we have

$$f(x) = 0 \iff m_1 = 0 \iff x = m_2 \in M_2.$$

(3) For every $x \in M$ we have $f(x) \in M_1$ and so, by (1), we have that $f[f(x)] = f(x)$. ◊

Definition A morphism $f : M \to M$ such that $f \circ f = f$ will be called *idempotent*.

By Theorem 6.12(3), projections are idempotent. In fact, as we shall now show, the converse is also true.

Theorem 6.13 *An R-morphism $f : M \to M$ is a projection if and only if it is idempotent, in which case $M = \operatorname{Im} f \oplus \operatorname{Ker} f$ and f is the projection on $\operatorname{Im} f$ parallel to $\operatorname{Ker} f$.*

Proof Suppose that f is idempotent. If $x \in \operatorname{Im} f \cap \operatorname{Ker} f$ then $x = f(y)$ for some $y \in M$ and $f(x) = 0$, so

$$x = f(y) = f[f(y)] = f(x) = 0.$$

Moreover, for every $x \in M$,

$$f(x - f(x)) = f(x) - f[f(x)] = 0,$$

Products and coproducts 71

and so $x - f(x) \in \operatorname{Ker} f$. Then from the identity $x = f(x) + x - f(x)$ we deduce that $M = \operatorname{Im} f \oplus \operatorname{Ker} f$.

Suppose now that $m = x + y$ where $x \in \operatorname{Im} f$ and $y \in \operatorname{Ker} f$. Then $x = f(z)$ for some $z \in M$ and $f(y) = 0$, so that

$$f(m) = f(x + y) = f(x) + 0 = f[f(z)] = f(z) = x.$$

In other words, f is the projection on $\operatorname{Im} f$ parallel to $\operatorname{Ker} f$.

As observed above, the converse is clear by Theorem 6.12(3). ◊

Corollary *If $f : M \to M$ is a projection then so is $\operatorname{id}_M - f$, in which case $\operatorname{Im} f = \operatorname{Ker}(\operatorname{id}_M - f)$.*

Proof Writing $f \circ f$ as f^2, we deduce from $f^2 = f$ that

$$(\operatorname{id}_M - f)^2 = \operatorname{id}_M - f - f + f^2 = \operatorname{id}_M - f.$$

Moreover, by Theorem 6.12(1),

$$x \in \operatorname{Im} f \iff x = f(x) \iff (\operatorname{id}_M - f)(x) = 0$$

and so $\operatorname{Im} f = \operatorname{Ker}(\operatorname{id}_M - f)$. ◊

We shall now show how the decomposition of an R-module into a finite direct sum of submodules can be expressed in terms of projections. As we shall see in due course, this result opens the door to a deep study of linear transformations and their representation by matrices.

Theorem 6.14 *A module M is a direct sum of submodules $M_1, \ldots, M_n$ if and only if there are non-zero morphisms $p_1, \ldots, p_n : M \to M$ (which are necessarily projections) such that*

(1) $\sum_{i=1}^{n} p_i = \operatorname{id}_M$;

(2) $(i \neq j) \; p_i \circ p_j = 0$.

Proof Suppose first that $M = \bigoplus_{i=1}^{n} M_i$. Then for each i we have $M = M_i \oplus \sum_{j \neq i} M_j$. Let p_i be the projection on M_i parallel to $\sum_{j \neq i} M_j$. Then, for $x \in M$ and $i \neq j$,

$$\begin{aligned} p_i[p_j(x)] \in p_i^{\to}(\operatorname{Im} p_j) &= p_i^{\to}(M_j) \quad \text{by Theorem 6.12} \\ &\subseteq p_i^{\to}\left(\sum_{j \neq i} M_j\right) \\ &= p_i^{\to}(\operatorname{Ker} p_i) \quad \text{by Theorem 6.12} \\ &= \{0\} \end{aligned}$$

and so $p_i \circ p_j = 0$ for $i \neq j$. Also, since every $x \in M$ can be written uniquely as $x = \sum_{i=1}^{n} x_i$ where $x_i \in M_i$ for $i = 1, \ldots, n$, and since $p_i(x) = x_i$ for each i, we observe that

$$x = \sum_{i=1}^{n} x_i = \sum_{i=1}^{n} p_i(x) = \left(\sum_{i=1}^{n} p_i\right)(x),$$

whence we deduce that $\sum_{i=1}^{n} p_i = \text{id}_M$.

Conversely, suppose that $p_1, \ldots, p_n$ satisfy (1) and (2). Then

$$p_i = p_i \circ \text{id}_M = p_i \circ \left(\sum_{i=1}^{n} p_j\right) = \sum_{i=1}^{n}(p_i \circ p_j) = p_i \circ p_i$$

so that each p_i is idempotent and hence a projection. By (1) we have

$$x = \text{id}_M(x) = \left(\sum_{i=1}^{n} p_i\right)(x) = \sum_{i=1}^{n} p_i(x) \in \sum_{i=1}^{n} \text{Im}\, p_i,$$

whence $M = \sum_{i=1}^{n} \text{Im}\, p_i$.

Moreover, if $x \in \text{Im}\, p_i \cap \sum_{j \neq i} \text{Im}\, p_j$ then $x = p_i(x)$ and $x = \sum_{j \neq i} x_j$ where $p_j(x_j) = x_j$ for every $j \neq i$. Consequently, by (2),

$$x = p_i(x) = p_i\left(\sum_{j \neq i} x_j\right) = p_i\left(\sum_{j \neq i} p_j(x_j)\right) = \sum_{j \neq i}(p_i \circ p_j)(x_j) = 0.$$

It follows that $M = \bigoplus_{i=1}^{n} \text{Im}\, p_i$. $\Diamond$

EXERCISES

6.1 Let M be an R-module and let $M_1, \ldots, M_n$ be submodules of M such that $M = \bigoplus_{i=1}^{n} M_i$. For $k = 1, \ldots, n$ let N_k be a submodule of M_k and let $N = \sum_{i=1}^{n} N_i$. Prove that

(1) $N = \bigoplus_{i=1}^{n} N_i$;

(2) $M/N \simeq \bigoplus_{i=1}^{n} M_i/N_i$.

6.2 Let M be an R-module of finite height. Prove that for every R-morphism $f : M \to N$ there exists $n \in \mathbb{N}$ such that
$$M = \operatorname{Im} f^n \oplus \operatorname{Ker} f^n.$$

[*Hint.* The ascending chain $\{0\} \subseteq \operatorname{Ker} f \subseteq \operatorname{Ker} f^2 \subseteq \ldots$ terminates after p steps, and the descending chain $M \supseteq \operatorname{Im} f \supseteq \operatorname{Im} f^2 \supseteq \ldots$ after q steps; consider $n = \max\{p, q\}$.]

6.3 If M, N are R-modules let $\operatorname{Mor}_R(M, N)$ be the set of R-morphisms $f : M \to N$. Show that $\operatorname{Mor}_R(M, N)$ is an abelian group which is an R-module whenever R is commutative.

Suppose now that R is commutative and that $A_1, \ldots, A_n$ are submodules of M such that $M = \bigoplus_{i=1}^{n} A_i$. For $j = 1, \ldots, n$ let L_j denote the set of R-morphisms $g : M \to N$ such that $\bigoplus_{i \neq j} A_i \subseteq \operatorname{Ker} g$. Prove that each L_j is an R-module and that $L_j \simeq \operatorname{Mor}_R(A_j, M)$.

[*Hint.* Observe that there is an R-isomorphism $\vartheta_j : A_j \to M/\bigoplus_{i \neq j} A_i$.

Now form the diagram

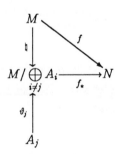

in which f_* is given by Theorem 3.4. Examine the mapping $f \mapsto f_* \circ \vartheta_j$.]

6.4 If M is an R-module and $(N_i)_{i \in I}$ is a family of R-modules, establish the following isomorphisms of abelian groups:

(a) $\operatorname{Mor}_R\left(\bigoplus_{i \in I} N_i, M\right) \simeq \underset{i \in I}{\times} \operatorname{Mor}_R(N_i, M)$;

(b) $\operatorname{Mor}_R\left(M, \underset{i \in I}{\times} N_i\right) \simeq \underset{i \in I}{\times} \operatorname{Mor}_R(M, N_i)$.

[*Hint.* (a) : Given $f : \bigoplus_{i \in I} N_i \to M$ let $\vartheta(f) = (f \circ \operatorname{in}_i)_{i \in I}$;

(b) : Given $f : M \to \underset{i \in I}{\times} N_i$ let $\zeta(f) = (\mathrm{pr}_i \circ f)_{i \in I}$.]

Deduce that

$$\mathrm{Mor}_R\left(\bigoplus_{i \in I} M_i, \underset{j \in J}{\times} N_j\right) \simeq \underset{(i,j) \in I \times J}{\times} \mathrm{Mor}_R(M_i, N_j).$$

Establish an abelian group isomorphism $\mathrm{Mor}_{\mathbb{Z}}(\mathbb{Z}, \mathbb{Z}) \simeq \mathbb{Z}$. Deduce that, for all positive integers n and m,

$$\mathrm{Mor}_{\mathbb{Z}}(\mathbb{Z}^n, \mathbb{Z}^m) \simeq \mathbb{Z}^{nm}.$$

6.5 A submodule M of $\underset{i \in I}{\times} M_i$ is said to be a *subdirect product* of the family $(M_i)_{i \in I}$ if, for every $i \in I$, the restriction $\mathrm{pr}_i^M : M \to M_i$ of the canonical projection pr_i is an R-epimorphism.

If N is an R-module and there is a family $(f_i)_{i \in I}$ of epimorphisms $f_i : N \to M_i$ such that $\bigcap_{i \in I} \mathrm{Ker}\, f_i = \{0\}$, prove that N is isomorphic to a subdirect product of $(M_i)_{i \in I}$.

Show that $\mathbb{Z}$ is isomorphic to a subdirect product of $(\mathbb{Z}/n\mathbb{Z})_{n > 1}$.

6.6 An R-morphism $f : M \to N$ is said to be *regular* if there is an R-morphism $g : N \to M$ such that $f \circ g \circ f = f$. Prove that $f : M \to N$ is regular if and only if

(1) $\mathrm{Ker}\, f$ is a direct summand of M;

(2) $\mathrm{Im}\, f$ is a direct summand of N.

[*Hint.* Use the canonical sequences

$$\begin{array}{ccccc} \mathrm{Ker}\, f & \xrightarrow{i_1} & M & \xrightarrow{\natural_1} & M/\mathrm{Ker}\, f \\ & & \zeta^{-1} \uparrow \downarrow \zeta & & \\ \mathrm{Im}\, f & \xrightarrow{i_2} & N & \xrightarrow{\natural_2} & N/\mathrm{Im}\, f \end{array}$$

and Theorem 6.11.]

6.7 Let $(M_i)_{i \in I}$ and $(N_i)_{i \in I}$ be families of R-modules. If, for every $i \in I$, $f_i : M_i \to N_i$ is an R-morphism define the *direct sum of the family* $(f_i)_{i \in I}$ to be the R-morphism $f : \bigoplus_{i \in I} M_i \to \bigoplus_{i \in I} N_i$ given by $f((m_i)_{i \in I}) = (f_i(m_i))_{i \in I}$. Prove that

(1) $\mathrm{Ker}\, f = \bigoplus_{i \in I} \mathrm{Ker}\, f_i$;

(2) $\operatorname{Im} f = \bigoplus_{i \in I} \operatorname{Im} f_i$.

If $(L_i)_{i \in I}$ is also a family of R-modules and $g_i : L_i \to M_i$ is an R-morphism for every $i \in I$, let g be the direct sum of the family $(g_i)_{i \in I}$. Prove that

$$\bigoplus_{i \in I} L_i \xrightarrow{g} \bigoplus_{i \in I} M_i \xrightarrow{f} \bigoplus_{i \in I} N_i$$

is an exact sequence if and only if, for every $i \in I$,

$$L_i \xrightarrow{g_i} M_i \xrightarrow{f_i} N_i$$

is an exact sequence.

6.8 An R-module M is said to be *indecomposable* if its only direct summands are $\{0\}$ and M. Show that the $\mathbb{Z}$-module $\mathbb{Q}_p/\mathbb{Z}$ is indecomposable.

6.9 Prove that an R-module M together with a family $(\pi_j)_{j \in I}$ of R-morphisms with $\pi_j : M \to M_j$ is a product of the family $(M_j)_{j \in I}$ if and only if there is a family $(i_j)_{j \in I}$ of R-morphisms $i_j : M_j \to M$ such that

(1) $\pi_k \circ i_j = \begin{cases} \operatorname{id}_{M_j} & \text{if } k = j, \\ 0 & \text{if } k \neq j; \end{cases}$

(2) for every $(x_j)_{j \in I} \in \underset{j \in I}{\times} M_j$ there is a unique $x \in M$ such that $\pi_j(x) = x_j$ for every $j \in I$.

6.10 If M is an R-module and $f : M \to M$ is an R-morphism prove that the following statements are equivalent:

(1) $M = \operatorname{Im} f + \operatorname{Ker} f$;

(2) $\operatorname{Im} f = \operatorname{Im}(f \circ f)$.

6.11 If M is an R-module and $p, q : M \to M$ are projections prove that

(1) $\operatorname{Im} p = \operatorname{Im} q$ if and only if $p \circ q = q$ and $q \circ p = p$;

(2) $\operatorname{Ker} p = \operatorname{Ker} q$ if and only if $p \circ q = p$ and $q \circ p = q$.

6.12 Let V be a vector space and let p, q be projections. Prove that $p + q$ is a projection if and only if $p \circ q = q \circ p = 0$, in which case $p + q$ is the projection on $\operatorname{Im} p + \operatorname{Im} q$ parallel to $\operatorname{Im} p \cap \operatorname{Im} q$.

6.13 The diagram of R-modules and R-morphisms

$$\cdots \to A_i \xrightarrow{f_i} B_i \xrightarrow{g_i} C_i \xrightarrow{h_i} A_{i+1} \xrightarrow{f_{i+1}} B_{i+1} \xrightarrow{g_{i+1}} C_{i+1} \to \cdots$$

with vertical maps $\alpha_i, \beta_i, \gamma_i, \alpha_{i+1}, \beta_{i+1}, \gamma_{i+1}$ to

$$\cdots \to A'_i \xrightarrow{f'_i} B'_i \xrightarrow{g'_i} C'_i \xrightarrow{h'_i} A'_{i+1} \xrightarrow{f'_{i+1}} B'_{i+1} \xrightarrow{g'_{i+1}} C'_{i+1} \to \cdots$$

is given to be commutative with exact rows. If each γ_i is an isomorphism, establish the exact sequence

$$\cdots \to A_i \xrightarrow{\varphi_i} A'_i \oplus B_i \xrightarrow{\vartheta_i} B'_i \xrightarrow{h_i \circ \gamma_i^{-1} \circ g'_i} A_{i+1} \to \cdots$$

where φ_i is given by

$$\varphi_i(a_i) = (\alpha_i(a_i), f_i(a_i))$$

and ϑ_i is given by

$$\vartheta_i(a'_i, b_i) = f'_i(a'_i) - \beta_i(b_i).$$

7 Free modules; bases

Definition Let R be a unitary ring and let S be a non-empty set. By a *free R-module on S* we shall mean an R-module F together with a mapping $f : S \to F$ such that, for every R-module M and every mapping $g : S \to M$, there is a unique R-morphism $h : F \to M$ such that the diagram

is commutative. We denote such an R-module by (F, f).

- Roughly speaking, a free module F on S allows us to 'trade in' a mapping from S for an R-morphism from F.

Theorem 7.1 *If (F, f) is a free module on S then F is injective and $\operatorname{Im} f$ generates F.*

Proof To show that f is injective, let $x, y \in S$ be such that $x \neq y$; we have to show that $f(x) \neq f(y)$. For this purpose, let M be an R-module having more than one element (e.g. R itself will do) and choose any mapping $g : S \to M$ such that $g(x) \neq g(y)$. Let $h : F \to M$ be the unique R-morphism such that $h \circ f = g$. Then since $h[f(x)] = g(x) \neq g(y) = h[f(y)]$ we must have $f(x) \neq f(y)$.

To show that $\operatorname{Im} f$ generates F, let A be the submodule of F that is generated by $\operatorname{Im} f$ and consider the diagram

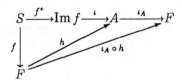

in which ι is the inclusion map from $\operatorname{Im} f$ to A, ι_A is that from A to F, and $f^+ : S \to \operatorname{Im} f$ is given by $f^+(x) = f(x)$ for every $x \in S$. Since F is free on S there is a unique R-morphism $h : F \to A$ such that $h \circ f = \iota_A \circ f^+$. The mapping $k : F \to F$ given by $k = \iota_A \circ h$ is then an R-morphism such that $k \circ f = \iota_A \circ \iota \circ f^+$. But since F is free on S there can be only one

R-morphism $\vartheta : F \to F$ such that $\vartheta \circ f = \iota_A \circ \iota \circ f^+$; and clearly $\vartheta = \mathrm{id}_F$ does just that. We must therefore have $\iota_A \circ h = k = \mathrm{id}_F$, from which we see that ι_A is surjective. We conclude that $A = F$ and consequently that Im f generates F. ◊

Theorem 7.2 [Uniqueness] *Let (F, f) be a free R-module on the non-empty set S. Then (F', f') is also a free R-module on S if and only if there is a unique R-morphism $j : F \to F'$ such that $j \circ f = f'$.*

Proof Suppose first that (F', f') is also free on S. Then there are R-morphisms $j : F \to F'$ and $k : F' \to F$ such that the diagrams

are commutative. Then since $k \circ j \circ f = k \circ f' = f$ we have the commutative diagram

But again since F is free on S only one morphism can complete this diagram in a commutative manner; and clearly id_F does just that. We conclude, therefore, that $k \circ j = \mathrm{id}_F$. In a similar manner we can show that $j \circ k = \mathrm{id}'_F$. It follows that j, k are mutually inverse R-isomorphisms.

Suppose, conversely, that the condition is satisfied. Then since $f = j^{-1} \circ f'$ we can use the fact that (F, f) is free on S to build, for any R-module M and any mapping $g : S \to M$, the diagram

$$S \xrightarrow{g} M$$

in which $h \circ j^{-1} \circ f' = h \circ f = g$. That (F', f') is also free on S will follow if we can show that if $t : F' \to M$ is an R-morphism such that $t \circ f' = g$

Free modules; bases 79

then $t = h \circ j^{-1}$. Now $t \circ f' = g$ is equivalent to $t \circ j \circ f = g$ which, by the uniqueness of h, gives $t \circ j = h$, whence $t = h \circ j^{-1}$. $\diamond$

We shall now settle the question concerning the existence of free modules.

Theorem 7.3 [Existence] *For every non-empty set S and every unitary ring R there is a free R-module on S.*

Proof Consider the set F of mappings $\vartheta : S \to R$ which are such that $\vartheta(s) = 0$ for 'almost all' $s \in S$, i.e. all but a finite number of $s \in S$. It is readily seen that F, equipped with the laws of composition given by

$$(\forall s \in S) \quad (\vartheta + \zeta)(s) = \vartheta(s) + \zeta(s);$$
$$(\forall \lambda \in R)(\forall s \in S) \quad (\lambda \vartheta)(s) = \lambda \vartheta(s),$$

is an R-module. Now define a mapping $f : S \to F$ by assigning to $s \in S$ the mapping $f(s) : S \to R$ given by

$$[f(s)](t) = \begin{cases} 1 & \text{if } t = s; \\ 0 & \text{if } t \neq s. \end{cases}$$

We shall show that (F, f) is a free R-module on S.

For this purpose, suppose that M is an R-module and that $g : S \to M$ is a mapping. Define a mapping $h : F \to M$ by the prescription

$$h(\vartheta) = \sum_{s \in S} \vartheta(s) g(s).$$

Note that each such sum is well defined since there is at most a finite number of terms different from zero. It is clear that h is an R-morphism. Moreover, since

$$(\forall s \in S) \quad h[f(s)] = \sum_{t \in S} [f(s)](t) \cdot g(t) = g(s),$$

we have $h \circ f = g$. To establish the uniqueness of h, we note first that for all $\vartheta \in F$ and all $t \in S$ we have

$$\vartheta(t) = \vartheta(t) \cdot 1_R = \sum_{s \in S} \vartheta(s) \cdot [f(s)](t) = \left(\sum_{s \in S} \vartheta(s) f(s)\right)(t)$$

and so

$$(\forall \vartheta \in F) \quad \vartheta = \sum_{s \in S} \vartheta(s) f(s).$$

Thus, if $h' : F \to M$ is also an R-morphism such that $h' \circ f = g$, then for all $\vartheta \in F$ we have

$$h'(\vartheta) = \sum_{s \in S} \vartheta(s) h'[f(s)] = \sum_{s \in S} \vartheta(s) g(s) = h(\vartheta).$$

Hence $h' = h$ as required. ◊

The above results show that, to within R-isomorphism, there is a unique free R-module on a non-empty set S. As a model of this, we may choose the R-module constructed in the above proof. We shall refer to this particular R-module by calling it henceforth *the* free R-module on S.

Definition We shall say that an R-module M is *free* if there is a free R-module (F, f) on some non-empty set S such that M is isomorphic to F.

Our immediate aim is to determine a simple criterion for a module to be free. For this purpose, we require the following notion.

Definition We shall say that a non-empty subset S of an R-module M is *linearly independent* (or *free*) if, for every finite sequence $x_1, \ldots, x_n$ of distinct elements of S,

$$\sum_{i=1}^{n} \lambda_i x_i = 0_M \Rightarrow \lambda_1 = \ldots = \lambda_n = 0_R.$$

Put another way, S is linearly independent if the only way of expressing 0 as a linear combination of distinct elements of S is the trivial way (in which all the coefficients are zero).

- Note that no linearly independent subset of an R-module M can contain the zero element of M; for we can fabricate the equality

$$1_R 0_M + 0_R x_1 + \ldots + 0_R x_n = 0_M$$

thus expressing 0_M s a linear combination of $\{0_M, x_1, \ldots, x_n\}$ in which not all the coefficients are zero.

- Note also that in a vector space V every singleton subset $\{x\}$ with $x \neq 0_V$ is linearly independent; for if we had $\lambda x = 0$ with $\lambda \neq 0$ then it would follow that

$$x = \lambda^{-1}\lambda x = \lambda^{-1} 0 = 0,$$

a contradiction. In contrast, if we consider a unitary ring R as an R-module then, if R has zero divisors, singleton subsets need not be linearly independent.

Definition By a *basis* of an R-module M we shall mean a linearly independent subset of M that generates M.

Free modules; bases

Theorem 7.4 *A non-empty subset S of an R-module M is a basis of M if and only if every element of M can be expressed in a unique way as a linear combination of elements of S.*

Proof If S is a basis of M then, since S generates M, every $x \in M$ can be expressed as a linear combination of elements of S. Suppose now that

$$x = \sum_{i=1}^{n} \lambda_i x_i = \sum_{j=1}^{m} \mu_j y_j$$

where $x_1, \ldots, x_n$ are distinct elements of S and $y_1, \ldots, y_m$ are distinct elements of M. Then we can form the equation

$$\sum_{i=1}^{n} \lambda_i x_i + \sum_{j=1}^{m} (-\mu_j) y_j = 0.$$

The linear independence of S now shows that for distinct x_i, y_j we have $\lambda_i = \mu_j = 0$, and when $x_i = y_j$ we have $\lambda_i = \mu_j$. Consequently x has a unique expression as a linear combination of elements of S.

Suppose, conversely, that this condition holds. Then clearly S generates M. Also, 0 can be expressed uniquely as a linear combination of elements of S, so that S is linearly independent. Thus S is a basis. ◊

- Observe from the above proof that in order to show that x has a unique expression as a linear combination of elements of S, there is no loss in generality in assuming that $m = n$ and $y_i = x_i$.

Example 7.1 If M is an R-module and n is a positive integer then the R-module R^n has the basis $\{e_1, \ldots, e_n\}$ where

$$e_i = (\underbrace{0, \ldots, 0}_{i-1}, 1, \ldots, 0).$$

We call this the *natural basis* of R^n.

Example 7.2 Consider the differential equation $D^2 f + 2Df + f = 0$. The reader will know that, loosely speaking, the general solution is $f : x \mapsto (A + Bx)e^{-x}$ where A, B are arbitrary constants. To be somewhat more precise, the solution space is the subspace $\text{LC}\{f_1, f_2\}$ of the vector space V of twice-differentiable functions $f : \mathbb{R} \to \mathbb{R}$ where $f_1(x) = e^{-x}$ and $f_2(x) = xe^{-x}$. Since $\{f_1, f_2\}$ is linearly independent (to see this, write $\lambda_1 f_1 + \lambda_2 f_2 = 0$ and differentiate), it therefore forms a basis for the solution space. Note that, in contrast, V does not have a finite basis.

Theorem 7.5 *Let (F, f) be the free R-module on S. Then $\text{Im } f$ is a basis of F.*

Proof Recall that (F, f) is the R-module constructed in the proof of Theorem 7.3. That $\operatorname{Im} f$ generates F is immediate from Theorem 7.1. Observe that, from the proof of Theorem 7.3, we have

$$(\forall \vartheta \in F) \qquad \vartheta = \sum_{s \in S} \vartheta(s) f(s).$$

Suppose now that we have

$$\sum_{i=1}^{m} \alpha_i f(x_i) = \sum_{i=1}^{m} \beta_i f(x_i)$$

where $f(x_1), \ldots, f(x_m)$ are distinct elements of $\operatorname{Im} f$. Consider the mappings $\vartheta, \zeta : S \to R$ given by the prescriptions

$$\vartheta(x) = \begin{cases} 0 & \text{if } (\forall i) \ x \neq x_i; \\ \alpha_i & \text{if } (\exists i) \ x = x_i. \end{cases}$$

$$\zeta(x) = \begin{cases} 0 & \text{if } (\forall i) \ x \neq x_i; \\ \beta_i & \text{if } (\exists i) \ x = x_i. \end{cases}$$

It is clear that $\vartheta, \zeta \in F$. Moreover,

$$\vartheta = \sum_{s \in S} \vartheta(s) f(s) = \sum_{i=1}^{m} \alpha_i f(x_i);$$

$$\zeta = \sum_{s \in S} \zeta(s) f(s) = \sum_{i=1}^{m} \beta_i f(x_i).$$

Consequently, $\vartheta = \zeta$ and hence $\alpha_i = \beta_i$. It now follows by Theorem 7.4 that $\operatorname{Im} f$ is a basis of F. $\Diamond$

Corollary *If (M, α) is a free R-module on S then $\operatorname{Im} \alpha$ is a basis of M.*

Proof Let (F, f) be the free R-module on S. Then by Theorem 7.2 there is a unique R-isomorphism such that the diagram

is commutative. Since then $\vartheta^{\to}(\operatorname{Im} f) = \operatorname{Im} \alpha$, we deduce from the fact that isomorphisms clearly carry bases to bases that $\operatorname{Im} \alpha$ is a basis of M. $\Diamond$

Free modules; bases 83

We can now establish a simple criterion for a module to be free.

Theorem 7.6 *An R-module is free if and only if it has a basis.*

Proof If M is a free R-module then M is isomorphic to the free R-module on some non-empty set. Since isomorphisms take bases to bases, it follows that M has a basis.

Conversely, suppose that S is a basis of M and let (F, f) be the free R-module on S. Then if $\iota_S : S \to M$ is the natural inclusion there is a unique R-morphism $h : F \to M$ such that $h \circ f = \iota_S$. We shall show that h is an isomorphism, whence the result will follow.

Now $\operatorname{Im} h$ is a submodule of M that contains S and so, since S generates M, we must have $\operatorname{Im} h = M$. Consequently, h is surjective. Also, we know (from the proof of Theorem 7.3 with $g = \iota_S$) that

$$h(\vartheta) = \sum_{s \in S} \vartheta(s) s$$

where $\vartheta(x) = 0$ for all but a finite number of elements $x_1, \ldots, x_n$ of S. If now $\vartheta \in \operatorname{Ker} h$ then we obtain

$$0 = \sum_{i=1}^{n} \vartheta(x_i) x_i$$

from which we deduce, since $\{x_1, \ldots, x_n\}$ is linearly independent, that $\vartheta(x_i) = 0$ for every x_i. This then implies that $\vartheta = 0$ and consequently h is also injective. ◊

Corollary *A free R-module is isomorphic to a direct sum of copies of R. More precisely, if $\{a_i \; ; \; i \in I\}$ is a basis of M then $M = \bigoplus_{i=1}^{n} Ra_i$ where $Ra_i \simeq R$ for every $i \in I$.*

Proof By Theorems 6.8 and 7.4, we have $M = \bigoplus_{i \in I} Ra_i$; and since each singleton $\{a_i\}$ is linearly independent, the R-morphism $f_i : R \to Ra_i$ given by $f_i(r) = ra_i$ is an R-isomorphism. ◊

The following important properties relate bases to R-morphisms.

Theorem 7.7 *Let M be a free R-module and let $A = \{a_i \; ; \; i \in I\}$ be a basis of M. If N is an R-module and if $(b_i)_{i \in I}$ is a family of elements of N then there is a unique R-morphism $f : M \to N$ such that*

$$(\forall i \in I) \quad f(a_i) = b_i.$$

Proof Since every element of M can be expressed uniquely as a linear combination of elements of A, we can define a mapping $f : M \to N$ by the prescription

$$f\left(\sum_{i=1}^{n} \lambda_i a_i\right) = \sum_{i=1}^{n} \lambda_i b_i.$$

It is readily verified that f is an R-morphism and that $f(a_i) = b_i$ for every $i \in I$. Suppose now that $g : M \to N$ is also an R-morphism such that $g(a_i) = b_i$ for every $i \in I$. Given $x \in M$ with say $x = \sum_{i=1}^{n} \lambda_i a_i$, we have

$$g(x) = g\left(\sum_{i=1}^{n} \lambda_i a_i\right) = \sum_{i=1}^{n} \lambda_i g(a_i) = \sum_{i=1}^{n} \lambda_i b_i = f(x)$$

and so $g = f$, whence the uniqueness of f follows. ◊

Corollary 1 *Let B be a non-empty subset of an R-module M and let $\iota_B : B \to M$ be the natural inclusion. Then B is a basis of M if and only if (M, ι_B) is a free R-module on B.*

Proof The necessity follows immediately from Theorem 7.7; and the sufficiency follows from the Corollary to Theorem 7.5. ◊

Corollary 2 *If M and N are free R-modules and if $f : M \to N$ is an R-morphism then f is an isomorphism if and only if, whenever $\{a_i \ ; \ i \in I\}$ is a basis of M, $\{f(a_i) \ ; \ i \in I\}$ is a basis of N.*

Proof It is enough to establish the sufficiency; and for this it is enough to observe from the proof of Theorem 7.7 that if $\{b_i \ ; \ i \in I\}$ is a basis then f is an isomorphism. ◊

Corollary 3 *Let $f : M \to N$ be an R-morphism. If M is free then f is completely determined by $f^{\to}(B)$ for any basis B of M.*

Proof This is immediate from Corollary 1; for by that result the restriction of f to B extends to a unique R-morphism from M to N, namely f itself. ◊

Corollary 4 *Let $f, g : M \to N$ be R-morphisms. If M is free and if $f(x) = g(x)$ for all x in some basis B of M then $f = g$.*

Proof The restriction of $f - g$ to B is the zero map and so, by Corollary 3, $f - g$ is the zero map from M to N. ◊

Free modules; bases   85

Concerning direct sums of free modules, the above results yield the following.

Theorem 7.8 *Let M be an R-module and let $(M_\lambda)_{\lambda \in I}$ be a family of submodules of M such that $M = \bigoplus_{\lambda \in I} M_\lambda$. If B_λ is a basis of M_λ for every $\lambda \in I$ then $\bigcup_{\lambda \in I} B_\lambda$ is a basis of M.*

Proof Given B_λ consider the diagram

$$\begin{array}{ccccc} B_\lambda & \xrightarrow{j_\lambda} & \bigcup_{\lambda \in I} B_\lambda & \xrightarrow{g} & N \\ {\scriptstyle i_\lambda}\downarrow & & \downarrow{\scriptstyle i} & & \\ M_\lambda & \underset{\mathrm{pr}_\lambda^\oplus}{\overset{\mathrm{in}_\lambda}{\rightleftarrows}} & \bigoplus_{\lambda \in I} M_\lambda = M & & \end{array}$$

in which i_λ, j_λ, i are the natural inclusions, and $g : \bigcup_{\lambda \in I} B_\lambda \to N$ is a mapping to an arbitrary R-module N.

Since B_λ is a basis of M_λ we have, by Corollary 1 of Theorem 7.7, that (M_λ, i_λ) is free on B_λ. There is therefore a unique R-morphism $\vartheta_\lambda : M_\lambda \to N$ such that $\vartheta_\lambda \circ i_\lambda = g \circ j_\lambda$. We now define a mapping $\zeta : \bigoplus_{\lambda \in I} M_\lambda \to N$ by the prescription

$$\zeta(y) = \sum_{\mu \in I} (\vartheta_\mu \circ \mathrm{pr}_\mu^\oplus)(y).$$

[Recall that $\mathrm{pr}_\mu^\oplus(y) = 0$ for all but a finite number of $\mu \in I$ so ζ is well defined.] For every $x \in B_\lambda$ we then have

$$\begin{aligned} (\zeta \circ i \circ j_\lambda)(x) &= \sum_{\mu \in I} (\vartheta_\mu \circ \mathrm{pr}_\mu^\oplus \circ i \circ j_\lambda)(x) \\ &= \sum_{\mu \in I} (\vartheta_\mu \circ \mathrm{pr}_\mu^\oplus \circ \mathrm{in}_\lambda \circ i_\lambda)(x) \\ &= (\vartheta_\lambda \circ i_\lambda)(x) \\ &= (g \circ j_\lambda)(x). \end{aligned}$$

Since the j_λ are natural inclusions and since $\{B_\lambda \,;\, \lambda \in I\}$ is a partition of $\bigcup_{\lambda \in I} B_\lambda$ [for if $\lambda, \mu \in I$ and $\lambda \neq \mu$ then necessarily $B_\lambda \cap B_\mu \subseteq M_\lambda \cap M_\mu = \{0\}$ by Theorem 6.9 and, since $\{0\}$ is not independent, $B_\lambda \cap B_\mu = \emptyset$], it follows that $\zeta \circ i = g$. The result will now follow from Corollary 1 of Theorem 7.7 if we can show that ζ is unique with respect to this property.

For this purpose, suppose that $k : \bigoplus_{\lambda \in I} M_\lambda \to N$ is also an R-morphism such that $k \circ i = g$. Then for every $\lambda \in I$ we have $k \circ i \circ j_\lambda = g \circ j_\lambda$ whence

$k \circ \mathrm{in}_\lambda \circ i_\lambda = g \circ j_\lambda$ and so $k \circ \mathrm{in}_\lambda = \vartheta_\lambda$ by the uniqueness of ϑ_λ. Then for every $y \in \bigoplus_{\lambda \in I} M_\lambda$ we have

$$k(y) = \sum_{\mu \in I}(k \circ \mathrm{in}_\mu \circ \mathrm{pr}_\mu^\oplus)(y) = \sum_{\mu \in I}(\vartheta_\mu \circ \mathrm{pr}_\mu^\oplus)(y) = \zeta(y)$$

and so $k = \zeta$. ◊

So far in this section we have restricted our attention to a non-empty set S. A free module on such a set is clearly never a zero module. Henceforth we shall make the convention that *the empty subset of an R-module M will be considered as a linearly independent subset of M*. This is not unreasonable since the condition that defined linear independence may be regarded as being satisfied 'vacuously' by $\emptyset$ simply because it has no elements.

- The courtesy of regarding $\emptyset$ as linearly independent yields the advantage of having $\emptyset$ as a basis for every zero module (recall Theorem 2.2), so that we can also regard a zero module as being free.

We shall now derive the important result that every vector space has a basis. This is a consequence of the following, the proof of which uses Zorn's axiom.

Theorem 7.9 *Let V be a vector space over a field F. If I is a linearly independent subset of V and if G is a set of generators of V with $I \subseteq G$, then there is a basis B of V such that $I \subseteq B \subseteq G$.*

Proof Let C be the collection of all linearly independent subsets A of V such that $I \subseteq A \subseteq G$. We note that $C \neq \emptyset$ since it clearly contains I. Let $T = \{A_j \ ; \ j \in J\}$ be a totally ordered subset of C and let $D = \bigcup_{j \in J} A_j$. Clearly, we have $I \subseteq D \subseteq G$. We shall show that D is linearly independent whence it will follow that $D \in C$ and is the supremum of T, so that C is inductively ordered. For this purpose, let $x_1, \ldots, x_n$ be distinct elements of D and suppose that $\sum_{i=1}^{n} \lambda_i x_i = 0$. Since every x_i belongs to some A_j and since T is totally ordered, there exists $A_k \in T$ such that $x_1, \ldots, x_n \in A_k$. Since A_k is linearly independent we deduce that $\lambda_1 = \ldots = \lambda_n = 0$ whence D is also linearly independent and consequently C is inductively ordered.

It follows by Zorn's axiom that C has a maximal element, B say. Now B is linearly independent (since it belongs to C); we shall show that it is also a set of generators of V whence it will be a basis with the required property. For this purpose, let W be the subspace generated by B and

Free modules; bases 87

let $x \in V$. Since G generates V we have $x = \sum_{i=1}^{n} \lambda_i g_i$ for some $\lambda_i, \ldots, \lambda_n \in F$ and $g_1, \ldots, g_n \in G$. Now if $x \notin W$ there exists some g_j such that $g_j \notin W$ (for otherwise every g_j would be in W and so x, being a linear combination of these elements, would also be in W, contradicting the hypothesis) whence $B \cup \{g_j\}$ is a linearly independent subset of G (for, if $\sum_{i=1}^{n} \lambda_i b_i + \mu g_j = 0$ with every $b_i \in B$ and $\mu \neq 0$ then $g_j = -\mu^{-1} \left(\sum_{i=1}^{n} \lambda_i b_i \right) \in W$, a contradiction, so that $\mu = 0$ and $\sum_{i=1}^{n} \lambda_i b_i = 0$ whence also every $\lambda_i = 0$). Since $g_j \notin B$, this contradicts the maximality of B in C. We conclude from this that we must have $x \in W$, whence $W = V$ and consequently B is a basis of V with $I \subseteq B \subseteq G$. ◊

Corollary 1 *Every vector space has a basis.*

Proof Take $I = \emptyset$ and $G = V$ in Theorem 7.9. ◊

Corollary 2 *Every linearly independent subset I of a vector space V can be extended to a basis of V.*

Proof Take $G = V$ in Theorem 7.9. ◊

The previous result leads to the following property of bases.

Theorem 7.10 *If B is a basis of M then B is both a minimal generating subset and a maximal linearly independent subset. In the case where V is a vector space, the following are equivalent*:

(1) *B is a basis of V;*
(2) *B is a minimal generating subset;*
(3) *B is a maximal linearly independent subset.*

Proof Suppose that B is not a minimal set of generators of M. Then there exists a set G of generators with $G \subset B$ and, for some $x \in B \setminus G$, the set $B \setminus \{x\}$ generates M. For such an element x we have $x = \sum_{i=1}^{n} \lambda_i x_i$ where $x_1, \ldots, x_n$ are distinct elements of $B \setminus \{x\}$. This can be written in the form

$$1_R x + \sum_{i=1}^{n} (-\lambda_i) x_i = 0$$

and contradicts the linear independence of B. Thus we deduce that B is a minimal generating subset.

Suppose now that $y \in M \setminus B$. Then there exist distinct elements $x_1, \ldots, x_n$ of B and $r_1, \ldots, r_n \in R$ such that $y = \sum_{i=1}^{n} r_i x_i$ whence

$$1_r y + \sum_{i=1}^{n} (-r_i) x_i = 0$$

which shows that $B \cup \{y\}$ is not linearly independent. Thus B is also a maximal linearly independent subset.

To show that (1),(2),(3) are equivalent in the case of a vector space V, it suffices to show that $(2) \Rightarrow (1)$ and $(3) \Rightarrow (1)$.

$(2) \Rightarrow (1)$: By Theorem 7.9 there is a basis $B^\star$ of V such that $\emptyset \subseteq B^\star \subseteq B$. Since $B^\star$ is also a generating subset, the minimality of B yields $B^\star = B$ whence B is a basis.

$(3) \Rightarrow (1)$: By Corollary 2 of Theorem 7.9 there is a basis $B^\star$ such that $B \subseteq B^\star$. Since B is also linearly independent the maximality of B yields $B = B^\star$ whence B is a basis. $\Diamond$

- Note that the implications $(2) \Rightarrow (1)$ and $(3) \Rightarrow (1)$ do not hold in general for R-modules. For example, if n is a positive integer then in the $\mathbb{Z}$-module $\mathbb{Z}/n\mathbb{Z}$ the set $\{1 + n\mathbb{Z}\}$ is a minimal generating subset but is not linearly independent (for we have $n(1 + n\mathbb{Z}) = 0 + n\mathbb{Z}$ with $n \neq 0$) and so is not a basis. In fact the $\mathbb{Z}$-module $\mathbb{Z}/n\mathbb{Z}$ has no basis. Likewise, in the $\mathbb{Z}$-module $\mathbb{Q}$ every element t with $t \neq 0$ and $t \neq 1/k$ where $k \neq 0$ is such that $\{t\}$ is a maximal linearly independent subset but is not a generating subset and so is not a basis. In fact the $\mathbb{Z}$-module $\mathbb{Q}$ has no basis.

We shall now establish a striking result concerning the cardinality of bases in a vector space. In the proof of this, the properties of infinite cardinals play an important role.

Theorem 7.11 *All bases of a vector space are equipotent.*

Proof Suppose first that the vector space V has an infinite basis B. Then we have $V = \bigoplus_{x \in B} Fx$ where $Fx = \{\lambda x \; ; \; \lambda \in F\}$ is the subspace generated by $\{x\}$. Now the mapping $\vartheta : F \to Fx$ given by $\vartheta(\lambda) = \lambda x$ is an isomorphism (recall Example 5.1 and Theorem 5.6), and so we have that $V = \bigoplus_{i \in I} V_i$ where $\operatorname{Card} I = \operatorname{Card} B$ and $V_i \simeq F$ for every $i \in I$.

Suppose now that $B^\star$ is any basis of V. For every $y \in B^\star$ let J_y denote the (finite) set of indices $i \in I$ such that the component of y in V_i is non-zero. Since $B^\star$ is a basis, we have $I = \bigcup_{y \in B^\star} J_y$. Now if $\operatorname{Card} B^\star$ were finite

Free modules; bases 89

we would have

$$\operatorname{Card} B = \operatorname{Card} I \le \sum_{y \in B^*} \operatorname{Card} J_y \le \operatorname{Card} B^* \cdot \max\{\operatorname{Card} J_y \; ; \; y \in B^*\}$$

and $\operatorname{Card} B$ would be finite, contradicting the hypothesis. We thus see that $\operatorname{Card} B^*$ is infinite. We now have

$$\operatorname{Card} B \le \sum_{y \in B^*} \operatorname{Card} J_y \le \operatorname{Card} B^* \cdot \aleph_0 \le (\operatorname{Card} B^*)^2 = \operatorname{Card} B^*.$$

Likewise we can show that $\operatorname{Card} B^* \le \operatorname{Card} B$. It now follows by the Schröder-Bernstein theorem that B^* is equipotent to B.

Suppose now that V has a finite basis. Then the above argument shows that all bases of V are finite. If $B = \{x_1, \ldots, x_n\}$ is a basis of V then, since each Fx_i is a simple F-module and since for $k = 1, \ldots, n$ we have

$$\bigoplus_{i=1}^{k} Fx_i \bigg/ \bigoplus_{i=1}^{k-1} Fx_i = \left(Fx_k \oplus \bigoplus_{i=1}^{k-1} Fx_i \right) \bigg/ \bigoplus_{i=1}^{k-1} Fx_i \simeq Fx_k,$$

we see that

$$\{0\} \subset Fx_1 \subset \bigoplus_{i=1}^{2} Fx_i \subset \bigoplus_{i=1}^{3} Fx_i \subset \ldots \subset \bigoplus_{i=1}^{n} Fx_i = V$$

is a Jordan-Hölder tower for V of height n, the number of elements in the basis B. The invariance of the height of such a tower now implies that all bases of V have the same number of elements. ◊

- The result of Theorem 7.11 is not true in general for free modules. Indeed, as we shall now show, if the ground ring R is 'bad' enough then it is possible for a free R-module to have bases of different cardinalities. Suppose that R is a unitary ring and consider the R-module $\bigoplus^{\mathbb{N}} R$ that consists of all finite sequences of elements of R, i.e. all mappings $f : \mathbb{N} \to R$ such that $f(i) = 0$ for all but finitely many $i \in \mathbb{N}$. As an R-module, $\bigoplus^{\mathbb{N}} R$ is free; its natural basis is $\{e_i \; ; \; i \in \mathbb{N}\}$ where

$$e_i(n) = \begin{cases} 1 & \text{if } n = i; \\ 0 & \text{otherwise.} \end{cases}$$

Consider now the ring $\operatorname{End} \bigoplus^{\mathbb{N}} R$ of group morphisms $f : \bigoplus^{\mathbb{N}} R \to \bigoplus^{\mathbb{N}} R$. Regarding $\operatorname{End} \bigoplus^{\mathbb{N}} R$ as an $\operatorname{End} \bigoplus^{\mathbb{N}} R$-module (in which the action is composition of mappings), we see that the singleton {id} is a basis for this module. However, consider now $f, g \in \text{End} \bigoplus^{\mathbb{N}} R$ given by

$$f(e_i) = \begin{cases} e_n & \text{if } i = 2n; \\ 0 & \text{if } i = 2n+1, \end{cases}$$

$$g(e_i)) = \begin{cases} 0 & \text{if } i = 2n; \\ e_n & \text{if } i = 2n+1. \end{cases}$$

Note that we have defined f, g only on the natural basis of $\bigoplus^{\mathbb{N}} R$; this is sufficient to describe f, g completely (Corollary 3 of Theorem 7.7). As can now be readily verified, every $\vartheta \in \text{End} \bigoplus^{\mathbb{N}} R$ can be expressed uniquely in the form

$$\vartheta = \alpha \circ f + \beta \circ g.$$

In fact, $\alpha, \beta \in \text{End} \bigoplus^{\mathbb{N}} R$ are given by $\alpha(e_i) = \vartheta(e_{2i})$ and $\beta(e_i) = \vartheta(e_{2i+1})$. Consequently we see that $\{f, g\}$ is also a basis for this module.

- Despite the above, it should be noted that if M, N are R-modules each having a basis of n elements then M and N are R-isomorphic. In fact they are each isomorphic to the R-module R^n. For example, if $\{a_1, \ldots, a_n\}$ is a basis of M then the mapping $f : M \to R$ given by $f\left(\sum_{i=1}^{n} \lambda_i a_i\right) = (\lambda_1, \ldots, \lambda_n)$ is an R-isomorphism. In particular, therefore, from the previous remark, if we denote the ring $\text{End} \bigoplus^{\mathbb{N}} R$ by A then the A-modules A and A^2 are A-isomorphic!

We shall now extend Theorem 7.11 to free R-modules where R is a commutative unitary ring. In the proof of this we shall make use of a result in ring theory known as *Krull's theorem*, namely that every commutative unitary ring has a maximal ideal.

Theorem 7.12 *Let R be a commutative unitary ring. If M is a free R-module then all bases of M are equipotent.*

Proof Let I be a maximal ideal of R. Then the quotient ring R/I is a field. Consider the subset IM of M consisting of all finite sums of the form $\sum_i \lambda_i x_i$ where $\lambda_i \in I$ and $x_i \in M$; in other words, IM is the set of all linear combinations of elements of M with coefficients in the maximal

Free modules; bases 91

ideal I. It is clear that IM is a submodule of M, so we can form the quotient module M/IM. Observing that

$$r - t \in I \Rightarrow (\forall x \in M) \; rx - tx = (r-t)x \in M$$
$$\Rightarrow (\forall x \in M) \; rx + IM = tx + IM;$$
$$x - y \in IM \Rightarrow (\forall t \in R) \; tx - ty = t(x-y) \in IM$$
$$\Rightarrow (\forall t \in R) \; tx + IM = ty + IM,$$

and consequently that

$$\left.\begin{array}{r}r - t \in I \\ x - y \in IM\end{array}\right\} \Rightarrow rx + IM = ty + IM,$$

we can define an action of R/I on M/IM by the prescription

$$(r + I) \cdot (x + IM) = rx + IM.$$

Clearly, this makes M/IM into a vector space over the field R/I.

Suppose now that $\{x_j \; ; \; j \in J\}$ is a basis of M. Then the mapping described by $x_j \mapsto x_j + IM$ is injective. For, suppose that

$$x_j + IM = x_k + IM$$

with $x_j \neq x_k$. Then $x_j - x_k \in IM$, which is impossible since I, being a maximal ideal of R, does not contain 1_R. We therefore deduce, on passing to quotients, that $(x_j + IM)_{j \in J}$ is a family of distinct elements of M/IM that generates M/IM. Now, all sums indicated being well defined, we have

$$\sum_j (r_j + I) \cdot (x_j + IM) = 0 + IM \Rightarrow \sum_j r_j x_j \in IM$$
$$\Rightarrow (\exists t_j \in I) \sum_j r_j x_j = \sum_j t_j x_j$$
$$\Rightarrow (\forall j) \; r_j = t_j \in I$$
$$\Rightarrow (\forall j) \; r_j + I = 0 + I$$

and so we see that $\{x_j + IM \; ; \; j \in J\}$ is indeed a basis of the R/I-vector space M/IM. Since this basis is equipotent ot the basis $\{x_j \; ; \; j \in J\}$, the result follows from Theorem 7.11. ◊

Because of the above results we can introduce the following terminology.

Definition If R is a commutative unitary ring (in particular, a field) and if M is a free R-module (in particular, a vector space) then by the *dimension* of M over R we shall mean the cardinality of any basis of M.

We denote this by dim M. In the case where dim M is finite we shall say that M is *finite-dimensional*.

EXERCISES

7.1 Let $f : M \to M$ be an R-morphism. Show that if f is a monomorphism then f is not a left zero divisor in the ring $\text{End}_R M$. If M is free, establish the converse : that if $f \in \text{End}_R M$ is not a left zero divisor then f is a monomorphism.

[*Hint.* Let $\{m_i \ ; \ i \in I\}$ be a basis of M and suppose that $\text{Ker}\, f \neq \{0\}$. Let $(n_i)_{i \in I}$ be a family of non-zero elements of $\text{Ker}\, f$ and let $g : M \to M$ be the unique R-morphism such that $g(m_i) = n_i$ for every $i \in I$ (Theorem 7.7). Observe that $\text{Im}\, g \subseteq \text{Ker}\, f$.]

7.2 Let p be a prime. Show that the $\mathbb{Z}$-module $\mathbb{Q}_p/\mathbb{Z}$ is not free.

[*Hint.* Show that the endomorphism on $\mathbb{Q}_p/\mathbb{Z}$ described by $x \mapsto px$ is neither a left zero divisor nor a monomorphism and use Exercise 7.1.]

7.3 Let $f : M \to M$ be an R-morphism. Show that if f is an epimorphism then f is not a right zero divisor in the ring $\text{End}_R M$. Give an example of a free $\mathbb{Z}$-module M and an $f \in \text{End}_{\mathbb{Z}} M$ such that f is neither a right zero divisor nor an epimorphism.

[*Hint.* Try multiplication by 2 on $\mathbb{Z}$.]

7.4 Let F be a field and let $q \in F[X]$ be of degree n. Show that if $\langle q \rangle$ denotes the ideal of $F[X]$ generated by q then $F[X]/\langle q \rangle$ can be made into a vector space over F. If $\natural : F[X] \to F[X]/\langle q \rangle$ is the natural epimorphism, show that

$$\{\natural(X^0), \natural(X^1), \ldots, \natural(X^{n-1})\}$$

is a basis for $F[X]/\langle q \rangle$.

7.5 By a *net* over the interval $[0, 1]$ of $\mathbb{R}$ we mean a finite sequence $(a_i)_{0 \le i \le n+1}$ such that

$$0 = a_0 < a_1 < \ldots < a_n < a_{n+1} = 1.$$

By a *step function* on $[0, 1[$ we mean a mapping $f : [0, 1[\to \mathbb{R}$ for which there is a net $(a_i)_{0 \le i \le n+1}$ over $[0, 1]$ and a finite sequence $(b_i)_{0 \le i \le n}$ of elements of $\mathbb{R}$ such that

$$(i = 0, \ldots, n)(\forall x \in [a_i, a_{i+1}[) \qquad f(x) = b_i.$$

Free modules; bases 93

Show that the set E of step functions on $[0, 1[$ is an $\mathbb{R}$-vector space and that a basis of E is the set $\{e_k \; ; \; k \in \mathbb{R}\}$ of functions $e_k :$ $[0, 1[\to \mathbb{R}$ given by

$$e_k(x) = \begin{cases} 0 & \text{if } 0 \leq x < k; \\ 1 & \text{if } k \leq x < 1. \end{cases}$$

By a *piecewise linear function* on $[0, 1[$ we mean a mapping $f :$ $[0, 1[\to \mathbb{R}$ for which there is a net $(a_i)_{0 \leq i \leq n+1}$ and finite sequences $(b_i)_{0 \leq i \leq n}, (c_i)_{0 \leq i \leq n}$ of elements of $\mathbb{R}$ such that

$$(i = 0, \ldots, n)(\forall x \in [a_i, a_{i+1}[) \quad f(x) = b_i x + c_i.$$

Show that the set F of piecewise linear functions on $[0, 1[$ is an $\mathbb{R}$-vector space and that a basis of F is the set $\{f_k \; ; \; k \in \mathbb{R}\}$ of functions $f_k : [0, 1[\to \mathbb{R}$ given by

$$f_k(x) = \begin{cases} 0 & \text{if } 0 \leq x < k; \\ x - k & \text{if } k \leq x < 1. \end{cases}$$

If G is the subset of F consisting of those piecewise linear functions g that are continuous with $g(0) = 0$, show that G and E are subspaces of F such that $F = G \oplus E$.

Show finally that the assignment $f \mapsto f^\star$, where

$$f^\star(x) = \int_0^x f(t)\, dt,$$

defines an isomorphism from E onto G.

[*Hint.* Draw pictures!]

7.6 For every positive integer n let E_n be the set of functions $f : \mathbb{R} \to \mathbb{R}$ given by a prescription of the form

$$f(x) = a_0 + \sum_{k=1}^{n}(a_k \cos kx + b_k \sin kx)$$

where $a_0, a_k, b_k \in \mathbb{R}$ for $k = 1, \ldots, n$. Show that E_n is a subspace of the R-vector space Map$(\mathbb{R}, \mathbb{R})$ of all mappings from $\mathbb{R}$ to itself. Prove that if $f \in E_n$ is the zero function then all the coefficients a_i, b_i are zero.

[*Hint.* Use induction; consider $D^2 f + n^2 f$ where D denotes the differentiation map.]

Deduce that the $2n+1$ functions

$$x \mapsto 1, \quad x \mapsto \cos kx, \quad x \mapsto \sin kx \; (k = 1, \ldots, n)$$

constitute a basis for E_n.

7.7 Let R be a commutative unitary ring and let I be an ideal of R. Prove that every linearly independent subset of the R-module I has at most one element.

[*Hint.* $xy - yx = 0$.]

Deduce that if I is finitely generated, but not principal (i.e. not generated by a singleton), then I has no basis.

7.8 Let R be a commutative unitary ring with the property that every ideal of R is a free R-module. Prove that R is a principal ideal domain.

[*Hint.* Use Exercise 7.7.]

7.9 Given $p, q \in \mathbb{C}$ with $q \neq 0$, let S be the set of sequences $a = (a_i)_{i \in \mathbb{N}}$ of complex numbers such that

$$(\forall n \in \mathbb{N}) \quad a_{n+2} + p a_{n+1} + q a_n = 0.$$

Show that S is a subspace of the vector space $\mathbb{C}^{\mathbb{N}}$. Show also that $S \simeq \mathbb{C}^2$.

[*Hint.* Consider $f : S \to \mathbb{C}^2$ given by $f(a) = (a_0, a_1)$.]

7.10 Given $\alpha, \beta \in \mathbb{R}$ with $\alpha \neq \beta$, show that the set of rational functions of the form

$$x \mapsto \frac{a_0 + a_1 x + \ldots + a_{r+s-1} x^{r+s-1}}{(x - \alpha)^r (x - \beta)^s}$$

is an R-vector space of dimension $r + s$.

7.11 Let $\mathbb{C}_n[X]$ be the $\mathbb{C}$-vector space of dimension $n + 1$ of complex polynomials of degree less than or equal to n. If $P_0, \ldots, P_n \in \mathbb{C}[X]$ are such that $\deg P_i = i$ for every i, prove that $\{P_0, \ldots, P_n\}$ is a basis of $\mathbb{C}_n[X]$.

[*Hint.* Show that X^k is a linear combination of $P_0, \ldots, P_n$.]

Suppose now that $P_0, \ldots, P_n$ are given by

$$P_0(X) = X^0; \quad (1 \leq k \leq n) \quad P_k(X) = X(X - 1) \cdots (X - k + 1).$$

Show that there is a unique $\mathbb{C}$-morphism $f : \mathbb{C}_n[X] \to \mathbb{C}_n[X]$ such that $f(X^k) = P_k$ for every k, and that f is a bijection.

[*Hint.* If $P(X) = \sum_{i=0}^{n} a_k X^k$ define $f(P) = \sum_{i=0}^{n} a_k P_k$.]

Free modules; bases 95

7.12 Let V be a finite-dimensional vector spce over a field F. Prove that V has precisely one basis if and only if either $V = \{0\}$, or $F \simeq \mathbb{Z}/2\mathbb{Z}$ and dim $V = 1$.

[*Hint.* If V has at least two elements then every singleton $\{x\}$ with $x \neq 0$ can be extended to a basis. Deduce that V is finite and, by considering the sum of all the non-zero elements, show that V has at most one non-zero element.]

7.13 Let V be the $\mathbb{R}$-vector space generated by the six functions

$$x \mapsto 1,\ x,\ x^2,\ e^x,\ xe^x,\ x^2 e^x.$$

Let W be the subspace generated by the first five of these functions. Describe the induced map $D_* : V/W \to V/W$ (Theorem 4.3) where D denotes the differentiation map.

7.14 Let F be a field of p elements and let V be a vector space of dimension n over F. Show that V has p^n elements. Deduce that V has $p^n - 1$ linearly independent singleton subsets. Use induction to show that if $1 \leq m \leq n$ then the number of linearly independent subsets of V consisting of m elements is

$$\frac{1}{m!} \prod_{t=0}^{m-1} (p^n - p^t).$$

Hence determine the number of bases of V.

7.15 Let R and S be unitary rings and let $f : S \to R$ be a 1-preserving ring morphism. If M is an R-module let $M_{[S]}$ denote M regarded as an S-module (as in Exercise 1.7). If $\{r_i\ ;\ i \in I\}$ is a set of generators (respectively, a linearly independent subset) of $R_{[S]}$ and if $\{m_j\ ;\ j \in J\}$ is a set of generators (respectively, a linearly independent subset) of M, prove that $\{r_i m_j\ ;\ (i,j) \in I \times J\}$ is a set of generators (respectively, a linearly independent subset) of $M_{[S]}$.

7.16 Give an example to show that a submodule of a free module need not be free.

[*Hint.* Let $R = \mathbb{Z} \times \mathbb{Z}$; show that the R-module R is free and consider the submodule $\mathbb{Z} \times \{0\}$.]

8 Groups of morphisms; projective modules

If M and N are R-modules then the set of R-morphisms $f : M \to N$ will be denoted by $\mathrm{Mor}_R(M, N)$.

- Since quite often the term R-*homomorphism* is used instead of R-morphism, the set $\mathrm{Mor}_R(M, N)$ is often denoted by $\mathrm{Hom}_R(M, N)$.

It is clear that $\mathrm{Mor}_R(M, N)$ forms an abelian group under the addition $(f, g) \mapsto f + g$ where

$$(\forall x \in M) \qquad (f + g)(x) = f(x) + g(x).$$

One is tempted to say that 'obviously' $\mathrm{Mor}_R(M, N)$ forms an R-module under the action $(\lambda, f) \mapsto \lambda f$ where

$$(\forall x \in M) \qquad (\lambda f)(x) = \lambda f(x).$$

However, this is not the case; for in general we have

$$(\lambda f)(rm) = \lambda f(rm) = \lambda r f(m) \neq r \lambda f(m) = r(\lambda f)(m).$$

Nevertheless, it is obvious from this that $\mathrm{Mor}_R(M, N)$ is an R-module whenever R is commutative (for then $\lambda r = r\lambda$). Concerning these groups of morphisms we have the following.

Theorem 8.1 *If $(N_i)_{i \in I}$ is a family of R-modules then for every R-module M there are abelian group isomorphisms*

(a) $\mathrm{Mor}_R\left(\bigoplus_{i \in I} N_i, M\right) \simeq \underset{i \in I}{\text{\Large X}} \mathrm{Mor}_R(N_i, M);$

(b) $\mathrm{Mor}_R\left(M, \underset{i \in I}{\text{\Large X}} N_i\right) \simeq \underset{i \in I}{\text{\Large X}} \mathrm{Mor}_R(M, N_i).$

Proof (a) Let $\vartheta \in \mathrm{Mor}_R\left(\bigoplus_{i \in I} N_i, M\right) \to \underset{i \in I}{\text{\Large X}} \mathrm{Mor}_R(N_i, M)$ be given by the prescription $\vartheta(f) = (f \circ \mathrm{in}_i)_{i \in I}$. It is clear that ϑ is an abelian group morphism. To show that ϑ is surjective, let $(g_i)_{i \in I} \in \underset{i \in I}{\text{\Large X}} \mathrm{Mor}_R(N_i, M)$. Then there is an R-morphism $\zeta : \bigoplus_{i \in I} N_i \to M$ such that every diagram

Groups of morphisms; projective modules

is commutative. Since then $\vartheta(\zeta) = (\zeta \circ \mathrm{in}_i)_{i \in I} = (g_i)_{i \in I}$ we see that ϑ is surjective. To show that ϑ is also injective, let $\alpha \in \mathrm{Ker}\,\vartheta$. Then $0 = \vartheta(\alpha) = (\alpha \circ \mathrm{in}_i)_{i \in I}$ and so every diagram

is commutative, in which 0 is the zero morphism. Since $\left(\bigoplus_{i \in I} N_i, (\mathrm{in}_i)_{i \in I}\right)$ is a coproduct of $(N_i)_{i \in I}$ and since the zero morphism from $\bigoplus_{i \in I} N_i$ to M also makes each such diagram commutative, we deduce that $\alpha = 0$, whence ϑ is also injective.

(b) This is dual to the proof of (a), and we leave the details to the reader. It suffices to replace $\bigoplus$ by $\bigtimes$, reverse the arrows, and define $\vartheta(f) = (\mathrm{pr}_i \circ f)_{i \in I}$. ◊

Corollary 1 *If R is commutative then each of the above $\mathbb{Z}$-isomorphisms is an R-isomorphism.* ◊

Corollary 2 *If $I = \{1, \ldots, n\}$, then there are $\mathbb{Z}$-isomorphisms*

$$\mathrm{Mor}_R\left(\bigoplus_{i=1}^n N_i, M\right) \simeq \bigoplus_{i=1}^n \mathrm{Mor}_R(N_i, M),$$

$$\mathrm{Mor}_R\left(M, \bigoplus_{i=1}^n N_i\right) \simeq \bigoplus_{i=1}^n \mathrm{Mor}_R(M, N_i).\ \diamond$$

We shall now focus our attention on natural ways of defining $\mathbb{Z}$-morphisms between $\mathbb{Z}$-modules of the form $\mathrm{Mor}_R(M, N)$.

Suppose that A, B are R-modules and that $f \in \mathrm{Mor}_R(A, B)$. If M is any R-module then we can define a mapping

$$\mathrm{Mor}_R(M, A) \xrightarrow{f_*} \mathrm{Mor}_R(M, B)$$

by the assignment
$$\vartheta \longrightarrow f_*(\vartheta) = f \circ \vartheta.$$

It is clear that f_* so defined is a $\mathbb{Z}$-morphism; we say that it is *induced* by f.

Likewise, we can define a $\mathbb{Z}$-morphism
$$\mathrm{Mor}_R(A, M) \xleftarrow{\;f^*\;} \mathrm{Mor}_R(B, M)$$
by the assignment
$$f^*(\vartheta) = \vartheta \circ f \longleftarrow \vartheta.$$

We also say that f^* is *induced* by f.

- A useful mnemonic in distinguishing between f_* and f^* is that *lower* star indicates composition on the *left* by f (unless, of course, you write mappings on the right, in which case forget it!).

Theorem 8.2 *Given R-morphisms* $A \xrightarrow{\;f\;} B \xrightarrow{\;g\;} C$ *we have*
(1) $(g \circ f)_* = g_* \circ f_*$;
(2) $(g \circ f)^* = f^* \circ g^*$.
If also $h \in \mathrm{Mor}_R(A, B)$ *and* $k \in \mathrm{Mor}_R(B, C)$ *then*
(3) $(f + h)_* = f_* + h_*$;
(4) $(g + k)^* = g^* + k^*$.

Proof (1) This is immediate on considering the composite assignment
$$\vartheta \longrightarrow f_*(\vartheta) = f \circ \vartheta \longrightarrow g_*(f \circ \vartheta) = g \circ f \circ \vartheta.$$

(2) This is immediate on considering the composite assignment
$$\vartheta \circ f \circ g = g^*(\vartheta \circ f) \longleftarrow \vartheta \circ f = f^*(\vartheta) \longleftarrow \vartheta.$$

(3) and (4) follow respectively from the fact that $(f + h) \circ \vartheta = (f \circ \vartheta) + (h \circ \vartheta)$ and $\vartheta \circ (g + k) = (\vartheta \circ g) + (\vartheta \circ k)$. ◊

- Property (1) of Theorem 8.2 is often referred to by saying that the assignment $f \mapsto f_*$ is *covariant*, and property (2) is referred to by saying that the assignment $f \mapsto f^*$ is *contravariant*.

Our main interest in the induced $\mathbb{Z}$-morphisms f_*, f^* is in examining what happens to short exact sequences of R-modules when we form the morphism groups of the terms in the sequence from, and to, some given R-module. More explicitly, we have the following results, in which we write 0_{AB} for the zero morphism from A to B.

Theorem 8.3 *Consider a short exact sequence*

$$0 \longrightarrow A' \xrightarrow{f} A \xrightarrow{g} A'' \longrightarrow 0$$

of R-modules and R-morphisms. If M is an arbitrary R-module then the induced sequences of $\mathbb{Z}$-modules and $\mathbb{Z}$-morphisms

(1) $\quad 0 \longrightarrow \mathrm{Mor}_R(M, A') \xrightarrow{f_*} \mathrm{Mor}_R(M, A) \xrightarrow{g_*} \mathrm{Mor}_R(M, A'');$

(2) $\quad \mathrm{Mor}_R(A', M) \xleftarrow{f^*} \mathrm{Mor}_R(A, M) \xleftarrow{g^*} \mathrm{Mor}_R(A'', M) \longleftarrow 0$

are exact.

Proof We shall show that (1) is exact. A similar argument will establish the exactness of (2).

We have to show that (a) f is a monomorphism; (b) $\mathrm{Im}\, f_* \subseteq \mathrm{Ker}\, g_*$; and (c) $\mathrm{Ker}\, g_* \subseteq \mathrm{Im}\, f_*$.

(a) : Given $\vartheta \in \mathrm{Ker}\, f_*$ we have $0_{MA} = f_*(\vartheta) = f \circ \vartheta$ whence $\vartheta = 0_{MA}$ since f is a monomorphism and so is left cancellable.

(b) : If $\vartheta \in \mathrm{Im}\, f_*$ then there exists $\vartheta' \in \mathrm{Mor}_R(M, A')$ such that $\vartheta = f_*(\vartheta') = f \circ \vartheta'$. Consequently

$$g_*(\vartheta) = g_*[f_*(\vartheta')] = (g \circ f)_*(\vartheta') = 0_{MA''}$$

since, by the exactness of the given sequence, $g \circ f = 0_{A'A''}$. Thus $\vartheta \in \mathrm{Ker}\, g_*$ and we have established (b).

(c) : If $\vartheta \in \mathrm{Ker}\, g_*$ then for every $x \in M$ we have

$$g[\vartheta(m)] = [g_*(\vartheta)](m) = 0_{MA''}$$

and so $\vartheta(m) \in \mathrm{Ker}\, g = \mathrm{Im}\, f$. Thus there exists $x' \in A'$ such that $\vartheta(m) = f(x')$; and since f is a monomorphism such an element x' is unique. We can therefore define a mapping $\vartheta' : M \to A'$ by setting $\vartheta'(m) = x'$. Clearly, ϑ' is an R-morphism and $\vartheta = f \circ \vartheta' = f_*(\vartheta') \in \mathrm{Im}\, f_*$. Thus we see that $\mathrm{Ker}\, g_* \subseteq \mathrm{Im}\, f_*$. ◊

- It is important to note that the induced sequences of Theorem 8.3 are not short exact in general, for the induced $\mathbb{Z}$-morphisms $g_\star$ and $f^\star$ need not be surjective. For example, consider the short exact sequence of $\mathbb{Z}$-modules and $\mathbb{Z}$-morphisms

$$0 \longrightarrow \mathbb{Z} \overset{\iota}{\longrightarrow} \mathbb{Q} \overset{\natural}{\longrightarrow} \mathbb{Q}/\mathbb{Z} \longrightarrow 0.$$

The induced $\mathbb{Z}$-morphism

$$\mathrm{Mor}_\mathbb{Z}(\mathbb{Z}/2\mathbb{Z}, \mathbb{Q}) \overset{\natural_\star}{\longrightarrow} \mathrm{Mor}_\mathbb{Z}(\mathbb{Z}/2\mathbb{Z}, \mathbb{Q}/\mathbb{Z})$$

cannot be surjective since

$$\mathrm{Mor}_\mathbb{Z}(\mathbb{Z}/2\mathbb{Z}, \mathbb{Q}) = 0 \neq \mathrm{Mor}_\mathbb{Z}(\mathbb{Z}/2\mathbb{Z}, \mathbb{Q}/\mathbb{Z}).$$

In fact, given $\vartheta \in \mathrm{Mor}_\mathbb{Z}(\mathbb{Z}/2\mathbb{Z}, \mathbb{Q})$ let $x = \vartheta(1 + 2\mathbb{Z})$. We have

$$2x = 2\vartheta(1 + 2\mathbb{Z}) = \vartheta(2 + 2\mathbb{Z}) = \vartheta(0 + 2\mathbb{Z}) = 0,$$

whence $x = 0$ and consequently $\vartheta = 0$. On the other hand, the mapping described by

$$0 + 2\mathbb{Z} \mapsto 0 + \mathbb{Z}, \quad 1 + 2\mathbb{Z} \mapsto \tfrac{1}{2} + \mathbb{Z}$$

is a non-zero element of $\mathrm{Mor}_\mathbb{Z}(\mathbb{Z}/2\mathbb{Z}, \mathbb{Q}/\mathbb{Z})$.

In a similar way, the induced $\mathbb{Z}$-morphism

$$\mathrm{Mor}_\mathbb{Z}(\mathbb{Z}, \mathbb{Z}) \overset{\iota^\star}{\longleftarrow} \mathrm{Mor}_\mathbb{Z}(\mathbb{Q}, \mathbb{Z})$$

cannot be surjective since

$$\mathrm{Mor}_\mathbb{Z}(\mathbb{Q}, \mathbb{Z}) = 0 \neq \mathrm{Mor}_\mathbb{Z}(\mathbb{Z}, \mathbb{Z}).$$

In fact, given $\vartheta \in \mathrm{Mor}_\mathbb{Z}(\mathbb{Q}, \mathbb{Z})$ suppose that $\vartheta(1) \neq 0$. Then for every non-zero $r \in \mathbb{Z}$ we have $\vartheta(1) = r\vartheta(1/r)$, whence r divides $\vartheta(1)$. But, by the fundamental theorem of arithmetic, $\vartheta(1)$ has only finitely many divisors. We deduce, therefore, that we must have $\vartheta(1) = 0$. For all non-zero $p, q \in \mathbb{Z}$ we then have

$$0 = p\vartheta(1) = p\vartheta\left(\frac{q}{q}\right) = pq\vartheta\left(\frac{1}{q}\right) = q\vartheta\left(\frac{p}{q}\right),$$

whence $\vartheta(p/q) = 0$ and so ϑ is the zero map. On the other hand, $\mathrm{id}_\mathbb{Z}$ is clearly a non-zero element of $\mathrm{Mor}_\mathbb{Z}(\mathbb{Z}, \mathbb{Z})$. Indeed, the groups $\mathrm{Mor}_\mathbb{Z}(\mathbb{Z}, \mathbb{Z})$ and $\mathbb{Z}$ are isomorphic; see Exercise 2.4.

Groups of morphisms; projective modules

Despite the above remark, we have the following situation.

Theorem 8.4 *Suppose that*

$$0 \longrightarrow A' \xrightarrow{f} A \xrightarrow{g} A'' \longrightarrow 0$$

is a split short exact sequence of R-modules and R-morphisms. Then for every R-module M there are induced split exact sequences of $\mathbb{Z}$-modules and $\mathbb{Z}$-morphisms

$$0 \longrightarrow \mathrm{Mor}_R(M, A') \xrightarrow{f_*} \mathrm{Mor}_R(M, A) \xrightarrow{g_*} \mathrm{Mor}_R(M, A'') \longrightarrow 0;$$

$$0 \longleftarrow \mathrm{Mor}_R(A', M) \xleftarrow{f^*} \mathrm{Mor}_R(A, M) \xleftarrow{g^*} \mathrm{Mor}_R(A'', M) \longleftarrow 0.$$

Proof We shall establish the first sequence, the second being similar. To show that (1) is exact, it suffices by Theorem 8.3 to show, using the fact that the original short exact sequence splits, that g_* is surjective. Suppose then that $\vartheta'' \in \mathrm{Mor}_R(M, A'')$ and let $\bar{g}$ be a splitting morphism associated with g. Consider the mapping $\vartheta = \bar{g} \circ \vartheta''$. Clearly, $\vartheta \in \mathrm{Mor}_R(M, A)$ and $g_*(\vartheta) = g \circ \vartheta = g \circ \bar{g} \circ \vartheta'' = \vartheta''$ and so g_* is indeed surjective. That (1) now splits follows from the fact that $g \circ \bar{g} = \mathrm{id}_{A''}$; for, by Theorem 8.2(1), we have $g_* \circ \bar{g}_* = (g \circ \bar{g})_* = (\mathrm{id}_{A''})_* = \mathrm{id}$ where this last identity map is that on $\mathrm{Mor}_R(M, A'')$. ◇

The above discussion leads in a natural way to the following notion.

Definition An R-module M is said to be *projective* if, for every short exact sequence

$$0 \longrightarrow A' \xrightarrow{f} A \xrightarrow{g} A'' \longrightarrow 0$$

of R-modules and R-morphisms, the induced $\mathbb{Z}$-morphism

$$\mathrm{Mor}_R(M, A) \xrightarrow{g_*} \mathrm{Mor}_R(M, A'')$$

is surjective.

We now derive an alternative characterisation of projective modules.

Theorem 8.5 *An R-module P is projective if and only if every diagram of R-modules and R-morphisms of the form*

$$\begin{array}{c} P \\ \downarrow{\vartheta''} \\ A \xrightarrow{g} A'' \longrightarrow 0 \end{array} \quad (exact)$$

can be extended to a commutative diagram

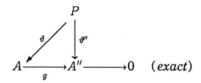

Proof Clearly, P is projective if and only if for every epimorphism g the induced morphism g_* is surjective; in other words, if and only if for every $\vartheta'' \in \mathrm{Mor}_R(P, A'')$ there exists $\vartheta \in \mathrm{Mor}_R(P, A)$ such that $\vartheta'' = g_*(\vartheta) = g \circ \vartheta$. $\Diamond$

- Roughly speaking, Theorem 8.5 says that P is projective if morphisms from P can be 'lifted' through epimorphisms. The morphism ϑ that completes the above diagram is called a *projective lifting* of ϑ''.

- Projective liftings are not unique in general.

- The characterisation in Theorem 8.5 is often taken as a definition of a projective module.

As for examples of projective modules, an abundance is provided by the following result.

Theorem 8.6 *Every free module is projective.*

Proof Suppose that M is a free R-module and consider the diagram

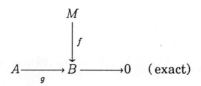

We have to establish the existence of an R-morphism $\vartheta : M \to A$ such that $g \circ \vartheta = f$. For this purpose, let S be a basis of M. Then, since g is surjective, for every $x \in S$ there exists $a \in A$ such that $f(x) = g(a)$. For each $x \in S$ choose once and for all an element $a_x \in A$ such that $f(x) = g(a_x)$. We can then define a mapping $\zeta : S \to A$ by the prescription $\zeta(x) = a_x$. Since M is free on S, we can extend ζ to a unique R-morphism $\vartheta : M \to A$ such that $\vartheta \circ \iota_S = \zeta$ where $\iota_S : S \to M$ is the natural inclusion. If $f_S : S \to B$ denotes the restriction of f to S, we then have $g \circ \vartheta \circ \iota_S = g \circ \zeta = f_S = f \circ \iota_S$ whence we see that the R-morphisms

Groups of morphisms; projective modules

$g \circ \vartheta$ and f coincide on the basis S. Applying Corollary 4 of Theorem 7.7, we deduce that $g \circ \vartheta = f$ and consequently that M is projective. ◊

- The converse of Theorem 8.6 is not true; see Exercise 8.2 for an example of a projective module that is not free. Thus the class of projective modules is larger than the class of free modules.

Our objective now is to obtain further useful characterisations of projective modules. For this purpose, we require the following result.

Theorem 8.7 *Every module is isomorphic to a quotient module of a free module.*

Proof Let M be an R-module and let S be a set of generators of M (e.g. the set M itself will do). Let F be a free R-module on S. Then the natural inclusion $\iota : S \to M$ extends to a unique R-morphism $h : F \to M$. Since $S = \iota^\to(S) \subseteq h^\to(F)$ and since S generates M, it follows that $h^\to(F) = M$. Thus h is an epimorphism and so $M = \operatorname{Im} h \simeq F/\operatorname{Ker} h$. ◊

Corollary *Every finitely generated module is isomorphic to a quotient module of a free module having a finite basis.* ◊

Theorem 8.8 *For an R-module P the following are equivalent:*
(1) *P is projective;*
(2) *every exact sequence $M \longrightarrow P \longrightarrow 0$ splits;*
(3) *P is a direct summand of a free R-module.*

Proof (1) $\Rightarrow$ (2) : Consider the diagram

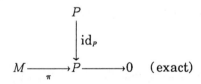

Since P is projective there is an R-morphism $f : P \to M$ such that $\pi \circ f = \operatorname{id}_P$; in other words, $M \longrightarrow P \longrightarrow 0$ splits.

(2) $\Rightarrow$ (3) : Let F be a free R-module with the set P as a basis and form the exact sequence

$$F \xrightarrow{\pi} P \longrightarrow 0$$

where π is an extension of id_P and so is an epimorphism. By the hypothesis, the short exact sequence

$$0 \longrightarrow \operatorname{Ker} \pi \xrightarrow{\iota} F \xrightarrow{\pi} P \longrightarrow 0$$

splits on the right. As we saw in Section 6, we then have $F \simeq \text{Ker}\,\pi \oplus P$. This direct sum module is therefore free and has P as a direct summand.

$(3) \Rightarrow (1)$: Consider the diagram

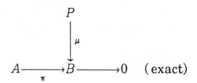

By hypothesis, there exists an R-module Q and a free R-module F such that $F = P \oplus Q$. Define $\mu' : F \to B$ by the prescription

$$(\forall x = p + q \in F) \quad \mu'(x) = \mu(p).$$

It is readily seen that μ' is an R-morphism. Now F, being free, is projective by Theorem 8.6 and so μ' can be lifted to an R-morphism $\mu'' : F \to A$ such that $\pi \circ \mu'' = \mu'$. It then follows that the restriction μ''_P of μ'' to P is an R-morphism such that $\pi \circ \mu''_P = \mu$, and hence that P is projective. $\Diamond$

We now make the observation that since a vector space V has a basis (Corollary 1 to Theorem 7.9), it is free (Theorem 7.6), whence it is projective (Theorem 8.6). Consequently, by Theorem 8.8, we can assert that

every short exact sequence of vector spaces splits.

This simple observation yields the following results.

Theorem 8.9 *Every subspace of a vector space is a direct summand.*

Proof The canonical short exact sequence

$$0 \longrightarrow W \xrightarrow{\iota_W} V \xrightarrow{\natural_W} V/W \longrightarrow 0$$

splits and so, by Theorem 6.11, $\text{Ker}\,\natural_W = W$ is a direct summand of V. $\Diamond$

Corollary 1 *If W is a subspace of V then*

$$\dim V = \dim W + \dim V/W.$$

Proof There exists a subspace W' such that $V = W \oplus W'$. By Theorem 7.8 we have $\dim V = \dim W + \dim W'$; and by Theorem 6.10 we have $\dim W' = \dim V/W$. $\Diamond$

Corollary 2 *If V, W are vector spaces and if $f : V \to W$ is a linear transformation then*

$$\dim V = \dim \operatorname{Im} f + \dim \operatorname{Ker} f.$$

Proof This follows from Corollary 1 and the first isomorphism theorem on taking $W = \operatorname{Ker} f$. ◊

- $\dim \operatorname{Im} f$ is often callled the *rank* of f and $\dim \operatorname{Ker} f$ is often called the *nullity* of f. Then Corollary 2 above can be expressed in the form :

$$\text{rank} + \text{nullity} = \text{dimension of departure space}.$$

This is sometimes also referred to as the *dimension theorem*.

Corollary 3 *If W is a subspace of V then $\dim W \leq \dim V$. Moreover, if V is of finite dimension then $\dim W = \dim V$ implies that $W = V$.*

Proof The first statement is clear. As for the second, if $\dim V$ is finite then, by Corollary 1, $\dim V = \dim W$ implies that $\dim V/W = 0$, whence $W = V$ since the only vector spaces of dimension 0 are the zero spaces. ◊

- The second part of Corollary 3 is not true when V is of infinite dimension, for infinite cardinals are not cancellable under addition. Likewise it does not hold in general for modules; for example, the free $\mathbb{Z}$-module $\mathbb{Z}$ has dimension 1 (since $\{1\}$ is a basis) as does every submodule $t\mathbb{Z}$ with $t \neq 0, 1$ (since $\{t\}$ is a basis).

Corollary 4 *If V, W are finite-dimensional vector spaces with $\dim V = \dim W$ and if $f : V \to W$ is a linear transformation then the following are equivalent :*

(1) f *is injective*;
(2) f *is surjective*;
(3) f *is bijective*.

Proof This is immediate from Corollaries 2 and 3; for f is injective if and only if $\dim \operatorname{Ker} f = 0$, which is the case if and only if $\dim W = \dim V = \dim \operatorname{Im} f$, which is equivalent to $W = \operatorname{Im} f$, i.e. to f being surjective. ◊

Using exact sequence, we can generalise Corollary 1 above as follows.

Theorem 8.10 *For an exact sequence*

$$0 \longrightarrow V_0 \xrightarrow{f_0} V_1 \xrightarrow{f_1} \cdots \xrightarrow{f_{n-1}} V_n \longrightarrow 0$$

of vector spaces and linear transformations we have

$$\sum_{k \text{ odd}} \dim V_k = \sum_{k \text{ even}} \dim V_k.$$

If, moreover, every V_i is of finite dimension then

$$\sum_{i=0}^{n}(-1)^n \dim V_i = 0.$$

Proof Clearly, $\dim V_0 = \dim \operatorname{Im} f_0$ and $\dim V_n = \dim \operatorname{Im} f_{n-1}$. Moreover, for $0 < k < n-2$ we have, by Corollary 2 of Theorem 8.9,

$$\dim V_{k+1} = \dim \operatorname{Im} f_{k+1} + \dim \operatorname{Ker} f_{k-1}$$
$$= \dim \operatorname{Im} f_{k+1} + \dim \operatorname{Im} f_k.$$

On summing separately the dimensions of the odd-numbered spaces and the even-numbered spaces in the sequence, we see that

$$\sum_{k \text{ odd}} \dim V_k = \sum_{k=0}^{n} \dim \operatorname{Im} f_k = \sum_{k \text{ even}} \dim V_k.$$

The second statement is an obvious consequence of the first. ◊

Corollary 1 *If A, B are subspaces of a vector space V then*

$$\dim(A+B) + \dim(A \cap B) = \dim(A \times B) = \dim A + \dim B.$$

Proof Consider the exact sequence

$$0 \longrightarrow A \xrightarrow{\vartheta} A \times B \xrightarrow{\pi} B \longrightarrow 0$$

where ϑ is given by $a \mapsto (a, 0)$ and π is given by $(a, b) \mapsto b$. Applying the above results, we obtain

$$\dim(A \times B) = \dim A + \dim B.$$

Consider now the sequence

$$0 \longrightarrow A \cap B \xrightarrow{\alpha} A \times B \xrightarrow{\beta} A + B \longrightarrow 0$$

where α is given by $x \mapsto (x, x)$ and β is given by $(a, b) \mapsto a - b$. It is clear that α is a monomorphism and that β is an epimorphism. Moreover, $\operatorname{Ker} \beta = \operatorname{Im} \alpha$ and so the sequence is exact. Applying the theorem, we obtain

$$\dim(A \times B) = \dim(A \cap B) + \dim(A + B). \quad \diamond$$

Corollary 2 *If V, W, X are finite-dimensional vector spaces and if $f : V \to W$ and $g : W \to X$ are linear transformations then*

$$\dim \operatorname{Im}(g \circ f) = \dim \operatorname{Im} f - \dim(\operatorname{Im} f \cap \operatorname{Ker} g)$$
$$= \dim(\operatorname{Im} f + \operatorname{Ker} g) - \dim \operatorname{Ker} g.$$

Proof By Corollary 1, we have

$$\dim(\operatorname{Im} f + \operatorname{Ker} g) + \dim(\operatorname{Im} f \cap \operatorname{Ker} g) = \dim \operatorname{Im} f + \dim \operatorname{Ker} g.$$

Rearranging this, we obtain

$$\dim \operatorname{Im} f - \dim(\operatorname{Im} f \cap \operatorname{Ker} g) = \dim(\operatorname{Im} f + \operatorname{Ker} g) - \dim \operatorname{Ker} g.$$

We now show that the left-hand side coincides with $\dim \operatorname{Im}(g \circ f)$. For this purpose, we note that if g_f denotes the restriction of g to the subspace $\operatorname{Im} f$ then

$$\operatorname{Ker} g_f = \dim \operatorname{Im} f \cap \operatorname{Ker} g \quad \text{and} \quad \dim \operatorname{Im} g_f = \dim(g \circ f).$$

We thus have

$$\dim \operatorname{Im}(g \circ f) = \dim \operatorname{Im} g_f = \dim \operatorname{Im} f - \dim \operatorname{Ker} g_f$$
$$= \dim \operatorname{Im} f - \dim(\operatorname{Im} f \cap \operatorname{Ker} g). \quad \diamond$$

Theorem 8.11 *If V and W are finite-dimensional vector spaces and if $f : V \to W$ is a linear transformation then, for every subspace V' of V,*

$$\dim f^{\to}(V') = \dim V' - \dim(\operatorname{Ker} f \cap V'),$$

and, for every subspace W' of W,

$$\dim f^{\leftarrow}(W') = \dim(\operatorname{Im} f \cap W') + \dim \operatorname{Ker} f.$$

Proof The first statement is immediate from Corollary 2 of Theorem 8.9 on observing that if f' is the restriction of f to V' then $\operatorname{Im} f' = f^{\to}(V')$ and $\operatorname{Ker} f' = \operatorname{Ker} f \cap V'$.

As for the second equality, note by the Corollary to Theorem 3.2 that

$$\operatorname{Im} f \cap W' = f^{\to}[f^{\leftarrow}(W')]$$

and so, by the first part of the theorem,

$$\dim(\operatorname{Im} f \cap W') = \dim f^{\leftarrow}(W') - \dim(\operatorname{Ker} f \cap f^{\leftarrow}(W')).$$

The result now follows from the fact that $f^{\leftarrow}$ is inclusion-preserving, so that $\operatorname{Ker} f = f^{\leftarrow}\{0\} \subseteq f^{\leftarrow}(W'). \quad \diamond$

We end the present section by returning to the problem of the commutative completion of certain triangles in the presence of projective modules.

Theorem 8.12 *Let P be a projective module and suppose that the diagram*

$$\begin{array}{ccc} & & P \\ & & \downarrow \vartheta \\ X \xrightarrow{\alpha} & Y \xrightarrow{\beta} & Z \end{array}$$

is such that the row is exact and $\beta \circ \vartheta = 0$. Then there is an R-morphism $\zeta : P \to X$ such that $\alpha \circ \zeta = \vartheta$.

Proof Since $\beta \circ \vartheta = 0$ we have $\operatorname{Im} \vartheta \subseteq \operatorname{Ker} \beta = \operatorname{Im} \alpha$. Applying the projectivity of P to the diagram

$$\begin{array}{ccc} & & P \\ & & \downarrow \vartheta^+ \\ X \xrightarrow{\alpha^+} & \operatorname{Im} \alpha & \to 0 \end{array}$$

in which ϑ^+, α^+ are given by $p \mapsto \vartheta(p), x \mapsto \alpha(x)$ we obtain the existence of an R-morphism $\zeta : P \to X$ such that $\alpha^+ \circ \zeta = \vartheta^+$. Since then

$$(\forall y \in P) \quad \alpha[\zeta(y)] = \alpha^+[\zeta(y)] = \vartheta^+(y) = \vartheta(y),$$

we have $\alpha \circ \zeta = \vartheta$. ◊

Theorem 8.13 *Consider the diagram*

of R-modules and R-morphisms. Suppose that C is projective. Then the following are equivalent :

(1) *there is an R-morphism $h : C \to B$ such that $f \circ h = g$;*
(2) *$\operatorname{Im} g \subseteq \operatorname{Im} f$.*

Proof $(1) \Rightarrow (2)$: as in Theorem 3.5.

(2) ⇒ (1) : Consider the diagram

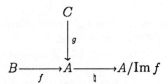

in which the row is exact. If (2) holds then $\operatorname{Im} g \subseteq \operatorname{Im} f = \operatorname{Ker} \natural$ so that $\natural \circ g = 0$ and (1) follows by Theorem 8.12. ◊

Theorem 8.14 *Consider the diagram*

of vector spaces and linear transformations. The following are equivalent :
(1) *there is a linear transformation* $h : B \to C$ *such that* $h \circ f = g$;
(2) $\operatorname{Ker} f \subseteq \operatorname{Ker} g$.

Proof (1) ⇒ (2) : as in Theorem 3.4.
(2) ⇒ (1) : If (2) holds then by Theorem 3.4 there is a linear transformation $h' : \operatorname{Im} f \to C$ such that $h'[f(x)] = g(x)$ for all $x \in A$. Let $\{f(e_i) \, ; \, i \in I\}$ be a basis of the subspace $\operatorname{Im} f$ of B. By Theorem 7.9 we can extend this to a basis

$$\{f(e_i) \, ; \, i \in I\} \cup \{x_j \, ; \, j \in J\}$$

of B. Choosing a family $(c_j)_{j \in J}$ of elements of C, we can define a linear transformation $h : B \to C$ by the prescription

$$(\forall i \in I) \; h[f(e_i)] = g(e_i), \qquad (\forall j \in J) \; h(x_j) = c_j.$$

Then clearly h is such that $h \circ f = g$. ◊

- Note that Theorem 8.14 does not hold in general for modules. The reason for this is essentially that if N is a submodule of a module M such that N has a basis then we cannot in general extend this to a basis of M. For example, $\{2\}$ is a basis of the submodule $2\mathbb{Z}$ of the free $\mathbb{Z}$-module $\mathbb{Z}$, but this cannot be extended to a basis of $\mathbb{Z}$.

EXERCISES

8.1 Let m and n be integers, each greater than 1. Show that the prescription $\vartheta(x + m\mathbb{Z}) = nx + nm\mathbb{Z}$ describes a $\mathbb{Z}$-morphism $\vartheta : \mathbb{Z}/m\mathbb{Z} \to \mathbb{Z}/nm\mathbb{Z}$. Prove that the $\mathbb{Z}$-module

$$\mathrm{Mor}_{\mathbb{Z}}(\mathbb{Z}/m\mathbb{Z}, \mathbb{Z}/nm\mathbb{Z})$$

is generated by $\{\vartheta\}$ and deduce that

$$\mathrm{Mor}_{\mathbb{Z}}(\mathbb{Z}/m\mathbb{Z}, \mathbb{Z}/nm\mathbb{Z}) \simeq \mathbb{Z}/m\mathbb{Z}.$$

[*Hint.* Show that $\mathrm{Ann}_{\mathbb{Z}}(\vartheta) = m\mathbb{Z}$ and use Exercise 4.1.]

8.2 Let R be a commutative unitary ring and let I be an ideal of R. If M is an R-module define

$$M_I = \{x \in M \ ; \ I \subseteq \mathrm{Ann}_R(x)\}.$$

Show that M_I is an R-module and prove that the assignment $f \mapsto f(1 + I)$ defines an R-isomorphism

$$\vartheta : \mathrm{Mor}_R(R/I, M) \to M_I.$$

Hence establish a $\mathbb{Z}$-isomorphism

$$\mathrm{Mor}_{\mathbb{Z}}(\mathbb{Z}/n\mathbb{Z}, \mathbb{Q}/\mathbb{Z}) \simeq \mathbb{Z}/n\mathbb{Z}.$$

[*Hint.* Consider $\zeta : \mathbb{Z} \to (\mathbb{Q}/\mathbb{Z})_{n\mathbb{Z}}$ given by $\zeta(m) = \frac{m}{n} + \mathbb{Z}$.]

8.3 If $(P_\alpha)_{\alpha \in I}$ is a family of R-modules each of which is projective prove that $\bigoplus_{\alpha \in I} P_\alpha$ is projective. Conversely, prove that if $\bigoplus_{\alpha \in I} P_\alpha$ is projective then so is each P_α.

[*Hint.* Working with the diagram

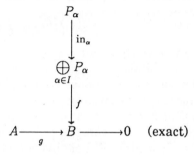

determine the appropriate fill-in maps.]

8.4 If p is a prime, establish a short exact sequence
$$0 \longrightarrow \mathbb{Z}/p\mathbb{Z} \longrightarrow \mathbb{Z}/p^2\mathbb{Z} \longrightarrow \mathbb{Z}/p\mathbb{Z} \longrightarrow 0$$
that does not split. Deduce that a submodule of a projective module need not be projective.

8.5 Let n be an integer greater than 1. For every divisor r of n consider the ideal $r(\mathbb{Z}/n\mathbb{Z})$ of the ring $\mathbb{Z}/n\mathbb{Z}$. Show how to construct an exact sequence
$$0 \longrightarrow \frac{n}{r}(\mathbb{Z}/n\mathbb{Z}) \longrightarrow \mathbb{Z}/n\mathbb{Z} \longrightarrow r(\mathbb{Z}/n\mathbb{Z}) \longrightarrow 0.$$
Prove that the following are equivalent:

1. the above short exact sequence splits;
2. h.c.f.$\{r, n/r\} = 1$;
3. the $\mathbb{Z}/n\mathbb{Z}$-module $r(\mathbb{Z}/n\mathbb{Z})$ is projective.

Hence give an example of a projective module that is not free.

[*Hint.* Consider $\mathbb{Z}/2\mathbb{Z}$ as a $\mathbb{Z}/6\mathbb{Z}$-module.]

8.6 Suppose that in the diagram of R-modules and R-morphisms

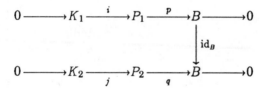

the rows are exact and P_1 is projective. Prove that there are R-morphisms $\beta : P_1 \to P_2$ and $\alpha : K_1 \to K_2$ such that the completed diagram is commutative.

[*Hint.* Use the projectivity of P_1 to construct β. As for α, observe that $q \circ \beta \circ i = 0$; use Theorem 3.7.]

Consider now the sequence
$$0 \longrightarrow K_1 \xrightarrow{\vartheta} P_1 \oplus K_2 \xrightarrow{\pi} P_2 \longrightarrow 0$$
in which $\vartheta(k) = (i(k), \alpha(k))$ and $\pi(p_1, k_2) = \beta(p_1) - j(k_2)$. Show that this sequence is exact. Deduce that if P_2 is also projective then $P_1 \oplus K_2 \simeq P_2 \oplus K_1$.

8.7 Show that every diagram of R-modules and R-morphisms of the form

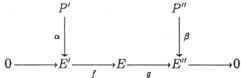

in which the row is exact and P', P'' are projective can be extended to a commutative diagram

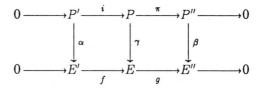

in which the top row is exact and P is also projective.

[*Hint.* Take $P = P' \oplus P''$. Let $\overline{\beta}$ be a projective lifting of β and let $j : P \to P'$ be a left-hand splitting morphism; consider the mapping $\gamma = (f \circ \alpha \circ j) = (\overline{\beta} \circ \pi)$.]

8.8 Let the diagram of R-modules and R-morphisms

$$\cdots \longrightarrow P_3 \xrightarrow{g_3} P_2 \xrightarrow{g_2} P_1 \xrightarrow{g_1} A$$
$$\downarrow k_0$$
$$\cdots \longrightarrow Q_3 \xrightarrow{h_3} Q_2 \xrightarrow{h_2} Q_1 \xrightarrow{h_1} B \longrightarrow 0$$

be such that both rows are exact and each P_i is projective. Prove that for every positive integer n there is an R-morphism $k_n : P_n \to Q_n$ such that $h_n \circ k_n = k_{n-1} \circ g_n$.

[*Hint.* Use Theorem 8.5 and induction.]

8.9 Let R be a commutative unitary ring. Let X and Y be R-modules with X projective. If A, B are submodules of X, Y respectively show that the set

$$\Delta_{A,B} = \{f \in \mathrm{Mor}_R(X, Y) \; ; \; f^\to(A) \subseteq B\}$$

is a submodule of the R-module $\mathrm{Mor}_R(X, Y)$. Show also that there is an R-isomorphism

$$\mathrm{Mor}_R(X/A, Y/B) \simeq \Delta_{A,B}/\Delta_{X,B}.$$

Groups of morphisms; projective modules

8.10 If M and N are R-modules consider the set $P(M,N)$ that consists of those R-morphisms $f : M \to N$ that 'factor through projectives' in the sense that there is a commutative diagram

in which P is projective. Show that $P(M,N)$ forms a subgroup of the group $\text{Mor}_R(M,N)$.

[*Hint.* If f factors through P and g factors through Q show that $f - g$ factors through $P \oplus Q$.]

If $[M,N]$ denotes the corresponding quotient group, prove that for every exact sequence

$$0 \longrightarrow N' \xrightarrow{f} N \xrightarrow{g} N'' \longrightarrow 0$$

there is an induced exact sequence of $\mathbb{Z}$-modules and $\mathbb{Z}$-morphisms

$$[M,N'] \xrightarrow{f'} [M,N] \xrightarrow{g'} [M,N''].$$

[*Hint.* The sequence

$$\text{Mor}_R(M,N') \xrightarrow{f_*} \text{Mor}_R(M,N) \xrightarrow{g_*} \text{Mor}_R(M,N'')$$

is exact. If $\alpha \in P(M,N')$ observe that $f_*(\alpha) \in P(M,N)$ and use Theorem 4.3 to produce f'; similarly for g'.]

8.11 An exact sequence of the form

$$\cdots \longrightarrow P_n \xrightarrow{f_n} P_{n-1} \xrightarrow{f_{n-1}} \cdots \xrightarrow{f_1} P_0 \longrightarrow 0$$

is said to *split* if there exist R-morphisms $g_i : P_i \to P_{i+1}$ such that

1. $f_1 \circ g_0 = \text{id}_{P_0}$;
2. $(\forall i \geq 1)\ g_{i-1} f_i + f_{i+1} g_i = \text{id}_{P+i}$.

Prove by induction that if each P_i is projective then the above sequence splits.

8.12 Let Δ_n be the ring of lower triangular $n \times n$ matrices $X = [x_{ij}]$ over a field F (so that $x_{ij} = 0$ if $i < j$). Let $A, B \in \Delta_n$ be given respectively by

$$a_{ij} = \begin{cases} 1 & \text{if } i = j+1; \\ 0 & \text{otherwise}, \end{cases} \qquad b_{ij} = \begin{cases} 1 & \text{if } i = j = 1; \\ 0 & \text{otherwise}. \end{cases}$$

If $\Theta_n = \{X \in \Delta_n \,;\, (i = 1, \ldots, n)\ x_{ii} = 0\}$ prove that

$$0 \longrightarrow \Delta_n B \xrightarrow{f} \Delta_n \xrightarrow{g} \Theta_n \longrightarrow 0,$$

where f is the natural inclusion and g is given by $g(X) = XA$, is a split exact sequence of Δ_n-modules. Deduce that Θ_n is a projective Δ_n-module.

8.13 Let V and W be finite-dimensional vector spaces over a field F. If $f : V \to W$ is a linear transformation let the rank of f be written $\rho(f)$. Prove that, for every $\lambda \in F$,

$$\rho(f) = \begin{cases} \rho(f) & \text{if } \lambda \neq 0; \\ 0 & \text{if } \lambda = 0. \end{cases}$$

Prove also that if $g : V \to W$ is also linear then

$$|\rho(f) - \rho(g)| \leq \rho(f + g) \leq \rho(f) + \rho(g).$$

[*Hint.* Establish the right-hand inequality from $\text{Im}(f+g) \subseteq \text{Im}\, f + \text{Im}\, g$. As for the left-hand inequality, write $f = (f + g) - g$ and apply the right-hand inequality together with the first part of the question.]

8.14 Let V, W, X be finite-dimensional vector spaces over a field F and let $f : V \to W, g : W \to X$ be linear transformations. Prove that

$$\rho(f) + \rho(g) - \dim V \leq \rho(g \circ f) \leq \min\{\rho(f), \rho(g)\}.$$

[*Hint.* For the left-hand inequality consider the restriction of g to $\text{Im}\, f$; and for the right-hand inequality us Corollary 2 of Theorem 8.11.]

9 Duality; transposition

In the previous section we noted that if M and N are R-modules then the abelian group $\text{Mor}_R(M,N)$ is not in general an R-module. Let us now examine the particular case where $N = R$. Noting that expressions of the form $rf(m)\lambda$ are meaningful for $r,\lambda \in R$ and $f \in \text{Mor}_R(M,R)$, we can give $\text{Mor}_R(M,R)$ the structure of a *right R-module* as follows : for every $f \in \text{Mor}_R(M,R)$ and every $\lambda \in R$ let $f\lambda : M \to R$ be given by the prescription $(f\lambda)(m) = f(m)\lambda$. Then by virtue of the equalities

$$(f\lambda)(rm) = f(rm)\lambda = rf(m)\lambda = r(f\lambda)(m),$$

the group morphism $f\lambda$ is indeed an R-morphism so is in $\text{Mor}_R(M,R)$.

Definition If M is an R-module then the *dual* of M is the *right R-module* $M^d = \text{Mor}_R(M,R)$. The elements of the dual module M^d are called *linear forms* (or *linear functionals*) on M.

It is clear that in an analogous way we can start with a right R-module N and form its dual, which will be a *left R-module*. In particular, if M is an R-module then we can form the dual of the right R-module M^d, thus obtaining the (left) R-module $(M^d)^d$. We denote this by M^{dd} and call it the *bidual* of M.

- Similarly, we can make $\text{Mor}_R(R,M)$ into a left R-module, but this turns out to be isomorphic to M; see Exercise 9.1.

Example 9.1 As in the remark following Theorem 8.3, we have

$$\mathbb{Q}^d = 0, \ (\mathbb{Z}/2\mathbb{Z})^d = 0, \ \mathbb{Z}^d \simeq \mathbb{Z}.$$

More generally, it can likewise be shown that $(\mathbb{Z}/n\mathbb{Z})^d = 0$ as a $\mathbb{Z}$-module. In contrast, note that $(\mathbb{Z}/n\mathbb{Z})^d \simeq \mathbb{Z}/n\mathbb{Z}$ as $\mathbb{Z}/n\mathbb{Z}$-modules; for the $\mathbb{Z}/n\mathbb{Z}$-module $\mathbb{Z}/n\mathbb{Z}$ is free, of dimension 1, and the assignment $f \mapsto f(1+n\mathbb{Z})$ yields a $\mathbb{Z}/n\mathbb{Z}$-isomorphism from $(\mathbb{Z}/n\mathbb{Z})^d$ to $\mathbb{Z}/n\mathbb{Z}$.

If M is an R-module and M^d is its dual then in what follows we shall write a typical element of M^d as x^d (remember that it is an R-morphism). We shall also use the following notation : given $x \in M$ and $y^d \in M^d$ we denote by $\langle x, y^d\rangle$ the element $y^d(x)$ of R. Then, for all $x, y \in M$, all $x^d, y^d \in M^d$, and all $\lambda \in R$,

(α) $\langle x+y, x^d\rangle = \langle x, x^d\rangle + \langle y, x^d\rangle$;
(β) $\langle x, x^d + y^d\rangle = \langle x, x^d\rangle + \langle x, y^d\rangle$;

(γ) $\langle \lambda x, x^d \rangle = \lambda \langle x, x^d \rangle$;
(δ) $\langle x, x^d \lambda \rangle = \langle x, x^d \rangle \lambda$.

Theorem 9.1 *Let M be a free R-module with basis $\{e_i \; ; \; i \in I\}$. For every $i \in I$ let $e_i^d : M \to R$ be the (unique) R-morphism such that*

$$e_i^d(e_j) = \begin{cases} 0 & \text{if } j \neq i; \\ 1 & \text{if } j = i. \end{cases}$$

Then $\{e_i^d \; ; \; i \in I\}$ is a linearly independent subset of M^d.

Proof It is clear that $e_i^d \in M^d$ for every $i \in I$ and that $e_i^d = e_j^d$ if and only if $i = j$. That $\{e_i^d \; ; \; i \in I\}$ is linearly independent now follows from the fact that if $\sum_{i=1}^{t} e_i^d \lambda_i = 0$ in M^d then for $j = 1, \ldots, n$ we have

$$0_R = \left(\sum_{i=1}^{t} e_i^d \lambda_i\right)(e_j) = \sum_{i=1}^{t}(e_i^d \lambda_i)(e_j) = \sum_{i=1}^{t} e_i^d(e_j)\lambda_i = \lambda_j. \quad \diamond$$

Corollary 1 *If $x = \sum_{i=1}^{n} x_i e_i$ then $e_i^d(x) = x_i$ for each i.*

Proof $e_i^d(x) = e_i^d\left(\sum_{j=1}^{n} x_j e_j\right) = \sum_{j=1}^{n} x_j e_i^d(e_j) = x_i. \quad \diamond$

- Because of Corollary 1, the R-morphisms e_i^d are often called the *coordinate forms* associated with the elements e_i.

Corollary 2 *If I is finite, say $i = \{1, \ldots, n\}$ then $\{e_1^d, \ldots, e_n^d\}$ is a basis of M^d.*

Proof Given $f \in M^d$ and $x = \sum_{i=1}^{n} x_i e_i \in M$ we have, using Corollary 1,

$$\left(\sum_{i=1}^{n} e_i^d f(e_i)\right)(x) = \sum_{i=1}^{n} e_i^d(x) f(e_i) = \sum_{i=1}^{n} x_i f(e_i) = f(x)$$

and consequently $f = \sum_{i=1}^{n} e_i^d f(e_i)$, which shows that $\{e_1^d, \ldots, e_n^d\}$ generates M^d. Since this set is linearly independent, it therefore forms a basis of M^d. $\diamond$

Definition If M is a free R-module and if $B = \{e_1, \ldots, e_n\}$ is a finite basis of M then by the basis *dual* to B we shall mean the basis $\{e_1^d, \ldots, e_n^d\}$ of M^d.

Duality; transposition

Let us now return to the identities (α) to (δ) above. It is clear from (β) and (δ) that, for every $x \in M$, the mapping $x^{dd} : M^d \to R$ given by the prescription

$$x^{dd}(x^d) = \langle x, x^d \rangle$$

is a linear form on the right R-module M^d and so is an element of the bidual module M^{dd} (hence our choice of the notation x^{dd}). Consider now the mapping $dd_M : M \to M^{dd}$ given by $x \mapsto x^{dd}$. It is quickly verified using (α) and (γ) that this is an R-morphism. We call it the *canonical R-morphism from M to M^{dd}*.

- The various notational conventions described above may be summarised in the identities

$$x^{dd}(x^d) = \langle x, x^d \rangle = x^d(x)$$

where $x \in M$, $x^d \in M^d$ and $x^{dd} \in M^{dd}$.

In general, the canonical R-morphism dd_M need not be either injective or surjective. However, we do have the following important result:

Theorem 9.2 *If M is a free R-module then the canonical morphism $dd_M : M \to M^{dd}$ is a monomorphism. If, moreover, M has a finite basis then dd_M is an isomorphism.*

Proof Let $\{e_i \; ; \; i \in I\}$ be a basis of M and let $\{e_i^d \; ; \; i \in I\}$ be the set of corresponding coordinate forms. Suppose that $x \in \operatorname{Ker} dd_M$ and that $x = \sum_{i \in J} x_i e_i$ where J is some finite subset of I. Then since x^{dd} is the zero element of M^{dd} we have $x^{dd}(y^d) = 0$ for all $y^d \in M^d$. In particular, for every $i \in J$ we have, by Corollary 1 of Theorem 9.1,

$$0 = x^{dd}(e_i^d) = \langle x, e_i^d \rangle = e_i^d(x) = x_i.$$

It follows that $x = 0$ and hence that d_M is a monomorphism.

Suppose now that I is finite, say $I = \{1, \ldots, n\}$. Since

$$e_i^{dd}(e_j) = \langle e_i, e_j^d \rangle = e_j^d(e_i) = \begin{cases} 1 & \text{if } i = j; \\ 0 & \text{if } i \neq j, \end{cases}$$

we see that the e_i^{dd} are the coordinate forms associated with the e_i^d and so, by Corollary 2 of Theorem 9.1, $\{e_1^{dd}, \ldots, e_n^{dd}\}$ is the basis of M^{dd} that is dual to the basis $\{e_1^d, \ldots, e_n^d\}$ of M^d. It now follows by the Corollary to Theorem 7.6 that dd_M is an R-isomorphism. $\diamond$

- In the case where M is a free R-module having a finite basis we shall henceforth agree to *identify* M and M^{dd}. We can do so, of course, only because of the isomorphism dd_M which is *canonical* in the sense that it is independent of any choice of bases. However, we shall not identify M and M^d in this case, despite the fact that for any (finite) basis of M the corresponding dual basis has the same cardinality so that M and M^d are also isomorphic (by the Corollary to Theorem 7.6). In fact, M and M^d are not canonically isomorphic. What we mean here by a canonical isomorphism is an R-isomorphism $\zeta : M \to M^d$ which is such that, for all $x, y \in M$ and all R-isomorphisms $f : M \to M$,

$$\langle x, \zeta(y) \rangle = \langle f(x), \zeta[f(y)] \rangle.$$

We refer the reader to Exercise 9.8 for specific details.

Definition Let M and N be R-modules. By the *transpose* of an R-morphism $f : M \to N$ we shall mean the induced R-morphism

$$M^d = \mathrm{Mor}_R(M, R) \longleftarrow \mathrm{Mor}_R(N, R) = N^d$$

given by the assignment

$$y^d \circ f \longleftarrow y^d.$$

We shall denote the transpose of f by f^t.

- It is clear from the definition that f^t is a $\mathbb{Z}$-morphism. That it is indeed an R-morphism follows from the equalities $[f^t(y^d\lambda)](x) = (y^d\lambda)[f(x)] = y^d[f(x)]\lambda = [(y^d \circ f)(x)]\lambda = [(y^d \circ f)\lambda](x) = [f^t(y^d)\lambda](x)$.

- In terms of the notation introduced previously, we have, for all $x \in M$ and all $y^d \in N^d$,

$$\langle f(x), y^d \rangle = \langle x, f^t(y^d) \rangle.$$

In fact, the left-hand side is $y^d[f(x)] = (y^d \circ f)(x) = [f^t(y^d)](x)$ which is the right-hand side.

The principal properties of transposition are as follows.

Theorem 9.3 (1) *For every R-module M, $(\mathrm{id}_M)^t = \mathrm{id}_{M^d}$;*
(2) *If $f, g \in \mathrm{Mor}_R(M, N)$ then $(f + g)^t = f^t + g^t$;*
(3) *If $f \in \mathrm{Mor}_R(M, N)$ and $g \in \mathrm{Mor}_R(N, P)$ then $(g \circ f)^t = f^t \circ g^t$.*

Proof (1) is obvious from the definition of transpose; and (2), (3) are special cases of Theorem 8.2(4),(2). ◊

Duality; transposition

Corollary *If $f : M \to N$ is an R-somorphism then so is $f^T : N^d \to M^d$; moreover, in this case, $(f^t)^{-1} = (f^{-1})^t$.*

Proof This is immediate from (1) and (3). ◊

We can, of course, apply the previous definition to the R-morphism $f^t : N^d \to M^d$, thereby obtaining its transpose, namely $(f^t)^t : M^{dd} \to N^{dd}$ given by $\vartheta \mapsto \vartheta \circ f^t$. We call $(f^t)^t$ the *bitranspose* of f and shall denote it henceforth by f^{tt}.

The connection between bitransposes and biduals can be summarised as follows.

Theorem 9.4 *For every R-morphism $f : M \to N$ there is the commutative diagram*

$$\begin{array}{ccc} M & \xrightarrow{f} & N \\ {\scriptstyle dd_M}\downarrow & & \downarrow{\scriptstyle dd_N} \\ M^{dd} & \xrightarrow{f^{tt}} & N^{dd} \end{array}$$

Proof For every $x \in M$ we have $(dd_N \circ f)(x) = [f(x)]^{dd_N}$ and $(f^{tt} \circ dd_M)(x) = f^{tt}(x^{dd_M})$. The result follows from the fact that, for all $x \in M$ and all $y^d \in N^d$,

$$[f(x)]^{dd_N}(y^d) = \langle f(x), y^d \rangle = \langle x, f^t(y^d) \rangle;$$

$$[f^{tt}(x^{dd_M})](y^d) = (x^{dd_M} \circ f^t)(y^d) = \langle x, f^t(y^d) \rangle. \quad \diamond$$

- Note that if M and N are free R-modules each having a finite basis and if $f : M \to N$ is an R-morphism then, on identifying M^{dd} with M and N^d with N, we obtain from Theorem 9.4 the equality $f^{tt} = f$. These identifications also happily reduce notational complexities to a reasonable level!

By way of applying Theorem 9.4, we can obtain the following generalisation of Theorem 9.2.

Theorem 9.5 *If P is a projective module then the canonical morphism dd_P is a monomorphism; moreover, if P is finitely generated then dd_P is an isomorphism.*

Proof If P is projective then, by Theorem 8.7, P is a direct summand of a free module F. Let $\iota_P : P \to F$ be the natural inclusion. Then by

Theorem 9.4 we have the commutative diagram

$$\begin{array}{ccc} P & \xrightarrow{\iota_P} & F \\ {\scriptstyle dd_P}\downarrow & & \downarrow{\scriptstyle dd_F} \\ P^{dd} & \xrightarrow{\iota_P^{tt}} & F^{dd} \end{array}$$

Since dd_F is injective by Theorem 9.2, we deduce that $\iota_P^{dd} \circ dd_P$ is injective whence so is dd_P.

Suppose now that P is also finitely generated. Then by the Corollary to Theorem 8.5 there is a free R-module F with a finite basis and an R-epimorphism $\pi : F \to P$. Since P is projective, Theorem 8.8 yields the split short exact sequence

$$0 \longrightarrow \operatorname{Ker} \pi \xrightarrow{\iota} F \xrightarrow{\pi} P \longrightarrow 0$$

in which ι is the natural inclusion. Applying Theorem 8.4(2) twice (with $M = R$) and Theorem 9.4, we obtain the commutative diagram

$$\begin{array}{ccccccccc} 0 & \longrightarrow & K & \xrightarrow{\iota} & K & \xrightarrow{\pi} & P & \longrightarrow & 0 \\ & & {\scriptstyle dd_K}\downarrow & & \downarrow{\scriptstyle dd_F} & & \downarrow{\scriptstyle dd_P} & & \\ 0 & \longrightarrow & K^{dd} & \xrightarrow{\iota^{tt}} & F^{dd} & \xrightarrow{\pi^{tt}} & P^{dd} & \longrightarrow & 0 \end{array}$$

in which $K = \operatorname{Ker} \pi$ and each row is split exact. Since dd_F is an isomorphism by Theorem 9.2, we deduce that $dd_P \circ \pi$ is surjective, whence so is dd_P. It now follows from the first part of the theorem that dd_P is an isomorphism. ◊

We now consider further relations between a module and its dual.

Definition If M is an R-module then $x \in M$ is said to be *annihilated* by $x^d \in M^d$ if $x^d(x) = \langle x, x^d \rangle = 0$.

It is clear from the equalities (β) and (γ) immediately preceding Theorem 9.1 that for every non-empty subset E on M the set of elements of M^d that annihilate every element of E forms a submodule of M^d. We denote this submodule of M^d by

$$E^{\square} = \{ x^d \in M^d \ ; \ (\forall x \in E) \ \langle x, x^d \rangle = 0 \}$$

and call this the *submodule of M^d that annihilates E*. In particular, we obviously have $\{0_M\}^{\square} = M^d$ and $M^{\square} = \{0_{M^d}\}$.

Duality; transposition

The connection between duality, transposition, and annihilation can be summarised as follows, in which $L(M)$ denotes the lattice of submodules of M.

Theorem 9.6 *For every R-morphism $f : M \to N$ there is the commutative diagram*

$$\begin{array}{ccc} L(M) & \xrightarrow{\square} & L(M^d) \\ {\scriptstyle f^\rightarrow} \downarrow & & \downarrow {\scriptstyle (f^t)^\leftarrow} \\ L(N) & \xrightarrow{\square} & L(N^d) \end{array}$$

Proof What we have to show is that, for every submodule A of M,

$$[f^\rightarrow(A)]^\square = (f^t)^\leftarrow(A^\square).$$

This follows immediately from the observation that

$$\begin{aligned} y^d \in [f^\rightarrow(A)]^\square &\iff (\forall x \in f^\rightarrow(A))\ 0 = \langle x, y^d \rangle \\ &\iff (\forall a \in A)\ 0 = \langle f(a), y^d \rangle = \langle a, f^t(y^d) \rangle \\ &\iff f^t(y^d) \in A^\square \\ &\iff y^d \in (f^t)^\leftarrow(A^\square). \quad \Diamond \end{aligned}$$

Corollary 1 *If $f : M \to N$ is an R-morphism then*

$$(\operatorname{Im} f)^\square = \operatorname{Ker} f^T.$$

Proof It suffices to take $A = M$ in the above. $\Diamond$

Corollary 2 *If A is a submodule of M then $(M/A)^d \simeq A^\square$.*

Proof Consider the natural short exact squence

$$0 \longrightarrow A \xrightarrow{\iota} M \xrightarrow{\natural} M/A \longrightarrow 0.$$

By Theorem 8.3 we have the induced short exact sequence

$$A^d \xleftarrow{\iota^t} M^d \xleftarrow{\natural^t} (M/A)^d \xleftarrow{} 0.$$

Then, by Corollary 1 and the exactness,

$$(M/A)^d \simeq \operatorname{Im} \natural^t = \operatorname{Ker} \iota^t = (\operatorname{Im} \iota)^\square = A^\square. \quad \Diamond$$

We now turn our attention to finite-dimensional vector spaces and derive some further properties of the dimension function.

Theorem 9.7 *If V is a finite-dimensional vector space and if W is a subspace of V then*

$$\dim W^\square = \dim V - \dim W.$$

Moreover, if we identify V with its bidual then $(W^\square)^\square = W$.

Proof If $W = V$ then the result is clear. Suppose then that $W \subset V$. Let $\dim V = n$ and note that by Corollary 3 of Theorem 8.9 we have $\dim W = m$ where $m < n$. Let $\{a_1, \ldots, a_m\}$ be a basis of W and extend this to a basis $\{a_1, \ldots, a_m, a_{m+1}, \ldots, a_n\}$ of V. Let $\{a_1^d, \ldots, a_n^d\}$ be the basis of V^d that is dual to this basis of V. If now $x = \sum_{i=1}^{n} a_i^d \lambda_i \in W^\square$ then for $j = 1, \ldots, m$ we have

$$0 = \langle a_j, x^d \rangle = \sum_{i=1}^{n} \langle a_j, a_i^d \rangle \lambda_i = \lambda_j.$$

It follows immediately that $\{a_{m+1}^d, \ldots, a_n^d\}$ is a basis for $W^\square$ and hence that

$$\dim W^\square = n - m = \dim V - \dim W.$$

As for the second statement, consider the subspace $(W^\square)^\square$ of $V^{dd} = V$. By definition, every element of W is annihilated by every element of $W^\square$ and so we have $W \subseteq (W^\square)^\square$. On the other hand, by the first part of the theorem,

$$\dim(W^\square)^\square = \dim V^d - \dim W^\square = n - (n - m) = m = \dim W.$$

We conclude from Corollary 3 of Theorem 8.9 that $(W^\square)^\square = W$. ◊

Corollary *The assignment $W \mapsto W^\square$ yields a bijection from the set of m-dimensional subspaces of V to the set of $(m - n)$-dimensional subspaces of V^d.* ◊

Theorem 9.8 *Let V and W be finite-dimensional vector spaces over the same field F. If $f : V \to W$ is a linear transformation then*
 (1) $(\operatorname{Im} f)^\square = \operatorname{Ker} f^t$;
 (2) $(\operatorname{Ker} f)^\square = \operatorname{Im} f^t$;
 (3) $\dim \operatorname{Im} f = \dim \operatorname{Im} f^t$;
 (4) $\dim \operatorname{Ker} f = \dim \operatorname{Ker} f^t$.

Proof (1) follows from Corollary 1 of Theorem 9.6; and (2) follows from (1), Theorem 9.7 and the remark preceding Theorem 9.5. As for (3) and (4), we observe that, by (1) and (2),

$$\begin{aligned}\dim \operatorname{Im} f^t &= n - \dim \operatorname{Ker} f^t \\ &= n - \dim(\operatorname{Im} f)^\square \\ &= n - (n - \dim \operatorname{Im} f) \\ &= \dim \operatorname{Im} f \\ &= n - \dim \operatorname{Ker} f,\end{aligned}$$

from which both (3) and (4) follow. ◊

We shall see the importance of Theorem 9.8 in the next section.

EXERCISES

9.1 Let M be an R-module. For every R-morphism $f : R \to M$ and every $\lambda \in R$ let $\lambda f : R \to M$ be given by the prescription $(\lambda f)(r) = f(r\lambda)$. Show that $\lambda f \in \text{Mor}_R(R, M)$ and deduce that $\text{Mor}_R(R, M)$ is an R-module. Show also that the mapping $\vartheta : \text{Mor}_R(R, M) \to M$ given by $\vartheta(f) = f(1_R)$ is an R-isomorphism.

9.2 Let $\mathbb{R}_n[X]$ denote the $(n+1)$-dimensional vector space of polynomials over $\mathbb{R}$ of degree less than or equal to n. If $t_1, \ldots, t_{n+1}$ are $n+1$ distinct real numbers and if for $i = 1, \ldots, n+1$ the mappings $\zeta_{t_i} : \mathbb{R}_n[X] \to \mathbb{R}$ are the corresponding *substitution morphisms*, given by $\zeta_{t_i}(p) = p(t_i)$ for each i, prove that

$$B = \{\zeta_{t_i} \; ; \; i = 1, \ldots, n+1\}$$

is a basis for the dual space $(\mathbb{R}[x])^d$. Determine a basis of $\mathbb{R}_n[X]$ of which B is the dual.

[*Hint*. Consider the *Lagrange polynomials*

$$P_j = \prod_{i \neq j} \frac{X - t_i}{t_j - t_i}$$

where $i, j = 1, \ldots, n+1$.]

9.3 Let R be a commutative unitary ring regarded as an R-module. Let m and n be positive integers. Given $f_1, \ldots, f_m \in (R^n)^d$, define $f : R^n \to R^m$ by the prescription

$$f(x) = (f_1(x), \ldots, f_m(x)).$$

Show that $f \in \text{Mor}_R(R^n, R^m)$. Show also that every element of $\text{Mor}_R(R^n, R^m)$ is of this form for some $f_1, \ldots, f_m \in (R^n)^d$.

[*Hint*. Consider $f_i = \text{pr}_i \circ f$.]

9.4 If $(M_i)_{i \in I}$ is a family of R-modules prove that

$$\left(\bigoplus_{i \in I} M_i \right)^d \simeq \underset{i \in I}{\times} M_i^d.$$

[*Hint*. Use Theorem 8.1(a).]

9.5 Prove that an R-morphism is surjective if and only if its transpose is injective.

9.6 Let V be a finite-dimensional vector space and let $(V_i)_{i \in I}$ be a family of subspaces of V. Prove that

$$\left(\bigcup_{i \in I} V_i\right)^\square = \bigcap_{i \in I} V_i^\square, \qquad \left(\bigcap_{i \in I} V_i\right)^\square = \sum_{i \in I} V_i^\square.$$

[*Hint*. Observe first that $X \subseteq Y$ implies $Y^\square \subseteq X^\square$.]

9.7 If V is a finite-dimensional vector space and W is a subspace of V prove that $V^d/W^\square \simeq W^d$.

[*Hint*. Show that $f - g \in W^\square$ if and only if the restrictions of f, g to W coincide.]

9.8 In this exerise we indicate a proof of the fact that if V is a vector space of dimension $n > 1$ over a field F then there is no canonical isomorphism $\zeta : V \to V^d$ except when $n = 2$ and $F \simeq \mathbb{Z}/2\mathbb{Z}$. By such an isomorphism ζ we mean one such that, for all $x, y \in V$ and all isomorphisms $f : V \to V$,

$$(\star) \qquad \langle x, \zeta(y) \rangle = \langle f(x), \zeta[f(y)] \rangle.$$

If ζ is such an isomorphism show that for $y \neq 0$ the subspace $\operatorname{Ker}\zeta(y) = \{\zeta(y)\}^\square$ is of dimension $n - 1$.

Suppose first that $n > 3$. If there exists $t \neq 0$ such that $t \in \operatorname{Ker}\zeta(t)$ let $\{t, x_1, \ldots, x_{n-2}\}$ be a basis of $\operatorname{Ker}\zeta(t)$, and extend this to form a basis $\{t, x_1, \ldots, x_{n-2}, z\}$ of V. Let $f : V \to V$ be the (unique) linear transformation such that

$$f(t) = t, \quad f(x_1) = z, \quad f(z) = x_1, \quad f(x_i) = x_i (i \neq 1).$$

Show that f is an isomorphism that does not satisfy $(\star)$. [Take $x = x_1, y = t$.] If, on the other hand, for every $t \neq 0$ we have $t \notin \operatorname{Ker}\zeta(t)$, let $\{x_1, \ldots, x_{n-1}\}$ be a basis of $\operatorname{Ker}\zeta(t)$ with $\{x_1, \ldots, x_n, t\}$ a basis of V. Show that

$$\{x_1 + x_2, x_2 + t, x_3, \ldots, x_{n-1}, t + x_1\}$$

is also a basis of V. Show also that $x_2 \in \operatorname{Ker}\zeta(x_1)$. [Assume the contrary and use Theorem 7.10.] Now show that if $f : V \to V$ is the (unique) linear transformation such that

$$f(x_1) = x_1 + x_2, \quad f(x_2) = x_2 + t, \quad f(t) = t + x_1, \quad f(x_i) = x_i (i \neq 1, 2)$$

Duality; transposition

then f is an isomorphism that does not satisfy $(\star)$. [Take $x = x_1, y = t$.] Conclude that we must have $n = 2$.

Suppose now that $|F| > 3$ and let $\lambda \in F$ be such that $\lambda \neq 0, 1$. If there exists $t \neq 0$ such that $t \in \text{Ker}\,\zeta(t)$ observe that $\{t\}$ is a basis of $\text{Ker}\,\zeta(t)$ and extend this to a basis $\{t, z\}$ of V. If $f : V \to V$ is the (unique) linear transformation such that $f(t) = t, f(z) = \lambda z$, show that f is an isomorphism that does not satisfy $(\star)$. [Take $x = z, y = t$.] If, on the other hand, for all $t \neq 0$ we have $t \notin \text{Ker}\,\zeta(t)$ let $\{z\}$ be a basis of $\text{Ker}\,\zeta(t)$ so that $\{z, t\}$ is a basis of V. If $f : V \to V$ is the (unique) linear transformation such that $f(z) = \lambda z, f(t) = t$, show that f is an isomorphism that does not satisfy $(\star)$. [Take $x = y = z$.] Conclude that we must have $|F| = 2$.

Now examine the F-vector space F^2 where $F \simeq \mathbb{Z}/2\mathbb{Z}$.

[*Hint.* Observe that the dual of F^2 is the set of linear transformations $f : F \times F \to F$. Since $|F^2| = 4$ there are $2^4 = 16$ laws of composition on F. Only four of these can be linear transformations from $F \times F$ to F; and each is determined by its action on the natural basis of F^2. Compute $(F^2)^d$ and determine a canonical isomorphism from F^2 onto $(F^2)^d$.]

9.9 Let E and F be vector spaces over a field K and let $u : E \to F$ be a linear transformation. If V is a subspace of E prove that

$$[u^\to(V)]^d \simeq (u^t)^\to(F^d)/[V^\square \cap (u^t)^\to(F^d)].$$

[*Hint.* Consider the mapping $\vartheta : (u^t)^\to(F^d) \to [u^\to(V)]^d$ described by sending $f^d \circ u$ to the restriction of f^d to $u^\to(V)$.]

10 Matrices; linear equations

In this section we introduce the notion of a *matrix* and illustrate the importance of some of the previous results in the study of *linear equations*. The reader will undoubtedly be familiar with several aspects of this, in perhaps a less general setting, and for this reason we shall be as brief as possible.

Definition Let S be a non-empty set. By an $m \times n$ *matrix* over S we shall mean a mapping $f : [1,m] \times [1,n] \to S$. We shall write the element $f(i,j)$ of S as x_{ij} and denote such a mapping by the array

$$\begin{bmatrix} x_{11} & x_{12} & \cdots & x_{1n} \\ x_{21} & x_{22} & \cdots & x_{2n} \\ \vdots & \vdots & \ddots & \vdots \\ x_{m1} & x_{m2} & \cdots & x_{mn} \end{bmatrix}$$

which consists of m rows and n columns, the entry x_{ij} appearing at the intersection of the i-th row and the j-th column. We shall often abbreviate this to $[x_{ij}]_{m \times n}$.

It is clear from the definition of equality for mappings that $[x_{ij}]_{m \times n} = [y_{ij}]_{p \times q}$ if and only if $m = p, n = q$ and $x_{ij} = y_{ij}$ for all i, j.

We shall denote the set of $m \times n$ matrices over a commutative unitary ring R by $\text{Mat}_{m \times n}(R)$. Clearly, this forms an R-module under the component-wise definitions

$$[x_{ij}]_{m \times n} + [y_{ij}]_{m \times n} = [x_{ij} + y_{ij}]_{m \times n}, \quad \lambda[x_{ij}]_{m \times n} = [\lambda x_{ij}]_{m \times n}.$$

Definition Let R be a commutative unitary ring and let M be a free R-module of dimension n. By an *ordered basis* of M we shall mean a sequence $(a_i)_{1 \le i \le n}$ of elements of M such that the set $\{a_1, \ldots, a_n\}$ is a basis of M. We shall often write an ordered basis as simply $(a_i)_n$.

- Note that every basis of n elements gives rise to $n!$ distinct ordered bases since there are $n!$ bijections on a set of n elements.

Suppose now that R is a commutative unitary ring and that M, N are free R-modules of dimensions m, n respectively. Suppose further that $(a_i)_m, (b_i)_n$ are given ordered bases of M, N and that $f : M \to N$ is an R-morphism. Then we know by Theorem 7.7 and its Corollary 3 that f is entirely determined by the mn scalars x_{ij} such that

$$(j = 1, \ldots, m) \quad f(a_j) = \sum_{i=1}^{n} x_{ji} b_i = x_{j1} b_1 + \ldots + x_{jn} b_n.$$

Matrices; linear equations

The $n \times m$ matrix $X = [x_{ji}]_{n \times m}$ is called the *matrix of f relative to the ordered bases* $(a_i)_m$ *and* $(b_i)_n$. We shall denote it by $\text{Mat}[f,(b_i)_n,(a_i)_m]$ or simply $\text{Mat}\, f$ if there is no confusion over the ordered bases.

- Note that it is an **n** × **m** matrix that represents an R-morphism from an **m**-dimensional module to an **n**-dimensional module. This conventional twist is deliberate and the reason for it will soon be clear.

Theorem 10.1 *Let R be a commutative unitary ring and let M, N be free R-modules of dimensions m, n referred respectively to ordered bases $(a_i)_m, (b_i)_n$. Then the mapping*

$$\vartheta : \text{Mor}_R(M, N) \to \text{Mat}_{n \times m}(R)$$

given by $\vartheta(f) = \text{Mat}[f,(b_i)_n,(a_i)_m]$ *is an R-isomorphism.*

Proof It suffices to note that if we have $A = \text{Mat}\, f$ and $B = \text{Mat}\, g$ then $A + B = \text{Mat}(f + g)$, and $\lambda A = \text{Mat}(\lambda f)$. ◊

Definition If $X = [x_{ij}]_{m \times n}$ and $Y = [y_{ij}]_{n \times p}$ are matrices over a commutative unitary ring R then we define the *product XY* to be the $m \times p$ matrix $[z_{ij}]_{m \times p}$ given by

$$z_{ij} = \sum_{k=1}^{n} x_{ik} y_{kj}.$$

The reason for this (rather strange) definition is made clear by the following result.

Theorem 10.2 *Let R be a commutative unitary ring and let M, N, P be free R-modules of dimensions m, n, p respectively. If $(a_i)_m, (b_i)_n, (c_i)_p$ are fixed ordered bases of M, N, P respectively and if $f \in \text{Mor}_R(M, N), g \in \text{Mor}_R(N, P)$ then*

$$\text{Mat}[g \circ f,(c_i)_p,(a_i)_m] = \text{Mat}[g,(c_i)_p,(b_i)_n]\, \text{Mat}[f,(b_i)_n,(a_i)_m].$$

Proof Let $\text{Mat}\, f = [x_{ij}]_{n \times m}$ and $\text{Mat}\, g = [y_{ij}]_{p \times n}$. Then for $j = 1, \ldots, m$ we have

$$(g \circ f)(a_j) = g\left(\sum_{i=1}^{n} x_{ij} b_i\right) = \sum_{i=1}^{n} x_{ij} g(b_i) = \sum_{k=1}^{p}\left(\sum_{i=1}^{n} x_{ij} y_{ki}\right) c_k$$

from which the result follows. ◊

- In a more succinct way, the equality of Theorem 10.1 can be written in the form $\text{Mat}(g \circ f) = \text{Mat}\, g\, \text{Mat}\, f$. It is precisely to have the product in this order that the above twist is adopted.

Corollary 1 *Matrix multiplication as defined above is associative.*

Proof Composition of morphisms is associative. ◊

Corollary 2 *If $(a_i)_n$ is an ordered basis of an n-dimensional module over a commutative unitary ring R then the mapping*

$$\vartheta : \text{Mor}_R(M, M) \to \text{Mat}_{n \times n}(R)$$

given by $\vartheta(f) = \text{Mat}[f, (a_i)_n, (a_i)_n]$ is an R-algebra isomorphism.

Proof It suffices to note that $\text{Mat}_{n \times n}(R)$ is an R-algebra, the identity element of which is the diagonal matrix

$$I_n = \begin{bmatrix} 1 & 0 & 0 & \cdots & 0 \\ 0 & 1 & 0 & \cdots & 0 \\ 0 & 0 & 1 & \cdots & 0 \\ \vdots & \vdots & \vdots & \ddots & \vdots \\ 0 & 0 & 0 & \cdots & 1 \end{bmatrix}$$

i.e. that given by $I_n = [\delta_{ij}]_{n \times n}$ where

$$\delta_{ij} = \begin{cases} 1 & \text{if } i = j; \\ 0 & \text{if } i \neq j, \end{cases}$$

and that $\text{Mat}[\text{id}_M, (a_i)_n, (a_i)_n] = I$. ◊

Corollary 3 $\text{Mat}[f, (a_i)_n, (a_i)_n]$ *is invertible in the ring $\text{Mat}_{n \times n}(R)$ if and only if f is an R-isomorphism.* ◊

The importance of invertible matrices is illustrated in the following result. This tells us how the matrix representing an R-morphism changes when we switch reference to a new ordered basis.

Theorem 10.3 *Let R be a commutative unitary ring and let M, N be free R-modules of dimensions m, n respectively.*

(1) Let $(a_i)_m, (a_i')_m$ be ordered bases of M and let $(b_i)_n, (b_i')_n$ be ordered bases of N. Given $f \in \text{Mor}_R(M, N)$, suppose that $\text{Mat}[f, (b_i)_n, (a_i)_m] = A$. Then

$$\text{Mat}[f, (b_i')_n, (a_i')_m] = Q^{-1}AP$$

where $P = \text{Mat}[\text{id}_M, (a_i)_m, (a_i')_m]$ and $Q = \text{Mat}[\text{id}_N, (b_i)_n, (b_i')_n]$.

(2) Conversely, if $(a_i)_m$ and $(b_i)_n$ are ordered bases of M and N respectively and if A, B are $n \times m$ matrices over R such that, for some invertible $m \times m$ matrix P and some invertible $n \times n$ matrix Q, $B = Q^{-1}AP$ then there

Matrices; linear equations

is an R-morphism $f : M \to N$ and ordered bases $(a'_i)_m, (b'_i)_n$ of M, N respectively such that $A = \text{Mat}[f, (b_i)_n, (a_i)_m]$ and $B = \text{Mat}[f, (b'_i)_n, (a'_i)_m]$.

Proof (1) Let $M; (a_i)_m$ denote that M is referred to the ordered basis $(a_i)_m$ and consider the commutative diagram

$$\begin{array}{ccc} M;(a_i)_m & \xrightarrow{f;A} & N;(b_i)_n \\ {\scriptstyle \text{id}_M;P} \uparrow & {\scriptstyle \text{id}_N;Q} \uparrow \downarrow {\scriptstyle \text{id}_N;Q^{-1}} & \\ M;(a'_i)_m & \xrightarrow[f;?]{} & N;(b'_i)_n \end{array}$$

in which, for example, $f; A$ indicates that A is the matrix of f relative to the ordered bases $(a_i)_m, (b_i)_n$. Note by Theorem 10.2 that $Q = \text{Mat}[\text{id}_N, (b_i)_n, (b'_i)_n]$ is invertible with $Q^{-1} = \text{Mat}[\text{id}_n, (b'_i)_n, (b_i)_n]$. Now by Theorem 10.2 again, we have

$Q^{-1}AP = \text{Mat}[\text{id}_N, (b'_i)_n, (b_i)_n] \text{Mat}[f, (b_i)_m, (a_i)_m] \text{Mat}[\text{id}_M, (a_i)_m, (a'_i)_m]$
$= \text{Mat}[\text{id}_N, (b'_i)_n, (b_i)_n] \text{Mat}[f, (b_i)_n, (a'_i)_m]$
$= \text{Mat}[f, (b'_i)_n, (a'_i)_m],$

so that the unknown ? in the diagram is indeed $Q^{-1}AP$.

(2) Let $P = [p_{ij}]_{m \times m}$ and $Q = [q_{ij}]_{n \times n}$ and define

$$a'_j = \sum_{i=1}^m p_{ij} a_i, \quad b'_j = \sum_{i=1}^n q_{ij} b_i.$$

Since P is invertible there is an R-isomorphism g with

$$P = \text{Mat}[g, (a_i)_m, (a_i)_m]$$

and since $a'_i = g(a_i)$ for each i we see that $(a'_i)_m$ is an ordered basis of M (for isomorphisms take bases to bases). It is now clear that

$$P = \text{Mat}[\text{id}_M, (a_i)_m, (a'_i)_m].$$

Similarly we see that $Q = \text{Mat}[\text{id}_N, (b_i)_n, (b'_i)_n]$. Now by Theorem 10.1 there exists $f \in \text{Mor}_R(M, N)$ with $A = \text{Mat}[f, (b_i)_n, (a_i)_m]$. Let $C = \text{Mat}[f, (b'_i)_n, (a'_i)_m]$; then by part (1) we have $C = Q^{-1}AP = B$. ◊

Definition If $(a_i)_m$ and $(a'_i)_m$ are ordered bases of M then the matrix $\text{Mat}[\text{id}_M, (a'_i)_m, (a_i)_m]$ is called the *transition matrix* (or *matrix representing the change of basis*) from $(a_i)_m$ to $(a'_i)_m$.

It is immediate from the previous result that two $m \times n$ matrices A, B represent the same R-morphism with respect to possibly different ordered bases if and only if there are invertible matrices P, Q (namely, transition matrices) with P of size $m \times m$ and Q of size $n \times n$ such that $B = Q^{-1}AP$. We describe this situation by saying that A and B are *equivalent*. It is clear that the relation of being equivalent is an equivalence relation on the set $\text{Mat}_{n \times m}(R)$. An important problem from both the theoretical and practical points of view is that of locating a particularly simple representative, or *canonical form*, in each equivalence class. In order to tackle this problem, we require the following notions.

Definition By the *transpose* of an $n \times m$ matrix $A = [a_{ij}]_{n \times m}$ we mean the $m \times n$ matrix $A^t = [a_{ji}]_{m \times n}$.

The reason for this choice of terminology is clear from the following result.

Theorem 10.4 *Let R be a commutative unitary ring and let M, N be free R-modules of dimensions m, n respectively. If $f : M \to N$ is an R-morphism and $(a_i)_m, (b_i)_n$ are ordered bases of M, N respectively with $\text{Mat}[f, (b_i)_m, (a_i)_m] = A$ then $\text{Mat}[f^t, (a_i^d)_m, (b_i^d)_n] = A^t$.*

Proof Let $\text{Mat}[f^t, (a_i^d)_m, (b_i^d)_n] = [b_{ij}]_{m \times n}$. Then we have

$$\begin{cases} \langle f(a_i), b_j^d \rangle = \Big\langle \sum_{t=1}^{n} a_{ti} b_t, b_j^d \Big\rangle = \sum_{t=1}^{n} a_{ti} \langle b_t, b_j^d \rangle = a_{ji}; \\ \qquad \| \\ \langle a_i, f^t(b_j^d) \rangle = \Big\langle a_i, \sum_{t=1}^{m} b_{tj} a_t^d \Big\rangle = \sum_{t=1}^{m} b_{tj} \langle a_i, a_t^d \rangle = b_{ij}, \end{cases}$$

from which the result follows. $\diamond$

Suppose now that $f : [1, m] \times [1, n] \to S$ is an $m \times n$ matrix. Then for every $p \in [1, n]$ the restriction $f_p : [1, m] \times \{p\} \to S$ is an $m \times 1$ matrix which we shall call the *p-th column matrix* of f. The *p-th row matrix* of f is defined similarly. The *p*-th column and row matrices of a matrix $A = [a_{ij}]_{m \times n}$ that represents f are then

$$\begin{bmatrix} a_{p1} & a_{p2} & \cdots & a_{pn} \end{bmatrix}, \quad \begin{bmatrix} a_{1p} \\ a_{2p} \\ \vdots \\ a_{mp} \end{bmatrix}.$$

Definition By the *column rank* of an $m \times n$ matrix A over a commutative unitary ring R we shall mean the dimension of the subspace of

Matrices; linear equations

$\text{Mat}_{m\times 1}(R)$ generated by the columns of A. Similarly, the *row rank* of A is the dimension of the subspace generated by the rows of A.

Theorem 10.5 *Let V and W be vector spaces of dimensions m, n respectively over a field F. Let $f : V \to W$ be a linear transformation amd let A be the matrix of f relative to fixed ordered bases $(a_i)_m, (b_i)_n$ respectively. Then the following coincide :*
  (1) *the column rank of A;*
  (2) $\dim \operatorname{Im} f$;
  (3) *the row rank of A;*
  (4) $\dim \operatorname{Im} f^t$.

Proof Let $A = [a_{ij}]_{n \times m}$ and recall that

$$(\star) \qquad (j = 1, \ldots, m) \quad f(a_j) = \sum_{i=1}^{n} a_{ij} b_i.$$

Now the mapping $\vartheta : \text{Mat}_{n \times 1}(F) \to W$ given by

$$\vartheta \begin{bmatrix} x_1 \\ x_2 \\ \vdots \\ x_n \end{bmatrix} = \sum_{i=1}^{n} x_i b_i$$

is clearly an isomorphism which, by virtue of $(\star)$, takes the j-th column matrix of A onto $f(a_j)$ and hence maps the subspace of $\text{Mat}_{n \times 1}(F)$ generated by the columns matrices of A onto $\operatorname{Im} f$. Thus we see that the column rank of A coincides with $\dim \operatorname{Im} f$. Likewise we can show that the column rank of A^t is $\dim \operatorname{Im} f^t$. Now since the column rank of A^t is clearly the same as the row rank of A, the result follows by Theorem 9.8(3). ◊

Corollary *If $g : V \to W$ is also a linear transformation and if the matrix of g relative to $(a_i)_m, (b_i)_n$ is B then the following statements are equivalent :*
  (1) *the subspace of $\text{Mat}_{n \times 1}(F)$ generated by the column matrices of A coincides with the subspace generated by the column matrices of B;*
  (2) $\operatorname{Im} f = \operatorname{Im} g$.

Proof It suffices to recall that ϑ in the above proof is an isomorphism that maps the column space of A onto $\operatorname{Im} f$. ◊

- Because of the equivalence of (a) and (c) in Theorem 10.5, we shall talk simply of the *rank* of a matrix over a field when referring to either the row rank or the column rank.

• In view of the definition of row and column rank, the result of Theorem 10.5 is really quite remarkable; for there is no obvious reason why we should expect the row rank of a matrix to coincide with the column rank. Indeed, the result holds only because the corresponding result for linear transformations [Theorem 9.8(3)] is a very natural one. In contrast, we note that Theorem 10.5, and hence Theorem 9.8, does not hold in general for R-modules; for an illustration of this we refer the reader to Exercise 10.7.

It is clear that, relative to the equivalence relation described above, the only matrix that is equivalent to the $m \times n$ zero matrix is the $m \times n$ zero matrix. As for the other equivalence classes, we shall now locate a particularly simple representative of each.

Theorem 10.6 *Let A be a non-zero $m \times n$ matrix over a field F. If A is of rank r then A is equivalent to a matrix of the form*

$$\begin{bmatrix} I_r & O_{r \times n-r} \\ O_{n-r \times r} & O_{n-r \times n-r} \end{bmatrix}.$$

Proof Let V and W be vector spaces over F of dimensions n and m respectively. Let $(A_i)_n, (b_i)_m$ be ordered bases of V, W and let $f : V \to W$ be a linear transformation such that $A = \text{Mat}[f, (b_i)_m, (a_i)_n]$. By Corollary 2 of Theorem 8.9 we have $\dim \text{Ker } f = n - r$, so there is an ordered basis $\alpha = (a'_i)_n$ of V such that $\{a'_{r+1}, \ldots, a'_n\}$ is a basis of $\text{Ker } f$. Since $\{f(a'_1), \ldots, f(a'_r)\}$ is then a basis of $\text{Im } f$, we can extend this to a basis

$$\{f(a'_1), \ldots, f(a'_r), b'_{r+1}, \ldots, b'_m\}$$

of W. Let $\beta = (\beta_i)_m$ be the ordered basis given by $\beta_i = f(a'_i)$ for $i = 1, \ldots, r$ and $\beta_{r+i} = b'_{r+i}$ for $i = 1, \ldots, m-r$. Then it is readily seen that the matrix of f relative to the ordered bases α, β is of the stated form, and this is equivalent to A by Theorem 10.3. ◊

Corollary 1 *Two $m \times n$ matrices over a field are equivalent if and only if they have the same rank.* ◊

Corollary 2 *If A is an $n \times n$ matrix over a field then the following statements are equivalent:*
 (1) *A is invertible;*
 (2) *A is of rank n;*
 (3) *A is equivalent to the identity matrix I_n.*

Proof This is immediate from Corollary 4 of Theorem 8.9. ◊

- The matrix exhibited in Theorem 10.6 is called the *canonical matrix of rank r*, or the *canonical form of A under the relation of equivalence*.

We now consider a particularly important application of matrices, namely to the solution of systems of linear equations over a field. Consider the following system of m equations in n unknowns:

$$\begin{cases} a_{11}x_1 + a_{12}x_2 + \ldots + a_{1n}x_n = b_1 \\ a_{21}x_1 + a_{22}x_2 + \ldots + a_{2n}x_n = b_2 \\ \vdots \quad\quad \vdots \quad\quad\quad\quad \vdots \quad\quad \vdots \\ a_{m1}x_1 + a_{m2}x_2 + \ldots + a_{mn}x_n = b_m \end{cases}$$

The $m \times n$ matrix $A = [a_{ij}]_{m \times n}$ is called the *coefficient matrix* of the system. By abuse of language we shall henceforth refer to such a system by means of its matrix representation

$$A \begin{bmatrix} x_1 \\ x_2 \\ \vdots \\ x_n \end{bmatrix} = \begin{bmatrix} b_1 \\ b_2 \\ \vdots \\ b_m \end{bmatrix}$$

which we shall abbreviate for convenience to $A[x_i]_n = [b_i]_m$. In this way we can represent the given system by a single matrix equation. The $m \times (n+1)$ matrix whose first n column matrices are those of A and whose $(n+1)$-th column matrix is $[b_i]_m$ is called the *augmented matrix* of the system.

Theorem 10.7 *The system of equations $A[x_i]_n = [b_i]_m$ over a field F has a solution if and only if the coefficient matrix and the augmented matrix have the same rank.*

Proof For $i = 1, \ldots, n$ let the i-th column matrix of A be A_i. Then the system can be written in the form $\sum_{i=1}^{n} A_i x_i = [b_i]_m$. It follows that a solution exists if and only if $[b_i]_m$ belongs to the subspace of $\mathrm{Mat}_{m \times 1}(F)$ generated by the column matrices of A, which is the case if and only if the (column) rank of A is the (column) rank of the augmented matrix. ◊

If f is a linear transformation then related to the equation $f(x) = y$ there is the equation $f(x) = 0$. The latter is called the *associated homogeneous equation*. The importance of this stems from the following result.

Theorem 10.8 *If x_0 is a solution of the linear equation $f(x) = y$ then the solution set of this equation is the set of elements of the form $x_0 + z$ where z is a solution of the associated homogeneous equation.*

Proof This is immediate from the observation that
$$f(x) = y = f(x_0) \iff f(x - x_0) = 0. \quad \Diamond$$

- If V and W are vector spaces and if $f : V \to W$ is linear then clearly the solutions of the equation $f(x) = 0$ constitute the subspace Ker f. We can therefore rephrase Theorem 10.8 in the form : *the solution set of $f(x) = y$ is either empty or is a coset of the solution space of $f(x) = 0$.*

Theorem 10.9 *Let A be an $m \times n$ matrix over a field F. Then the solution space of $A[x_i]_n = [0]_m$ is of dimension $n -$ rank A.*

Proof Let $f_A : \text{Mat}_{n \times 1}(F) \to \text{Mat}_{m \times 1}(F)$ be the linear transformation given by $f_A([x_i]_n) = A[x_i]_n$. Then it is readily verified that, relative to the natural ordered bases of Mat)$n \times 1(F)$ and $\text{Mat}_{m \times 1}(F)$, the matrix of f_A is precisely A. The solution space of $A[x_i]_n = [0]_m$ is now Ker f_A and its dimension is $n - \dim\text{Im } f_A$ which, by Theorem 10.5, is $n -$ rank A. $\Diamond$

In the elementary solution of linear equations, as is often illustrated by means of worked examples in introductory courses on linear algebra, there are three basic operations involved, namely

1. interchanging two equations;
2. multiplying an equation by a non-zero scalar;
3. forming a new equation by adding to one a multiple of another.

The same operations can be performed on the rows of the coefficient matrix of the system and are called *elementary row operations*. When the coefficients are elements of a given field, these operations are 'reversible' in the sense that we can always perform the 'inverse' operation and obtain the system we started off with. Consequently, these operations do not alter the solution set of the system.

Definition By an *elementary matrix* of size $m \times m$ over a field we shall mean a matrix that has been obtained from the identity matrix I_m by applying to it a single elementary row operation.

The importance of elementary matrices stems from the following result.

Theorem 10.10 *Let A and B be $m \times n$ matrices over a field F. If B is obtained from A by means of a single elementary row operation then*

Matrices; linear equations 135

$B = PA$ where P is the elementary matrix obtained by applying the same elementary row operation to I_m.

Proof We shall make use of the *Kronecker symbol* δ_{ij} given by

$$\delta_{ij} = \begin{cases} 1 & \text{if } i = j; \\ 0 & \text{if } i \ne j. \end{cases}$$

Recall that $I_m = [\delta_{ij}]_{m \times m}$.

Suppose first that $P = [p_{ij}]$ is obtained from I_m by interchanging the i-th and j-th rows of I_m (with, of course, $i \ne j$). Then the r-th row of P is given by

$$(t = 1, \ldots, m) \qquad p_{rt} = \begin{cases} \delta_{it} & \text{if } r = j; \\ \delta_{jt} & \text{if } r = i; \\ \delta_{rt} & \text{otherwise.} \end{cases}$$

Consequently,

$$[PA]_{rs} = \sum_{t=1}^{m} p_{rt} a_{ts} = \begin{cases} a_{is} & \text{if } r = j; \\ a_{js} & \text{if } r = i; \\ a_{rs} & \text{otherwise.} \end{cases}$$

Thus we see that PA is the matrix obtained from A by interchanging the i-th and j-th rows of A.

Suppose now that P is obtained from I_m by multiplying the i-th row of I_m by λ. Then the r-th row of P is given by

$$(t = 1, \ldots, m) \qquad p_{rt} = \begin{cases} \lambda \delta_{it} & \text{if } r = i; \\ \delta_{rt} & \text{otherwise.} \end{cases}$$

Consequently,

$$[PA]_{rs} = \sum_{t=1}^{m} p_{rt} a_{ts} = \begin{cases} \lambda a_{is} & \text{if } r = i; \\ a_{rs} & \text{otherwise,} \end{cases}$$

whence we see that PA is obtained from A by multiplying the i-th row of A by λ.

Finally, suppose that P is obtained from I_m by adding λ times the i-th row to the j-th row (with $i \ne j$). Then, if E_{ij}^{λ} denotes the $m \times m$ matrix that has λ in the (i, j)-th position and 0 elsewhere, we have

$$[PA]_{rs} = [A + E_{ij}^{\lambda} A]_{rs} = \begin{cases} a_{rs} & \text{if } r \ne i; \\ a_{is} + \lambda a_{js} & \text{if } r = i. \end{cases}$$

Thus PA is obtained from A by adding λ times the i-th row of A to the j-th row of A. ◊

Since to every elementary row operation there corresponds a unique 'inverse' operation that restores the *status quo*, it is clear the every elementary matrix is invertible.

We shall now introduce a second equivalence relation on the set of $m \times n$ matrices over a field.

Definition We say that two $m \times n$ matrices over a field are *row equivalent* if one can be obtained from the other by means of a finite sequence of elementary row operations.

Clearly, this concept of row equivalence defines an equivalence relation on the set $\text{Mat}_{m \times n}(F)$. It can be variously characterised as follows.

Theorem 10.11 *Let A and B be $m \times n$ matrices over a field F. Then the following statements are equivalent :*

(1) *A and B are row equivalent*;
(2) *$A[x_i]_n = [0]_m$ if and only if $B[x_i]_n = [0]_m$*;
(3) *there is an invertible matrix P such that $B = PA$*;
(4) *the subspace of $\text{Mat}_{1 \times n}(F)$ generated by the row matrices of A coincides with that generated by the row matrices of B.*

Proof (1) $\Leftrightarrow$ (2) : Row operations do not alter solution sets.

(1) $\Rightarrow$ (3) : This is clear from Theorem 10.10 and the fact that every product of invertible matrices is invertible.

(3) $\Rightarrow$ (2) : Suppose that $B = PA$ where P is invertible. Then $AX = 0$ implies $BX = PAX = P0 = 0$, and the converse implication follows similarly using the fact that $A = P^{-1}B$.

(4) $\Leftrightarrow$ (2) : Clearly, (4) holds if and only if the subspace of $\text{Mat}_{n \times 1}(F)$ generated by the column matrices od A^t coincides with the subspace generated by the column matrices of B^t. Now if $f_A, f_B : \text{Mat}_{n \times 1}(F) \to \text{Mat}_{m \times 1}(F)$ are given by $f_A([x_i]_n) = A[x_i]_n$ and $f_B([x_i]_n) = B[x_i]_n$ then the matrices of f_A, f_B relative to the natural bases are A, B respectively. Taking dual spaces and transposes in the Corollary of Theorem 10.5, we thus see that (4) holds if and only if $\text{Im } f_A^t = \text{Im } f_B^t$ which, by Theorem 9.8(2), is equivalent to $(\text{Ker } f_A)^\square = (\text{Ker } f_B)^\square$ which, by Theorem 9.7, is equivalent to $\text{Ker } f_A = \text{Ker } f_B$ which is equivalent to (2). $\Diamond$

Corollary 1 *Row-equivalent matrices are equivalent.* $\Diamond$

Corollary 2 *If A is an $n \times n$ matrix over a field F then the following statements are equivalent :*

(1) *A is row-equivalent to I_n;*
(2) *$A[x_i]_n = [0]_n$ has a unique solution, namely $[0]_n$;*
(3) *A is invertible;*

Matrices; linear equations

(4) *A* is a product of elementary matrices. ◊

It is clear from the above that the only $m \times n$ matrix that is row-equivalent to the zero $m \times n$ matrix is the zero $m \times n$ matrix. As for the other equivalence classes, we shall now locate a particularly simple representative, or *canonical form*, in each.

Definition By a *row-echelon* (or *stairstep*) matrix we shall mean a matrix of the form

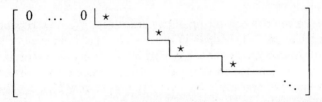

in which all the entries 'under the stairs' are zero, all the 'corner entries' (those marked ⋆) are non-zero, and all other entries are arbitrary.

- Note that the 'stairstep' descends one row at a time and that a 'step' may traverse several columns.

By a *Hermite matrix* we shall mean a row-echelon matrix in which every corner entry is 1 and every entry lying above a corner entry is 0.

Thus a Hermite matrix is of the typical form

$$\begin{bmatrix} 0 & \cdots & 0 & 1 & ? & ? & ? & 0 & 0 & ? & ? & ? & 0 & ? & ? & 0 & \cdots \\ & & & & & & & 1 & 0 & ? & ? & ? & 0 & ? & ? & 0 & \cdots \\ & & & & & & & & 1 & ? & ? & ? & 0 & ? & ? & 0 & \cdots \\ & & & & & & & & & & & & 1 & ? & ? & 0 & \cdots \\ & & & & & & & & & & & & & & & 1 & \cdots \\ & & & & & & & & & & & & & & & & \ddots \end{bmatrix}$$

Theorem 10.12 *Every non-zero $m \times n$ matrix A over a field F is row-equivalent to a unique $m \times n$ Hermite matrix.*

Proof Reading from the left, the first non-zero column of A contains at least one non-zero entry. A suitable permutation of the rows yields a row-equivalent matrix of the form

$$B = \begin{bmatrix} 0 & \cdots & 0 & b_{11} & b_{12} & \cdots & b_{1p} \\ 0 & \cdots & 0 & b_{21} & b_{22} & \cdots & b_{2p} \\ \vdots & & \vdots & \vdots & \vdots & & \vdots \\ 0 & \cdots & 0 & b_{m1} & b_{m2} & \cdots & b_{mp} \end{bmatrix}$$

in which $b_{11} \neq 0$. Now for $i = 2, \ldots, m$ subtract from the i-th row $b_{i1} b_{11}^{-1}$ times the first row. This yields a row-equivalent matrix of the form

$$C = \begin{bmatrix} 0 & \cdots & 0 & b_{11} & b_{12} & \cdots & b_{1p} \\ 0 & \cdots & 0 & 0 & c_{22} & \cdots & c_{2p} \\ \vdots & & \vdots & \vdots & & & \vdots \\ 0 & \cdots & 0 & 0 & c_{m2} & \cdots & c_{mp} \end{bmatrix}$$

in which we see the beginning of the stairstep. We now repeat the process using the $(m-1) \times (p-1)$ matrix $C' = [c_{ij}]$. It is clear that, continuing in this way, we eventually obtain a row-echelon matrix Z which, by its construction, is row-equivalent to A. Since every corner entry of Z is non-zero we can multiply every non-zero row of Z by the inverse of the corner entry in that row. This produces a row-echelon matrix Z' every corner entry of which is 1. We now subtract suitable multiples of every non-zero row from those rows lying above it to reduce to zero the entries lying above the corner entries. This then produces a Hermite matrix that is row-equivalent to A.

To establish the uniqueness of this Hermite matrix, it is clearly sufficient to show that if A, B are $m \times n$ Hermite matrices that are row-equivalent then $A = B$. This we proceed to do by induction on the number of columns.

It is clear that the only $m \times 1$ Hermite matrix is $[1\ 0\ \ldots\ 0]^t$, so the result is trivial in the case where $n = 1$. Suppose, by way of induction, that all row-equivalent Hermite matrices of size $m \times (n-1)$ are identical and let A, B be row-equivalent Hermite matrices of size $m \times n$. By Theorem 10.11 there is an invertible matrix P such that $B = PA$. Let A^+, B^+ be the $m \times (n-1)$ matrices consisting of the first $n-1$ columns of A, B. Then clearly $B^+ = PA^+$ and so A^+, B^+ are row-equivalent, by Theorem 10.11 again. By the induction hypothesis, therefore, we have $A^+ = B^+$. The result will now follow if we can show that the n-th columns of A and B coincide.

For this purpose, we note that since A, B are row-equivalent they have the same rank, namely the number of corner entries. If rank A = rank $B = r$ then we have either rank $A^+ = r$ or rank $A^+ = r - 1$. In the latter case the n-th columns of A and B consist of a corner entry 1 in the r-th row and zero entries elsewhere, whence these columns are equal. In the former case, let $i \in [1, r]$. Then by Theorem 10.11 we have

$$(\star) \qquad [b_{i1}\ \ldots\ b_{in}] = \sum_{k=1}^{r} \lambda_k [a_{k1}\ \ldots\ a_{kn}].$$

Matrices; linear equations

In particular, for the matrix $A^+ (= B^+)$ we have

$$[a_{i1} \ \ldots \ a_{i,n-1}] = \sum_{k=1}^{r} \lambda_k [a_{k1} \ \ldots \ a_{k,n-1}]$$

whence, since the first r rows matrices of A^+ form a linearly independent subset of $\text{Mat}_{1 \times (n-1)}(F)$, we obtain $\lambda_i = 1$ and $\lambda_k = 0$ for $k \neq i$. It now follows from $(\star)$ that

$$[b_{i1} \ \ldots \ b_{in}] = [a_{i1} \ \ldots \ a_{in}]$$

whence $b_{in} = a_{in}$. Thus the n-th columns of A, B coincide. $\diamond$

Interesting points about about the above proof are firstly the uniqueness of the Hermite form, and secondly that the proof describes a systematic procedure for solving systems of linear equations. By way of illustration, consider the following system whose coefficient matrix is in $\text{Mat}_{4 \times 4}(\mathbb{R})$:

$$\begin{cases} x + y + z + t = 4 \\ x + \lambda y + z + t = 4 \\ x + y + \lambda z + (3-\lambda)t = 6 \\ 2x + 2y + 2z + \lambda t = 6. \end{cases}$$

Applying the procedure for reducing the augmented matrix to echelon form, we obtain the matrix

$$Z = \begin{bmatrix} 1 & 1 & 1 & 1 & 4 \\ 0 & \lambda - 1 & 0 & 0 & 0 \\ 0 & 0 & \lambda - 1 & 2 - \lambda & 2 \\ 0 & 0 & 0 & \lambda - 2 & -2 \end{bmatrix}.$$

Now if $\lambda \neq 1, 2$ the rank of the coefficient matrix is clearly 4, as is that of the augmented matrix, and so a solution exists by Theorem 10.7; moreover, in this case a solution is unique, for the coefficient matrix is of rank 4 and so the solution space of the associated homogeneous system is of dimension $4 - 4 = 0$ (Theorem 10.9) whence the uniqueness follows by Theorem 10.8. To determine the solution in this case, we transform the above matrix to the unique associated Hermite matrix, obtaining

$$H = \begin{bmatrix} 1 & 0 & 0 & 0 & 4 + \frac{2}{\lambda - 2} \\ 0 & 1 & 0 & 0 & 0 \\ 0 & 0 & 1 & 0 & 0 \\ 0 & 0 & 0 & 1 & -\frac{2}{\lambda - 2} \end{bmatrix}.$$

The unique solution of the corresponding system of equations may now be read off, namely

$$x = 4 + \frac{2}{\lambda - 2}, \ y = 0, \ z = 0, \ t = -\frac{2}{\lambda - 2}.$$

Consider now the case $\lambda = 2$. Here the matrix Z becomes

$$\begin{bmatrix} 1 & 1 & 1 & 1 & 4 \\ 0 & 1 & 0 & 0 & 0 \\ 0 & 0 & 1 & 0 & 2 \\ 0 & 0 & 0 & 0 & -2 \end{bmatrix}.$$

In this case the rank of the coefficient matrix is 3 whereas that of the augmented matrix is 4. In this case, therefore, the solution set is empty.

Finally, consider the case where $\lambda = 1$. Here the matrix Z becomes

$$\begin{bmatrix} 1 & 1 & 1 & 1 & 4 \\ 0 & 0 & 0 & 0 & 0 \\ 0 & 0 & 0 & 1 & 2 \\ 0 & 0 & 0 & -1 & -2 \end{bmatrix},$$

the corresponding Hermite matrix being

$$\begin{bmatrix} 1 & 1 & 1 & 0 & 2 \\ 0 & 0 & 0 & 1 & 2 \\ 0 & 0 & 0 & 0 & 0 \\ 0 & 0 & 0 & 0 & 0 \end{bmatrix}.$$

Here the coefficient matrix and the augmented matrix are each of rank 2, so a solution exists. The dimension of the solution space of the associated homogeneous system is $4 - 2 = 2$ and, using Theorem 10.8, we see that (x, y, z, t) is a solution of the given system (when $\lambda = 1$) if and only if there exist $\alpha, \beta \in \mathbb{R}$ such that

$$(x, y, z, t) = (2, 0, 0, 2) + \alpha(-1, 1, 0, 0) + \beta(-1, 0, 1, 0).$$

We close this chapter with a brief mention of another equivalence relation, this time on the set of $n \times n$ matrices over a field F. Suppose that V is a vector space of dimension n over F and that $f : V \to V$ is linear. Under what conditions do two $n \times n$ matrices A, B represent f relative to different ordered bases? It is clear from Theorem 10.3 that this is the case if and only if there is an invertible $n \times n$ matrix P such that $B = P^{-1}AP$. When this is so, we shall say that A and B are *similar*. Similar matrices are evidently equivalent. The relation of being similar is clearly an equivalence relation on the set of $n \times n$ matrices over F. Again an important problem from both the theoretical and practical points of view is that of locating a particularly simple representative, or *canonical form*, in each equivalence class. In the exercises for this

Matrices; linear equations

chapter we indicate some of these canonical forms. In general, however, we require high-powered techniques to tackle this problem and we shall do so later.

EXERCISES

10.1 An $n \times n$ matrix A is called *symmetric* if $A^t = A$, and *skew-symmetric* if $A^t = -A$. Show that the set of symmetric $n \times n$ matrices over a field F is a subspace of $\text{Mat}_{n \times n}(F)$, as does the set of skew-symmetric $n \times n$ matrices. If F is not of characteristic 2, show that $\text{Mat}_{n \times n}(F)$ is the direct sum of these subspaces.

10.2 Define *elementary column operations* on a matrix in a similar way to elementary row operations. Call A, B *column-equivalent* if one can be obtained from the other by a finite sequence of column operations. Prove that A, are column-equivalent if and only if there is an invertible matrix Q such that $A = BQ$. Deduce that A, B are equivalent if and only if one can be obtained from the other by a finite sequence of row and column operations.

10.3 For the real matrix

$$\begin{bmatrix} 1 & 2 & 3 & -2 \\ 2 & -2 & 1 & 3 \\ 3 & 0 & 4 & 1 \end{bmatrix}$$

compute the canonical matrix N that is equivalent to A. Compute also invertible matrices P, Q such that $N = PAQ$.

10.4 Prove that an $n \times n$ matrix over a field is invertible if and only if ithe corresponding Hermite matrix is I_n. Use this fact to obtain a practical method of determining whether or not a given $n \times n$ matrix A has an inverse and, when it does, of computing A^{-1}.

10.5 Show that the set

$$K = \left\{ \begin{bmatrix} a & b \\ -b & a \end{bmatrix} \; ; \; a, b \in \mathbb{Z}_3 \right\}$$

is a subfield of the ring $\text{Mat}_{2 \times 2}(\mathbb{Z}_3)$. Show that the multiplicative group of non-zero elements of K is cyclic, of order 8, and generated by the element

$$\begin{bmatrix} 1 & 2 \\ 1 & 1 \end{bmatrix}.$$

10.6 Let $A_1, \ldots, A_n$ and $B_1, \ldots, B_m$ be R-modules. For every morphism $f : \bigoplus_{i=1}^{n} A_i \to \bigoplus_{j=1}^{m} B_j$ define
$$f_{ji} = \mathrm{pr}_j^B \circ f \circ \mathrm{in}_i^A$$
where pr_j^B and in_i^A are the obvious natural morphisms. If M is the set of $m \times n$ matrices $[\vartheta_{ji}]$ where each $\vartheta_{ji} \in \mathrm{Mor}_R(A_i, B_j)$, show that the mapping
$$\zeta : \mathrm{Mor}_R\left(\bigoplus_{i=1}^{n} A_i, \bigoplus_{j=1}^{m} B_j\right) \to M$$
described by $\zeta(f) = [f_{ji}]$ is an abelian group isomorphism, so that f is uniquely determined by the $m \times n$ matrix $[f_{ji}]$. Show also that the composite R-morphism
$$\bigoplus_{i=1}^{n} A_i \xrightarrow{f} \bigoplus_{j=1}^{m} B_j \xrightarrow{g} \bigoplus_{k=1}^{p} C_k$$
is represented by the matrix product $[g_{kj}][f_{ji}]$. Hence establish, for every R-module A, a ring isomorphism
$$\mathrm{Mor}_R(A^n, A^n) \simeq \mathrm{Mat}_{n \times n}[\mathrm{Mor}_R(A, A)].$$

10.7 Consider the matrix
$$A = \begin{bmatrix} 1 & 2 & 3 \\ 0 & 3 & 2 \end{bmatrix}$$
whose entries are in the ring $\mathbb{Z}_{30}$. Show that the row rank of A is 2 whereas the column rank of A is 1.

10.8 If $A, B \in \mathrm{Mat}_{n \times n}(F)$ are similar, prove that so also are A^m, B^m for every positive integer m.

10.9 Let V be a vector space of dimension n over a field F that is not of characteristic 2. Suppose that $f : V \to V$ is a linear transformation such that $f^2 = \mathrm{id}_V$. Prove that
$$V = \mathrm{Im}(\mathrm{id}_V + f) \oplus \mathrm{Im}(\mathrm{id}_V - f).$$
Deduce that an $n \times n$ matrix A over F is such that $A^2 = I_n$ if and only if A is similar to a matrix of the form
$$\left[\begin{array}{c|c} I_p & 0 \\ \hline 0 & -I_{n-p} \end{array}\right].$$

[*Hint*. Use Theorem 7.8.]

Suppose now that F is of characteristic 2 and that $f^2 = \text{id}_V$. If $g = \text{id}_V + f$ show that

$$x \in \text{Ker}\, g \iff x = f(x).$$

Show also that $g^2 = 0$. Deduce that an $n \times n$ matrix A over F is such that $A^2 = I_n$ if and only if A is similar to a matrix of the form

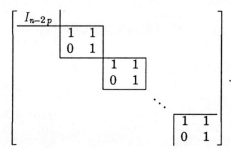

[*Hint*. Let f represent A relative to some fixed ordered basis of V and let g be as above. Note that $\text{Im}\, g \subseteq \text{Ker}\, g$. Let $\{g(c_1), \ldots, g(c_p)\}$ be a basis of $\text{Im}\, g$ extended to a basis $\{b_1, \ldots, b_{n-2p}, g(c_1), \ldots, g(c_p)\}$ of $\text{Ker}\, g$. Show that

$$\{b_1, \ldots, b_{n-2p}, g(c_1), c_1, g(c_2), c_2, \ldots, g(c_p), c_p\}$$

is a basis of V. Compute the matrix of f relative to this ordered basis.]

10.10 Let V be a vector space of dimension n over a field F and let $f : V \to V$ be a linear transformation such that $f^2 = 0$. Show that if $\text{Im}\, f$ is of dimension r then $2r \le n$. Suppose now that W is a subspace of V such that $V = \text{Ker}\, f \oplus W$. Show that W is of dimension r and that if $\{w_1, \ldots, w_r\}$ is a basis of W then $\{f(w_1), \ldots, f(w_r)\}$ is a linearly independent subset of $\text{Ker}\, f$. Deduce that $n - 2r$ elements $x_1, \ldots, x_{n-2r}$ can be chosen in $\text{Ker}\, f$ such that

$$\{w_1, \ldots, w_r, f(w_1), \ldots, f(w_r), x_1, \ldots, x_{n-2r}\}$$

is a basis of V. Deduce that a non-zero matrix A over F is such that $A^2 = 0$ if and only if A is similar to a matrix of the form

$$\begin{bmatrix} 0_r & 0 & \\ I_r & 0 & \\ 0 & 0 & \end{bmatrix}.$$

10.11 Let V be a vector space of dimension n over a field F. A linear transformation $f : V \to V$ (respectively, an $n \times n$ matrix A over F) is said to be *nilpotent of index* p if there is an integer $p > 1$ such that $f^{p-1} \neq 0$ and $f^p = 0$ (respectively, $A^{p-1} \neq 0$ and $A^p = 0$). Show that if f is nilpotent of index p and if $x \neq 0$ is such that $f^{p-1}(x) \neq 0$ then $\{x, f(x), \ldots, f^{p-1}(x)\}$ is a linearly independent subset of V. Hence show that f is nilpotent of index n if and only if there is an ordered basis $(a_i)_n$ of V such that

$$\mathrm{Mat}[\,f, (a_i)_n, (a_i)_n] = \left[\begin{array}{c|c} 0 & 0 \\ \hline I_{n-1} & 0 \end{array}\right].$$

Deduce that an $n \times n$ matrix A over F is nilpotent of index n if and only if A is similar to the above matrix.

10.12 Let V be a finite-dimensional vector space over $\mathbb{R}$ and let $f : V \to V$ be a linear transformation such that $f^2 = -\mathrm{id}_V$. Extend the action $\mathbb{R} \times V \to V$ to an action $\mathbb{C} \times V \to V$ by defining, for all $x \in V$ and all $\alpha + i\beta \in \mathbb{C}$,

$$(\alpha + i\beta)x = \alpha x - \beta f(x).$$

Show that in this way V becomes a vector space ove $\mathbb{C}$. Use the identity

$$\sum_{t=1}^{r}(\lambda_t - i\mu_t) v_t = \sum_{t=1}^{r} \lambda_t v_t + \sum_{t=1}^{r} \mu_t f(v_t)$$

to show that if $\{v_1, \ldots, v_r\}$ is a linearly independent subset of the $\mathbb{C}$-vector space V then $\{v_1, \ldots, v_r, f(v_1), \ldots, f(v_r)\}$ is a linearly independent subset of the $\mathbb{R}$-vector space V. Deduce that the dimension of V as a $\mathbb{C}$-vector space is finite, n say, and that as an $\mathbb{R}$-vector space V has a basis of the form

$$\{v_1, \ldots, v_n, f(v_1), \ldots, f(v_n)\},$$

so that the dimension of V as an $\mathbb{R}$-vector space is $2n$. If $(a_i)_{2n}$ is the ordered basis of V given by $a_i = v_i$ and $a_{n+i} = f(v_i)$ for $i = 1, \ldots, n$ show that

$$\mathrm{Mat}[\,f(a_i)_{2n}, (a_i)_{2n}] = \left[\begin{array}{c|c} 0 & -I_n \\ \hline I_n & 0 \end{array}\right].$$

Deduce that a $2n \times 2n$ matrix A over $\mathbb{R}$ is such that $A^2 = -I_{2n}$ if and only if A is similar to the above matrix.

10.13 Let V be a vector space of dimension 4 over $\mathbb{R}$. Let $\{b_1, b_2, b_3, b_4\}$ be a basis of V and, writing each $x \in V$ as $x = x_1 b_1 + x_2 b_2 + x_3 b_3 + x_4 b_4$, let

$$V_1 = \{x \in V \ ; \ x_3 = x_2 \text{ and } x_4 = x_1\};$$
$$V_2 = \{x \in V \ ; \ x_3 = -x_2 \text{ and } x_4 = -x_1\}.$$

Show that

(1) V_1 and V_2 are subspaces of V;

(2) $\{b_1 + b_4, b_2 + b_3\}$ is a basis of V_1 and $\{b_1 - b_4, b_2 - b_3\}$ is a basis of V_2;

(3) $V = V_1 \oplus V_2$;

(4) relative to the ordered bases $B = (b_i)_{1 \leq i \leq 4}$ and $C = (c_i)_{1 \leq i \leq 4}$ where $c_1 = b_1 + b_4, c_2 = b_2 + b_3, c_3 = b_2 - b_3, c_4 = b_1 - b_4$,

$$\text{Mat}[\text{id}_V, C, B] = \frac{1}{2} \begin{bmatrix} 1 & 0 & 0 & 1 \\ 0 & 1 & 1 & 0 \\ 0 & 1 & -1 & 0 \\ 1 & 0 & 0 & -1 \end{bmatrix}.$$

A 4×4 matrix $M = [m_{ij}]$ over $\mathbb{R}$ is called *centro-symmetric* if

$$(i, j = 1, \ldots, 4) \qquad m_{ij} = m_{5-i, 5-j}.$$

If M is centro-symmetric, prove that M is similar to a matrix of the form

$$\begin{bmatrix} \alpha & \beta & 0 & 0 \\ \gamma & \delta & 0 & 0 \\ \hline 0 & 0 & \epsilon & \zeta \\ 0 & 0 & \eta & \vartheta \end{bmatrix}.$$

10.14 Show that when a, b, c, d are positive real numbers the following system of equations has no solution :

$$\begin{cases} x + y + z + t = a \\ x - y - z + t = b \\ -x - y + z + t = c \\ -3x + y - 3z - 7t = d. \end{cases}$$

10.15 If $\alpha, \beta, \gamma \in \mathbb{R}$ show that the system of equations

$$\begin{cases} 2x + y + z = -6\beta \\ \alpha x + 3y + 2z = 2\beta \\ 2x + y + (\gamma + 1)z = 4 \end{cases}$$

has a unique solution except when $\gamma = 0$ and $\gamma = 6$. If $\gamma = 0$ show that there is only one value of β for which a solution exists and find the solution set in this case. Discuss the situation when $\gamma = 6$.

10.16 Given the real matrices

$$\begin{bmatrix} 3 & 2 & -1 & 5 \\ 1 & -1 & 2 & 2 \\ 0 & 5 & 7 & \vartheta \end{bmatrix}, \quad \begin{bmatrix} 0 & 3 \\ 0 & -1 \\ 0 & 6 \end{bmatrix}$$

prove that the matrix equation $AX = B$ has a solution if and only if $\vartheta = -1$. Find the solution set in this case.

11 Inner product spaces

In our discussion of vector spaces the ground field F has been arbitrary and its properties have played no significant rôle. In the present section we shall restrict F to be $\mathbb{R}$ or $\mathbb{C}$, the results obtained depending heavily on the properties of these fields.

Definition Let V be a vector space over $\mathbb{C}$. By an *inner product* on V we shall mean a mapping $V \times V \to \mathbb{C}$, described by $(x,y) \mapsto \langle x\,|\,y\rangle$, such that, $\overline{\alpha}$ denoting the complex conjugate of $\alpha \in \mathbb{C}$,

(1) $\langle x + x'\,|\,y\rangle = \langle x\,|\,y\rangle + \langle x'\,|\,y\rangle$;
(2) $\langle x\,|\,y + y'\rangle = \langle x\,|\,y\rangle + \langle x\,|\,y'\rangle$;
(3) $\langle \alpha x\,|\,y\rangle = \alpha\langle x\,|\,x\rangle$;
(4) $\langle y\,|\,x\rangle = \overline{\langle x\,|\,y\rangle}$;
(5) $\langle x\,|\,x\rangle \geq 0$ with equality if and only if $x = 0$.

By a *complex inner product space* we mean a $\mathbb{C}$-vector space V together with an inner product on V. By a *real inner product space* we mean an $\mathbb{R}$-vector space together with an inner product (this being defined as in the above, but with the bars denoting complex conjugates omitted). By an *inner product space* we shall mean either a complex inner product space or a real inner product space.

- The notation $\langle x\,|\,y\rangle$ is not to be confused with the notation $\langle x,y\rangle$ used in Chapter 9. We shall see the relationship between these notations later.

Note from the above definition that we have

$$(\forall x \in V) \quad \langle x\,|\,0\rangle = 0 = \langle 0\,|\,x\rangle.$$

In fact, this follows from (1), (2), (3) on taking $x' = -x$ and $y' = -y$.

Note also that we have

$$\langle x\,|\,\alpha y\rangle = \overline{\alpha}\langle x\,|\,y\rangle.$$

In fact, this follows from (3) and (4).

Finally, note that in a complex inner product space V we have $\langle x\,|\,x\rangle \in \mathbb{R}$ for all $x \in V$. This is immediate from (4).

Example 11.1 $\mathbb{C}^n$ is a complex inner product space under the *standard inner product* defined by

$$\langle (z_1,\ldots,z_n)\,|\,(w_1,\ldots,w_n)\rangle = \sum_{i=1}^{n} z_i\overline{w_i}.$$

Example 11.2 $\mathbb{R}^n$ is a real inner product space under the *standard inner product* defined by

$$\langle (x_1,\ldots,x_n) \mid (y_1,\ldots,y_n)\rangle = \sum_{i=1}^{n} x_i y_i.$$

In the cases where $n = 2, 3$ this inner product is sometimes called the *dot product*, a terminology that is popular when dealing with the geometric applications of vectors.

Example 11.3 Let $a, b \in \mathbb{R}$ be such that $a < b$ and let V be the $\mathbb{R}$-vector space of continuous functions $f : [a,b] \to \mathbb{R}$. Define a mapping $V \times V \to \mathbb{R}$ by the prescription

$$(f,g) \mapsto \langle f \mid g\rangle = \int_a^b f(x)g(x)\,dx.$$

Then this defines an inner product on V.

Definition Let V be an inner product space. For every $x \in V$ we define the *norm* of x to be the non-negative real number

$$\|x\| = \sqrt{\langle x \mid x\rangle}.$$

Given $x, y \in V$ we define the *distance* between x and y to be

$$d(x,y) = \|x - y\|.$$

It is clear from the above that $\|x\| = 0$ if and only if $x = 0$.

- In the real inner product space $\mathbb{R}^2$, for $x = (x_1, x_2)$ we have $\|x\|^2 = x_1^2 + x_2^2$ so that $\|x\|$ is the distance from x to the origin. Likewise, for $x = (x_1, x_2)$ and $y = (y_1, y_2)$ we have $\|x - y\|^2 = (x_1 - y_1)^2 + (x_2 - y_2)^2$, whence we see the connection between the concept of distance and the theorem of Pythagoras.

Concerning norms, the following result is fundamental.

Theorem 11.1 *Let V be an inner product space. Then, for all $x, y \in V$ and every scalar λ, we have*

(1) $\|\lambda x\| = |\lambda|\,\|x\|$;
(2) [Cauchy-Schwartz inequality] $|\langle x \mid y\rangle| \leq \|x\|\,\|y\|$;
(3) [Triangle inequality] $\|x + y\| \leq \|x\| + \|y\|$.

Proof (1) $\|\lambda x\|^2 = \langle \lambda x \mid \lambda x\rangle = \lambda\overline{\lambda}\langle x \mid x\rangle = |\lambda|^2\,\|x\|^2.$

Inner product spaces 149

(2) The result is trivial if $x = 0$. Suppose then that $x \neq 0$ so that $||x|| \neq 0$. Let $z = y - \dfrac{\langle y\,|\,y\rangle}{||x||^2}\,x$; then noting that $\langle z\,|\,x\rangle = 0$ we have

$$\begin{aligned}
0 \le ||z||^2 &= \left\langle y - \frac{\langle y\,|\,x\rangle}{||x||^2}\,\bigg|\,y - \frac{\langle y\,|\,x\rangle}{||x||^2}\,x\right\rangle \\
&= \langle y\,|\,y\rangle - \frac{\langle y\,|\,x\rangle}{||x||^2}\langle x\,|\,y\rangle \\
&= \langle y\,|\,y\rangle - \frac{|\langle x\,|\,y\rangle|^2}{||x||^2}
\end{aligned}$$

from which (2) follows.

(3) Using (2), we have

$$\begin{aligned}
||x+y||^2 = \langle x+y\,|\,x+y\rangle &= \langle x\,|\,x\rangle + \langle x\,|\,y\rangle + \langle y\,|\,x\rangle + \langle y\,|\,y\rangle \\
&= ||x||^2 + \langle x\,|\,y\rangle + \overline{\langle x\,|\,y\rangle} + ||y||^2 \\
&= ||x||^2 + 2\,\mathrm{Re}\langle x\,|\,y\rangle + ||y||^2 \\
&\le ||x||^2 + 2\,\langle x\,|\,y\rangle + ||y||^2 \\
&\le ||x||^2 + 2\,||x||\,||y|| + ||y||^2 \\
&= (||x|| + ||y||)^2
\end{aligned}$$

from which (3) follows. ◊

Definition Let V be an inner product space. Then $x, y \in V$ are said to be *orthogonal* if $\langle x\,|\,x\rangle = 0$. A non-empty subset S of V is called an *orthogonal subset* of V if, for all $x, y \in S$ with $x \neq y$, x and y are orthogonal. An orthogonal subset S in which $||x|| = 1$ for every x is called an *orthonormal subset*.

- In the real inner product space $\mathbb{R}^2$, if $x = (x_1, x_2)$ and $y = (y_1, y_2)$ then $\langle x\,|\,y\rangle = 0$ if and only if $x_1 y_1 + x_2 y_2 = 0$. Geometrically, this is equivalent to saying that the lines joining x and y to the origin are mutually perpendicular. In a general inner product space it is often convenient to think of an orthonormal set as a set of mutually perpendicular vectors each of length 1.

Example 11.4 With respect to the standard inner products, the natural bases of $\mathbb{R}^n$ and $\mathbb{C}^n$ are orthonormal sets.

An important property of orthonormal subsets is the following.

Theorem 11.2 *Orthonormal subsets are linearly independent.*

Proof Let S be an orthonormal subset of the inner product space V. Suppose that $x_1, \ldots, x_n$ are distinct elements of S and that $\lambda_1, \ldots, \lambda_n$

are scalars such that $\sum_{i=1}^{n} \lambda_i x_i = 0$. Then for each i we have

$$\begin{aligned}
\lambda_i = \lambda_i 1 = \lambda_i \langle x_i \,|\, x_i \rangle &= \sum_{i=1}^{n} \lambda_k \langle x_k \,|\, x_i \rangle \\
&= \Big\langle \sum_{i=1}^{n} \lambda_k x_k \,\Big|\, x_i \Big\rangle \\
&= \langle 0 \,|\, x_i \rangle \\
&= 0. \quad \diamondsuit
\end{aligned}$$

Theorem 11.3 *If V is an inner product space and if $\{e_1, \ldots, e_n\}$ is an orthonormal subset of V then*

[Bessel's inequality] $\quad (\forall x \in V) \quad \sum_{i=1}^{n} |\langle x \,|\, e_k \rangle|^2 \leq \|x\|^2.$

Moreover, if W is the subspace generated by $\{e_1, \ldots, e_n\}$ then the following are equivalent :

(1) $x \in W$;

(2) $\sum_{k=1}^{n} |\langle x \,|\, e_k \rangle|^2 = \|x\|^2$;

(3) $x = \sum_{k=1}^{n} \langle x \,|\, e_k \rangle e_k$;

(4) $(\forall y \in V) \; \langle x \,|\, y \rangle = \sum_{k=1}^{n} \langle x \,|\, e_k \rangle \langle e_k \,|\, y \rangle.$

Proof For Bessel's inequality let $z = x - \sum_{k=1}^{n} \langle x \,|\, e_k \rangle e_k$ and observe that

$$0 \leq \langle z \,|\, z \rangle = \langle x \,|\, x \rangle - \sum_{k=1}^{n} \langle x \,|\, e_k \rangle \overline{\langle x \,|\, e_k \rangle} = \|x\|^2 - \sum_{k=1}^{n} |\langle x \,|\, e_k \rangle|^2.$$

(2) $\Rightarrow$ (3) is now immediate.

(3) $\Rightarrow$ (4) : If $x = \sum_{k=1}^{n} \langle x \,|\, e_k \rangle e_k$ then for all $y \in V$ we have

$$\langle x \,|\, y \rangle = \Big\langle \sum_{k=1}^{n} \langle x \,|\, e_k \rangle e_k \,\Big|\, y \Big\rangle = \sum_{k=1}^{n} \langle x \,|\, e_k \rangle \langle e_k \,|\, y \rangle.$$

(4) $\Rightarrow$ (2) : This follows on taking $y = x$.
(3) $\Rightarrow$ (1) : This is obvious.

Inner product spaces

(1) $\Rightarrow$ (3) : If $x = \sum_{k=1}^{n} \lambda_k e_k$ then for $j = 1, \ldots, n$ we have

$$\lambda_j = \sum_{k=1}^{n} \lambda_k \langle e_k \mid e_j \rangle = \Big\langle \sum_{k=1}^{n} \lambda_k e_k \,\Big|\, e_j \Big\rangle = \langle x \mid e_j \rangle. \quad \diamond$$

Definition An *orthonormal basis* is an orthonormal subset that is a basis.

We know that every vector space has a basis. We shall now show that every finite-dimensional inner product space has an orthonormal basis. In so doing, we shall obtain a practical procedure for constructing such a basis.

Theorem 11.4 *Let V be an inner product space and for every $x \in V$ let $x^* = x/\|x\|$. If $\{x_1, \ldots, x_n\}$ is a linearly independent subset of V, define recursively*

$y_1 = x_1^*;$
$y_2 = (x_2 - \langle x_2 \mid y_1 \rangle y_1)^*;$
$\vdots$
$y_k = \Big(x_k - \sum_{i=1}^{k-1} \langle x_k \mid y_i \rangle y_i\Big)^*.$

Then $\{y_1, \ldots, y_k\}$ is an orthonormal set that generates the same subspace as $\{x_1, \ldots, x_k\}$.

Proof It is readily seen that $y_i \neq 0$ for every i and that y_i is a linear combination of $x_1, \ldots, x_i$. It is also clear that x_i is a linear combination of $y_1, \ldots, y_i$ for every i. Thus $\{x_1, \ldots, x_k\}$ and $\{y_1, \ldots, y_k\}$ generate the same subspace. It now suffices to prove that $\{y_1, \ldots, y_k\}$ is an orthogonal subset of V; and this we do by induction. For $k = 1$ the result is trivial. Suppose by way of induction that $\{y_1, \ldots, y_{t-1}\}$ is orthogonal where $t > 1$. Then, writing

$$\Big\|x_t - \sum_{i=1}^{t-1} \langle x_t \mid y_i \rangle y_i\Big\| = \alpha_t$$

we see that

$$\alpha_t y_t = x_t - \sum_{i=1}^{t-1} \langle x_t \mid y_i \rangle y_i$$

and so, for $j < t$,

$$\alpha_t \langle y_t \mid y_j \rangle = \langle x_t \mid y_j \rangle - \sum_{i=1}^{t-1} \langle x_t \mid y_i \rangle \langle y_i \mid y_j \rangle$$
$$= \langle x_t \mid y_j \rangle - \langle x_t \mid y_j \rangle$$
$$= 0.$$

As $\alpha_t \neq 0$, we deduce that $\langle y_t | y_j \rangle = 0$ for $j < t$, so $\{y_1, \ldots, y_t\}$ is orthogonal. $\Diamond$

Corollary *Every finite-dimensional inner product space has an orthonormal basis.*

Proof Simply apply Theorem 11.4 to a basis. $\Diamond$

- The construction of $\{y_1, \ldots, y_k\}$ from $\{x_1, \ldots, x_k\}$ in Theorem 11.4 is often called the *Gram-Schmidt orthonormalisation process*.

Theorem 11.5 *If V is an inner product space and $\{e_1, \ldots, e_n\}$ is an orthonormal basis of V then*

(1) $(\forall x \in V) \quad x = \sum_{k=1}^{n} \langle x | e_k \rangle e_k;$

(2) $(\forall x \in V) \quad \|x\|^2 = \sum_{k=1}^{n} |\langle x | e_k \rangle|^2;$

(3) $(\forall x \in V) \quad \langle x | y \rangle = \sum_{k=1}^{n} \langle x | e_k \rangle \langle e_k | y \rangle.$

Proof Since V is generated by the orthonormal subset $\{e_1, \ldots, e_n\}$, the result is immediate from Theorem 11.3. $\Diamond$

- The identity (1) of Theorem 11.5 is often referred to as the *Fourier expansion* of x relative to the orthonormal basis $\{e_1, \ldots, e_n\}$, the scalars $\langle x | e_k \rangle$ being called the *Fourier coefficients* of x. The identity (3) is called *Parseval's identity*.

Just as a linearly independent subset of a vector space can be extended to form a basis, so can an orthonormal subset (which by Theorem 11.2 is linearly independent) be extended to form an orthonormal basis. This is the content of the following result.

Theorem 11.6 *Let V be an inner product space of dimension n. If, for $k < n$, $\{x_1, \ldots, x_k\}$ is an orthonormal subset of V then there exist $x_{k+1}, \ldots, x_n \in V$ such that $\{x_1, \ldots, x_n\}$ is an orthonormal basis of V.*

Proof Let W be the subspace generated by $\{x_1, \ldots, x_k\}$. By Theorem 11.2, this set is a basis for W and so can be extended to form a basis

$$\{x_1, \ldots, x_k, x_{k+1}, \ldots, x_n\}$$

of V. Applying the Gram-Schmidt orthonormalisation process to this basis we obtain an orthonormal basis of V. The theorem now follows

Inner product spaces

on noting that, since $\{x_1, \ldots, x_k\}$ is orthonormal to start with, the first k terms of the new basis are precisely $x_1, \ldots, x_k$; for, referring to the formulae of Theorem 11.4, we have

$$y_1 = x_1^\star = x_1;$$
$$y_2 = (x_2 - \langle x_2 \mid x_1\rangle x_1)^\star = x_2^\star = x_2;$$
$$\vdots$$
$$y_k = \left(x_k - \sum_{i=1}^{k-1}\langle x_k \mid x_i\rangle x_i\right)^\star = x_k^\star = x_k. \quad \diamond$$

An isomorphism from one vector space to another carries bases to bases. The corresponding situation for inner product spaces is as follows.

Definition Let V and W be inner product spaces over the same field. Then $f : V \to W$ is called an *inner product isomorphism* if it is a vector space isomorphism that preserves inner products, in the sense that

$$(\forall x, y \in V) \qquad \langle f(x) \mid f(y)\rangle = \langle x \mid y\rangle.$$

Theorem 11.7 *Let V and W be finite-dimensional inner product spaces over the same field. Let $\{e_1, \ldots, e_n\}$ be an orthonormal basis of V. Then a linear mapping $f : V \to W$ is an inner product isomorphism if and only if $\{f(e_1), \ldots, f(e_n)\}$ is an orthonormal basis of W.*

Proof $\Rightarrow$: If f is an inner product isomorphism then $\{f(e_1), \ldots, f(e_n)\}$ is a basis of W. It is also orthonormal since

$$\langle f(e_i) \mid f(e_j)\rangle = \langle e_i \mid e_j\rangle = \begin{cases} 1 & \text{if } i = j; \\ 0 & \text{if } i \neq j. \end{cases}$$

$\Leftarrow$: Suppose that $\{f(e_1), \ldots, f(e_n)\}$ is an orthonormal basis of W. Then f carries a basis to a basis and so is a vector space isomorphism. Now for all $x \in V$ we have, using the Fourier expansion of x relative to $\{e_1, \ldots, e_n\}$,

$$\langle f(x) \mid f(e_j)\rangle = \left\langle f\left(\sum_{i=1}^n \langle x \mid e_i\rangle e_i\right) \Big| f(e_j)\right\rangle$$
$$= \left\langle \sum_{i=1}^n \langle x \mid e_i\rangle f(e_i) \,\Big|\, f(e_j)\right\rangle$$
$$= \sum_{i=1}^n \langle x \mid e_i\rangle, \langle f(e_i) \mid f(e_j)\rangle$$
$$= \langle x \mid e_j\rangle$$

and similarly $\langle f(e_j) \,|\, f(x)\rangle = \langle e_j \,|\, x\rangle$. It now follows by Parseval's identity applied to both V and W that

$$\begin{aligned}\langle f(x) \,|\, f(y)\rangle &= \sum_{i=1}^{n} \langle f(x) \,|\, f(e_j)\rangle \langle f(e_j) \,|\, f(y)\rangle \\ &= \sum_{i=1}^{n} \langle x \,|\, e_j\rangle \langle e_j \,|\, y\rangle \\ &= \langle x \,|\, y\rangle\end{aligned}$$

and consequently f is an inner product isomorphism. $\diamond$

We now pass to the consideration of the dual of an inner product space. For this purpose, we require the following notion.

Definition Let V and W be F-vector spaces where F is either $\mathbb{R}$ or $\mathbb{C}$. A mapping $f : V \to W$ is called a *conjugate transformation* if

$$(\forall x, y \in V)(\forall \lambda \in F) \qquad f(x+y) = f(x) + f(y), \quad f(\lambda x) = \overline{\lambda} f(x).$$

If, furthermore, f is a bijection then we say that it is a *conjugate isomorphism*.

- Note that when $F = \mathbb{R}$ a conjugate transformation is an ordinary linear transformation.

Theorem 11.8 *Let V be a finite-dimensional inner product space. Then there is a conjugate isomorphism $\vartheta_V : V \to V^d$. This is given by $\vartheta_V(x) = x^d$ where*

$$(\forall x \in V) \qquad x^d(x) = \langle x \,|\, x^d\rangle.$$

Proof It is clear that for every $y \in V$ the assignment $x \mapsto \langle x \,|\, y\rangle$ is linear and so defines an element of V^d. We shall write this element of V^d as y^d, so that we have $y^d(x) = \langle x \,|\, y\rangle$. Recalling the notation $\langle\,,\,\rangle$ introduced in Chapter 9, we therefore have the identities

$$\langle x \,|\, y\rangle = y^d(x) = \langle x, y^d\rangle.$$

Consider now the mapping $\vartheta_V : V \to V^d$ given by $\vartheta_V(x) = x^d$. For all $x, y, z \in V$ we have

$$\begin{aligned}\langle x, (y+z)^d\rangle &= \langle x \,|\, y+z\rangle \\ &= \langle x \,|\, y\rangle + \langle x \,|\, z\rangle \\ &= \langle x, y^d\rangle + \langle x, z^d\rangle \\ &= \langle x, y^d + z^d\rangle,\end{aligned}$$

Inner product spaces   155

from which we deduce that $(y+z)^d = y^d + z^d$ and hence that $\vartheta_V(y+z) = \vartheta_V(x) + \vartheta_V(z)$. Likewise,

$$\langle x, (\lambda y)^d \rangle = \langle x \mid \lambda y \rangle = \overline{\lambda} \langle x \mid x \rangle = \overline{\lambda} \langle x, y^d \rangle = \langle x, \overline{\lambda} y^d \rangle,$$

so that $(\lambda y)^d = \overline{\lambda} y^d$ and consequently $\vartheta_V(\lambda y) = \overline{\lambda} \vartheta_V(y)$. Thus we see that ϑ_V is a conjugate transformation.

That ϑ_V is injective follows from the fact that

$$x \in \operatorname{Ker} \vartheta_V \Rightarrow \langle x \mid x \rangle = \langle x, x^d \rangle = \langle x, 0 \rangle = 0$$
$$\Rightarrow x = 0.$$

To show that ϑ_V is also surjective, let $f \in V^d$. Let $\{e_1, \ldots, e_n\}$ be an orthonormal basis of V and let $x = \sum_{i=1}^{n} \overline{f(e_i)} e_i$. Then for $j = 1, \ldots, n$ we have

$$x^d(e_j) = \langle e_j \mid x \rangle = \left\langle e_j \mid \sum_{i=1}^{n} \overline{f(e_i)} e_i \right\rangle = \sum_{i=1}^{n} f(e_i) \langle e_j \mid e_i \rangle = f(e_j).$$

Since x^d and f thus coincide on the basis $\{e_1, \ldots, e_n\}$ we conclude that $f = x^d = \vartheta_V(x)$, so that ϑ_V is surjective. $\Diamond$

- Note from the above that we have the identity

$$(\forall x, y \in V) \qquad \langle x \mid y \rangle = \langle x, \vartheta_V(y) \rangle.$$

Since ϑ_V is a bijection, we also have the following identity (obtained by writing $\vartheta_V^{-1}(y)$ in place of y) :

$$(\forall x, y \in V) \qquad \langle x \mid \vartheta_V^{-1}(y) \rangle = \langle x, y \rangle.$$

The above result gives rise to the important notion of the *adjoint* of a linear transformation which we shall now describe.

Theorem 11.9 *Let V and W be finite-dimensional inner product spaces over the same field. If $f : V \to W$ is a linear transformation then there is a unique linear transformation $f^* : W \to V$ such that*

$$(\forall x \in V)(\forall y \in W) \qquad \langle f(x) \mid y \rangle = \langle x \mid f^*(y) \rangle.$$

Proof Using the above notations we have

$$\langle f(x) \mid y \rangle = \langle f(x), y^d \rangle = \langle x, f^t(y^d) \rangle$$
$$= \langle x \mid \vartheta_V^{-1}[f^t(y^d)] \rangle$$
$$= \langle x \mid (\vartheta_V^{-1} \circ f^t \circ \vartheta_W)(y) \rangle.$$

It follows immediately that $f^* = \vartheta_V^{-1} \circ f^t \circ \vartheta_W$ is the only linear transformation with the stated property. $\Diamond$

- Note that f^* can be alternatively characterised as the unique linear transformation from W to V such that the diagram

is commutative.

Definition The linear transformation f^* of Theorem 11.9 is called the *adjoint* of f.

Immediate properties of adjoints are listed in the following result.

Theorem 11.10 *Let V, W, X be finite-dimensional inner product spaces over the same field. Let $f, g : V \to W$ and $h : W \to X$ be linear transformations. Then*

(1) $(f + g)^* = f^* + g^*$;
(2) $(\lambda f)^* = \overline{\lambda} f^*$;
(3) $(h \circ f)^* = f^* \circ h^*$;
(4) $(f^*)^* = f$.

Proof (1) is immediate from $f^* = \vartheta_V^{-1} \circ f^t \circ \vartheta_W$ and the fact that $(f+g)^t = f^t + g^t$.

(2) Since $\langle (\lambda f)(x) \mid y \rangle = \lambda \langle f(x) \mid y \rangle = \lambda \langle x \mid f^*(y) \rangle = \langle x \mid \overline{\lambda} f^*(y) \rangle$, we have, by the uniqueness of adjoints, $(\lambda f)^* = \overline{\lambda} f^*$.

(3) Since $\langle h[f(x)] \mid y \rangle = \langle f(x) \mid h^*(y) \rangle = \langle x \mid f^*[h^*(y)] \rangle$ we have, again by the uniqueness of adjoints, $(h \circ f)^* = f^* \circ h^*$.

(4) Taking complex conjugates in Theorem 11.9, we obtain

$$(\forall y \in W)(\forall x \in V) \quad \langle f^*(y) \mid x \rangle = \langle y \mid f(x) \rangle.$$

It follows by the uniqueness of adjoints that $(f^*)^* = f$. ◇

Theorem 11.11 *Let V, W be inner product spaces over the same field and suppose that $\dim V = \dim W$. Then if $f : V \to W$ ia linear transformation the following are equivalent :*

(1) *f is an inner product isomorphism;*
(2) *f is a vector space isomorphism with $f^{-1} = f^*$;*
(3) *$f \circ f^* = \mathrm{id}_W$;*

Inner product spaces

(4) $f^* \circ f = \text{id}_V$.

Proof (1) $\Rightarrow$ (2) : If (1) holds then f^{-1} exists and
$$\langle f(x) | y \rangle = \langle f(x) | f[f^{-1}(y)] \rangle = \langle x | f^{-1}(y) \rangle.$$
It follows by the uniqueness of adjoints that $f^* = f^{-1}$.
(2) $\Rightarrow$ (3),(4) : these are obvious.
(3),(4) $\Rightarrow$ (1) : If, for example, (4) holds then f is injective whence it is bijective (by Corollary 4 of Theorem 8.9) and so $f^{-1} = f^*$. Then
$$\langle f(x) | f(y) \rangle = \langle x | f^*[f(y)] \rangle = \langle x | y \rangle$$
and so f is an inner product isomorphism. $\diamond$

Definition Let V be an inner product space. For every non-empty subset E of V we define the *orthogonal complement* of E in V to be the set
$$E^\perp = \{y \in V \; ; \; (\forall x \in E) \; \langle x | y \rangle = 0\}.$$

It is clear that $E^\perp$ is a subspace of V. The significance of this subspace is illustrated in the following result.

Theorem 11.12 *Let V be an inner product space. If W is a subspace of finite dimension then*
$$V = W \oplus W^\perp.$$

Proof Let $\{e_1, \ldots, e_m\}$ be an orthonormal basis of W, noting that this exists since W is of finite dimension. Given $x \in V$, let
$$x' = \sum_{i=1}^{m} \langle x | e_i \rangle e_i$$
and define $x'' = x - x'$. Then $x' \in W$ and for $j = 1, \ldots, m$ we have
$$\begin{aligned}\langle x'' | e_j \rangle &= \langle x - x' | e_j \rangle = \langle x | e_j \rangle - \langle x' | e_j \rangle \\ &= \langle x | e_j \rangle - \sum_{i=1}^{n} \langle x | e_i \rangle \langle e_i | e_j \rangle \\ &= \langle x | e_j \rangle - \langle x | e_j \rangle \\ &= 0.\end{aligned}$$

It follows from this that $x'' \in W^\perp$. Consequently $x = x' + x'' \in W + W^\perp$ and so $V = W + W^\perp$. Now if $x \in W \cap W^\perp$ then $\langle x | x \rangle = 0$ whence $\|x\| = 0$ and so $x = 0$. Thus we have $V = W \oplus W^\perp$. $\diamond$

Corollary *If V is a finite-dimensional inner product space and W is a subspace of V then $W = W^{\perp\perp}$ and*

$$\dim W^\perp = \dim W - \dim W.$$

Proof Since $V = W \oplus W^\perp$ it is clear from Theorem 7.8 that $\dim V = \dim W + \dim W^\perp$ and likewise $\dim V = \dim W^\perp + \dim W^{\perp\perp}$. It is also clear that $W \subseteq W^{\perp\perp}$. Since now

$$\dim W^{\perp\perp} = \dim V - \dim W^\perp = \dim V - (\dim V - \dim W) = \dim W,$$

it follows by Corollary 3 of Theorem 8.9 that $W^{\perp\perp} = W$. ◊

We end the present section by considering how the matrices of f and $f^\star$ are related.

Definition If $A = [a_{ij}]_{m \times n}$ is an $m \times n$ matrix over $\mathbb{C}$ then by its *adjoint* (or *conjugate transpose*) we mean the $n \times m$ matrix $A^\star$ the (i,j)-th entry of which is $\overline{a}_{ji}$.

Theorem 11.13 *Let V and W be finite-dimensional inner product spaces over the same field and let $(d_i)_n, (e_i)_m$ be ordered orthonormal bases of V, W. If $f : V \to W$ is a linear transformation such that*

$$\mathrm{Mat}[\,f, (e_i)_m, (d_i)_n\,] = A$$

then

$$\mathrm{Mat}[\,f^\star, (d_i)_n, (e_i)_m\,] = A^\star.$$

Proof For $j = 1, \ldots, n$ we have

$$f(d_j) = \sum_{i=1}^{m} \langle f(d_j) \mid e_i \rangle e_i$$

and so if $A = [a_{ij}]$ we have $a_{ij} = \langle f(d_j) \mid e_i \rangle$. Since likewise

$$f^\star(e_j) = \sum_{i=1}^{n} \langle f^\star(e_j) \mid d_i \rangle d_i$$

and since

$$\overline{a}_{ij} = \overline{\langle f(d_j) \mid e_i \rangle} = \langle e_i \mid f(d_j) \rangle = \langle f^\star(e_i) \mid d_j \rangle,$$

it follows that the matrix representing $f^\star$ is $A^\star$. ◊

Definition If V is a finite-dimensional inner product space and $f : V \to V$ is a linear transformation then we say that f is *self-adjoint* if $f = f^\star$. Likewise, an $n \times n$ matrix A will be called *self-adjoint* if $A = A^\star$.

Inner product spaces

- In the case where the ground field is $\mathbb{C}$ the term *hermitian* is often used instead of self-adjoint; and when the ground field is $\mathbb{R}$ the term *symmetric* is often used.

Definition If V is a finite-dimensional inner product space then an inner product isomorphism $f : V \to V$ is called a *unitary transformation*. An $n \times n$ matrix A is *unitary* if, relative to some ordered orthonormal basis, it represents a unitary transformation.

- When the ground field is $\mathbb{R}$ the term *orthogonal* is often used instead of unitary.

- Note by Theorem 11.11 that $f : V \to V$ is unitary if and only if f^{-1} exists and is $f^\star$; and that A is unitary if and only if A^{-1} exists and is $A^\star$. In particular, when the ground field is $\mathbb{R}$, A is orthogonal if and only if A^{-1} exists and is A^t.

If V is an inner product space of dimension n let $(d_i)_n$ and $(e_i)_n$ be ordered orthonormal bases of V. If U is an $n \times n$ matrix over the ground field of V then it is clear that U is unitary if and only if, relative to $(d_i)_n$ and $(e_i)_n$, U represents a unitary transformation (inner product isomorphism) $f : V \to V$. It is readily seen that if A, B are $n \times n$ matrices over the ground field of V then A, B represent the same linear transformation with respect to different ordered orthonormal bases of V if and only if there is a unitary matrix U such that $B = U^\star A U = U^{-1} A U$. We describe this by saying that A and B are *unitarily similar*. When the ground field is $\mathbb{R}$, in which case we have $B = U^t A U = U^{-1} A U$, we often use the term *orthogonally similar*.

It is clear that the relation of being unitarily (respectively, orthogonally) similar is an equivalence relation on the set of $n \times n$ matrices over $\mathbb{C}$ (respectively, $\mathbb{R}$). Just as with ordinary similarity, the problem of locating canonical forms is important. We shall deal with this problem later when we have all the necessary machinery to hand.

EXERCISES

11.1 If V is a real inner product space prove that
$$(\forall x, y \in V) \qquad ||x+y||^2 = ||x||^2 + ||y||^2 + 2\langle x \,|\, y \rangle$$
and interpret this geometrically when $V = \mathbb{R}^2$.

11.2 If V is an inner product space, establish the *parallelogram identity*
$$(\forall x, y \in V) \qquad ||x+y||^2 + ||x-y||^2 = 2||x||^2 + 2||y||^2$$
and interpret it geometrically when $V = \mathbb{R}^2$.

11.3 Write down the Cauchy-Schwartz inequality for the inner product spaces of Examples 11.1, 11.2, 11.3.

11.4 Show that equality holds in the Cauchy-Schwartz inequality if and only if $\{x,y\}$ is linearly dependent.

11.5 Let x,y be non-zero elements of the real inner product space $\mathbb{R}^3$. If ϑ is the angle $x0y$, prove that
$$\cos\vartheta = \frac{\langle x\mid y\rangle}{\|x\|\,\|y\|}.$$

11.6 Show that, in the real inner product space $\mathbb{R}^3$, Parseval's identity reduces to
$$\cos(\vartheta_1 - \vartheta_2) = \cos\vartheta_1\cos\vartheta_2 + \sin\vartheta_1\sin\vartheta_2.$$

11.7 Show that $V = \text{Mat}_{n\times n}(\mathbb{C})$ is an inner product space under the definition $\langle A\mid B\rangle = \text{tr}(AB^\star)$ where $\text{tr}\,A = \sum_{i=1}^n a_{ii}$ is the *trace* of A. Interpret the Cauchy-Schwartz inequality in this inner product space. Show further that if $E_{p,q} \in V$ is the $n\times n$ matrix whose (p,q)-th entry is 1 and all other entries are 0 then $\{E_{p,q}\ ;\ p,q = 1,\ldots,n\}$ is an orthonormal basis of V.

11.8 Use the Gram-Schmidt orthonormalisation process to construct an orthonormal basis for the subspace of $\mathbb{R}^4$ generated by
$$\{(1,1,0,1),(1,-2,0,0),(1,0,-1,2)\}.$$

11.9 Consider the inner product space of Example 11.3 with $a=0, b=1$. Find an orthonormal basis for the subspace generated by $\{f_1,f_2\}$ where $f_1 : x \mapsto 1$ and $f_2 : x \mapsto x$.

11.10 Let V be the complex inner product space of Exercise 11.7. For every $M \in V$ let $f_M : V \to V$ be given by $f_M(A) = MA$. Show that the adjoint of f_M is $f_{M^\star}$.

11.11 Let V be a finite-dimensional inner product space. If $f : V \to V$ is linear, prove that f is self-adjoint if and only if $\langle f(x)\mid x\rangle$ is real for all $x \in V$.

11.12 Let V and W be finite-dimensional inner product spaces over the same field and suppose that $\dim V = \dim W$. If $f : V \to W$ is linear, prove that the following are equivalent :

Inner product spaces

(1) f is an inner product isomorphism;
(2) f preserves inner products $[\langle f(x) | f(y) \rangle = \langle x | y \rangle]$;
(3) f preserves norms $[\|f(x)\| = \|x\|]$;
(4) f preserves distances $[d(f(x), f(y)) = d(x, y)]$.

11.13 Let V and W be subspaces of a finite-dimensional inner product space. Prove that

$$(V \cap W)^\perp = V^\perp + W^\perp, \quad (V + W)^\perp = V^\perp \cap W^\perp.$$

11.14 If V is a finite-dimensional inner product space and $f : V \to V$ is a linear transformation, prove that

$$\operatorname{Im} f^* = (\operatorname{Ker} f)^\perp, \quad \operatorname{Ker} f^* = (\operatorname{Im} f)^\perp.$$

11.15 Let V be a complex inner product space. For all $x, y \in V$ let $f_{x,y} : V \to V$ be given by

$$f_{x,y}(z) = \langle z | y \rangle x.$$

Prove that $f_{x,y}$ is linear and that

(a) $(\forall x, y, z \in V) \quad f_{x,y} \circ f_{y,z} = \|y\|^2 f_{x,z}$;
(b) the adjoint of $f_{x,y}$ is $f_{y,x}$.

11.16 If V is an inner product space then a linear transformation $f : V \to V$ is said to be *normal* if $f \circ f^* = f^* \circ f$. Prove that if $x \neq 0$ and $y \neq 0$ then the linear transformation $f_{x,y}$ of Exercise 11.5 is

(a) normal if and only if there exists $\lambda \in \mathbb{C}$ such that $x = \lambda y$;
(b) self-adjoint if and only if there exists $\lambda \in \mathbb{R}$ such that $x = \lambda y$.

11.17 Let W be a finite-dimensional subspace of a real inner product space V. Given $x \in V$, let $x = x_1 + x_2$ where $x_1 \in W$ and $x_2 \in W^\perp$. Show that $\|x\|^2 = \|x_1\|^2 + \|x_2\|^2$. Show also that if $y \in W$ then

$$\|x - y\|^2 = \|x - x_1\|^2 + \|x_1 - y\|^2.$$

Deduce that x_1 is the element of W that is 'nearest' to x.
[*Hint*. Observe that if $y \in W$ then $\|x - y\|^2 \geq \|x - x_1\|^2$.]
If $\{e_1, \ldots, e_n\}$ is an orthonormal subset of V prove that

$$\left\| x - \sum_{i=1}^{n} \lambda_i e_i \right\|$$

attains its smallest value when $\lambda_i = \langle x | e_i \rangle$ for $i = 1, \ldots, n$.

11.18 Let V be the real vector space of continuous functions $f : [-\pi, \pi] \to \mathbb{R}$. Show that V is an inner product space with respect to

$$\langle f | g \rangle = \frac{1}{\pi} \int_{-\pi}^{\pi} f(x) g(x) \, dx.$$

Prove that

$$B = \{x \mapsto \tfrac{1}{\sqrt{2}}, \ x \mapsto \sin nx, \ x \mapsto \cos nx \ ; \ n = 1, 2, 3, \ldots\}$$

is an orthonormal subset of V. If W_n is the subspace of dimension $2n+1$ generated by

$$\{x \mapsto \tfrac{1}{\sqrt{2}}, \ x \mapsto \sin kx, \ x \mapsto \cos kx \ ; \ k = 1, \ldots, n\},$$

prove that the element f_n of W_n that is nearest a given element f of V is given by

$$f_n(x) = \tfrac{1}{2} a_0 + \sum_{k=1}^{n} (a_k \cos kx + b_k \sin kx)$$

where

$$a_0 = \frac{1}{\pi} \int_{-\pi}^{\pi} f(x) \, dx, \quad a_k = \frac{1}{\pi} \int_{-\pi}^{\pi} f(x) \cos kx \, dx,$$

$$b_k = \frac{1}{\pi} \int_{-\pi}^{\pi} f(x) \sin kx \, dx.$$

12 Injective modules

Our principal concern now will be the relationship between modules and rings. Our motivation lies in questions such as : do rings R exist such that every R-module is projective? If so, what do they look like? But before considering such questions, we deal with the notion that is dual to that of a projective module.

Definition An R-module M is said to be *injective* if, for every short exact sequence

$$0 \longrightarrow A' \xrightarrow{f} A \xrightarrow{g} A'' \longrightarrow 0$$

of R-modules and R-morphisms, the induced $\mathbb{Z}$-morphism

$$\mathrm{Mor}_R(A', M) \xleftarrow{f^*} \mathrm{Mor}_R(A, M)$$

is surjective.

The following characterisation of injective modules is dual to that for projective modules.

Theorem 12.1 *An R-module I is injective if and only if every diagram of the form*

$$0 \longrightarrow A' \xrightarrow{f} A \quad (exact)$$
$$\vartheta' \downarrow $$
$$I $$

can be extended to a commutative diagram

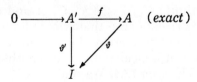

Proof I is injective if and only if, for every monomorphism f the induced morphism f^* is surjective; in other words, if and only if for every

$\vartheta' \in \mathrm{Mor}_R(A', I)$ there exists $\vartheta \in \mathrm{Mor}_R(A, I)$ such that $\vartheta' = f^*(\vartheta) = \vartheta \circ f$. ◊

We shall now find some examples of injective modules and derive a characterisation that is dual to Theorem 8.8. We warn the reader, however, that the proofs involved are considerably harder. We first establish the following useful result.

Theorem 12.2 *An R-module I is injective if and only if, for every left ideal L of R and every R-morphism $f : L \to I$, there is an R-morphism $\vartheta : R \to I$ such that the diagram*

is commutative, where ι_L is the natural inclusion.

Proof Necessity follows immediately from the definition of an injective module. To establish sufficiency, suppose that we have the situation

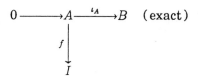

where, for convenience and without loss of generality, we assume that A is a submodule of B with ι_A the natural inclusion. Let F be the set of all pairs (A', f') where A' is a submodule of B with $A \subseteq A' \subseteq B$ and $f' : A' \to I$ extends f, in the sense that $f' \circ j = f$ where $j : A \to A'$ is the natural inclusion. It is clear that $(A, f) \in F$, so that $F \neq \emptyset$. Now order F by writing $(A', f') \leq (A'', f'')$ if and only if $A' \subseteq A''$ and f'' extends f'. It is readily seen that F is inductively ordered, so we can apply Zorn's axiom and thereby choose a maximal element (A_0, f_0) in F. Now if $A_0 = B$ it is clear that there is nothing more to prove : I will be injective by Theorem 12.1. We shall now show that in fact $A_0 = B$.

Suppose, by way of obtaining a contradiction, that $A_0 \neq B$ and let $x \in B \setminus A_0$. If we let $L = \{r \in R \ ; \ rx \in A_0\}$ then it is readily seen that L is a left ideal of R. Now let $h : L \to I$ be given by $h(r) = f_0(rx)$. Then clearly h is an R-morphism and, by the hypothesis, there is an

Injective modules

R-morphism $t : R \to I$ that extends h. Let $A_1 = A_0 + Rx$ and define $f_1 : A_1 \to I$ by the prescription

$$f_1(a_0 + rx) = f_0(a_0) + rt(1).$$

[Note that f is well defined; for if $a_0 + rx = a_0' + r'x$ then $(r - r')x = a_0' - a_0 \in A_0$ and so, applying f_0,

$$f_0(a_0' - a_0) = f_0[(r - r')x] = h(r - r') = t(r - r') = (r - r')t(1),$$

whence $f_0(a_0') + r't(1) = f_0(a_0) + rt(1)$.] Now it is clear that f_1 is an R-morphism and, on taking $r = 0$ in its definition, we have $f_1(a_0) = f_0(a_0)$ for every $a_0 \in A_0$, so that f_1 extends f_0. The pair (A_1, f_1) therefore belongs to F, contradicting the maximality of (A_0, f_0). This contradiction shows that $A_0 = B$ as asserted. ◊

Corollary *An R-module I is injective if and only if, for every left ideal L of R, the induced $\mathbb{Z}$-morphism*

$$\mathrm{Mor}_R(L, I) \xleftarrow{i^*} \mathrm{Mor}_R(R, I)$$

is surjective. ◊

We shall now determine precisely when a $\mathbb{Z}$-module is injective.

Definition An additive abelian group A is said to be *divisible* if, for every positive integer m, we have $mA = A$; in other words, if for every $x \in A$ and every positive integer m there exists $x' \in A$ such that $x = mx'$.

Example 12.1 The group $\mathbb{Q}/\mathbb{Z}$ is divisible. In fact, given any positive integer m we have $q + \mathbb{Z} = m(\frac{q}{m} + \mathbb{Z})$.

Theorem 12.3 *A $\mathbb{Z}$-module is injective if and only if it is divisible.*

Proof $\Rightarrow$: Suppose that the $\mathbb{Z}$-module A is divisible and let L be a left ideal of $\mathbb{Z}$ with $f : L \to A$ a $\mathbb{Z}$-morphism. Then L is necessarily of the form $m\mathbb{Z}$ for some non-negative $m \in \mathbb{Z}$. If $m \neq 0$ then, since A is divisible, there exists $a \in A$ such that $f(m) = ma$; and if $m = 0$ (in which case $L = \{0\}$) we have $f(m) = f(0) = 0 = m0$. Thus, in either case, there exists $a \in A$ such that $f(m) = ma$. For every $x \in L$ we then have, for some $z \in \mathbb{Z}$,

$$f(x) = f(mz) = zf(m) = zma = xa.$$

The $\mathbb{Z}$-morphism $t_a : \mathbb{Z} \to A$ given by $t_a(r) = ra$ therefore extends f and so A is injective by Theorem 12.2.

⇐ : Suppose that the $\mathbb{Z}$-module A is injective. Let $a \in A$ and let m be a positive integer. Define $f : m\mathbb{Z} \to A$ by the prescription $f(mz) = za$. Clearly, f is a $\mathbb{Z}$-morphism. By Theorem 12.2, there exists $b \in A$ such that $f(x) = xb$ for every $x \in m\mathbb{Z}$. In particular (taking $z = 1$ in the above) we have $a = f(m) = mb$. Thus we see that A is divisible. ◊

It is immediate from Theorem 12.3 and Example 12.1 that the $\mathbb{Z}$-module $\mathbb{Q}/\mathbb{Z}$ is injective. This module will play an important rôle in what follows and we note the following properties that it enjoys :

(α) *if C is a non-zero cyclic group then* $\mathrm{Mor}_\mathbb{Z}(C, \mathbb{Q}/\mathbb{Z}) \neq 0$.

In fact, suppose that C is generated by $\{x\}$. We can define a non-zero $\mathbb{Z}$-morphism $f : C \to \mathbb{Q}/\mathbb{Z}$ as follows : if C is infinite then we let $f(x)$ be any non-zero element of $\mathbb{Q}/\mathbb{Z}$, and if C is finite, say with x of order m, then we define $f(x) = \frac{1}{m} + \mathbb{Z}$.

(β) *if G is a non-zero abelian group then for every non-zero $x \in G$ there is a $\mathbb{Z}$-morphism $g : G \to \mathbb{Q}/\mathbb{Z}$ such that $g(x) \neq 0$.*

In fact, if C is the subgroup generated by $\{x\}$ then, as we have just seen, there is a $\mathbb{Z}$-morphism $f : C \to \mathbb{Q}/\mathbb{Z}$ with $f(x) \neq 0$; and since $\mathbb{Q}/\mathbb{Z}$ is injective f can be extended to a $\mathbb{Z}$-morphism $g : G \to \mathbb{Q}/\mathbb{Z}$.

(γ) *if G is a non-zero abelian group and if H is a proper subgroup of G then for every $a \in G \setminus H$ there is a $\mathbb{Z}$-morphism $h : G \to \mathbb{Q}/\mathbb{Z}$ such that $h^\to(H) = \{0\}$ and $h(a) \neq 0$.*

In fact, by (β) there is a $\mathbb{Z}$-morphism $g : G/H \to \mathbb{Q}/\mathbb{Z}$ such that $g(a + H) \neq 0 + \mathbb{Z}$ so it suffices to consider $h = g \circ \natural_H$.

We have seen earlier (Theorem 8.7) that for every R-module M there is a free, hence projective, R-module P and an R-epimorphism $f : P \to M$. Our aim now is to establish the dual of this, namely that there is an injective R-module I and an R-monomorphism $g : M \to I$; in other words, that every R-module M can be 'embedded' in an injective R-module I.

For this purpose, we require the assistance of a particular right R-module associated with M. Given an R-module M, consider its *character group*

$$M^+ = \mathrm{Mor}_\mathbb{Z}(M, \mathbb{Q}/\mathbb{Z}).$$

We can make M^+ into a *right* R-module by defining an action $M^+ \times R \to M^+$ as follows : with every $f \in M^+$ and every $r \in R$ associate the mapping $fr : M \to \mathbb{Q}/\mathbb{Z}$ given by the prescription $(fr)(x) = f(rx)$. It is readily seen that every fr so defined is a $\mathbb{Z}$-morphism and that M^+ thus has the structure of a right R-module. We call this right R-module the *character module* of M and denote it also by M^+.

Injective modules

We can, of course, repeat the above process on M^+ and form its character module, which we denote by M^{++}. Note that M^{++} so constructed is a *left* R-module, the action being $(r, f^+) \mapsto ef^+$ where $(rf^+)(g) = f^+(gr)$.

Consider now the mapping $\iota_M : M \to M^{++}$ defined as follows: for every $x \in M$ let $\iota_M(x) : M^+ \to \mathbb{Q}/\mathbb{Z}$ be given by

$$[\iota_M(x)](f) = f(x).$$

It is clear that ι_M so defined is a $\mathbb{Z}$-morphism. That it is an R-morphism follows from the identities

$$[r\iota_M(x)])(f) = [\iota_M(x)](fr) = (fr)(x) = f(rx) = [\iota_M(rx)](f).$$

We now show that ι_M is in fact an R-monomorphism. Suppose, by way of obtaining a contradiction, that $x \in M \setminus \{0\}$ is such that $\iota_M(x)$ is the zero of M^{++}. Then for every $f \in M^+$ we have $f(x) = [\iota_M(x)](f) = 0$. But, by the observation (β) above, there exists $f \in M^+$ such that $f(x) \neq 0$. This contradiction shows that $\text{Ker}\,\iota_M = \{0\}$ and so ι_M is a monomorphism which we shall call the *canonical R-monomorphism* from M to M^{++}.

Using the above notion of character module we can establish the following important result, which serves the purpose of providing an abundance of examples of injective modules.

Theorem 12.4 *If F is a free R-module then F^+ is injective.*

Proof Suppose that we have a diagram

$$0 \longrightarrow A' \xrightarrow{i} A \quad (\text{exact})$$
$$\downarrow\vartheta$$
$$F^+$$

Using the fact that $\mathbb{Q}/\mathbb{Z}$ is injective, we can construct the diagram

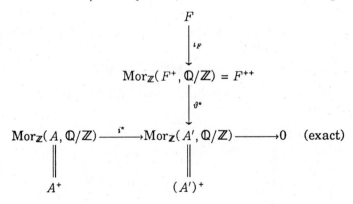

in which ϑ^* is the R-morphism induced by ϑ, and i^* is that induced by i. Now since F is free it is projective and so there is an R-morphism $\zeta : F \to A^+$ in the above diagram such that $i^* \circ \zeta = \vartheta^* \circ \iota_F$. We now let $\zeta^* : A^{++} \to F^+$ be the R-morphism induced by ζ and consider the diagram

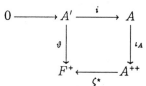

Our objective now is to show that this diagram is commutative; the R-morphism $\zeta^* \circ \iota_A$ will then be a completing morphism for the original diagram and F^+ will be injective as required.

To show that this latter diagram is commutative, we have to show that, for every $x \in A'$,

$$(\zeta^* \circ \iota_A \circ i)(x) = \vartheta(x).$$

Now each side of this equality is an element of F^+, so let t be an element of F. Then we have

$$\begin{aligned}
\left[(\zeta^* \circ \iota_A \circ i)(x)\right](t) &= [(\iota_A \circ i)(x) \circ \zeta](t) && [\text{definition of } \zeta^*]\\
&= [(\iota_A \circ i)(x)][\zeta(t)]\\
&= [\zeta(t)][i(x)] && [\text{definition of } \iota_A]\\
&= [\zeta(t) \circ i](x)\\
&= [(i^* \circ \zeta)(t)](x)\\
&= [(\vartheta^* \circ \iota_F)(t)](x)\\
&= [\iota_F(t) \circ \vartheta](x) && [\text{definition of } \vartheta^*]\\
&= [\iota_F(t)][\vartheta(x)]\\
&= [\vartheta(x)](t) && [\text{definition of } \iota_F]
\end{aligned}$$

from which the required equality follows immediately. ◊

Corollary *Every R-module can be embedded in an injective R-module.*

Proof We begin by observing that by Theorem 8.7 (for *right* modules) there is an exact sequence

$$F \xrightarrow{\pi} M^+ \longrightarrow 0$$

where F is a free right R-module. Then by Theorem 8.3 there is the induced exact sequence

$$\text{Mor}_{\mathbb{Z}}(F, \mathbb{Q}/\mathbb{Z}) \xleftarrow{\pi^*} \text{Mor}_{\mathbb{Z}}(M^+, \mathbb{Q}/\mathbb{Z}) \longleftarrow 0.$$
$$\| \qquad\qquad\qquad\qquad \|$$
$$F^+ \qquad\qquad\qquad\qquad M^{++}$$

Injective modules

Thus $\pi^* \circ \iota_M : M \to F^+$ is a monomorphism. Since the (left) R-module F^+ is injective, the result follows. ◊

We can now establish the analogue of Theorem 8.8.

Theorem 12.5 *For an R-module I the following are equivalent:*
(1) *I is injective;*
(2) *every exact sequence $0 \longrightarrow X \longrightarrow I$ splits;*
(3) *if $\alpha : I \to X$ is a monomorphism then $\operatorname{Im} \alpha$ is a direct summand of X.*

Proof (1) $\Rightarrow$ (2) : If I is injective then for every diagram

$$\begin{array}{ccccc} 0 & \longrightarrow & I & \stackrel{i}{\longrightarrow} & X \quad (\text{exact}) \\ & & \downarrow \text{id}_I & & \\ & & I & & \end{array}$$

there is an R-morphism $f : X \to I$ such that $f \circ i = \text{id}_I$. Such an R-morphism f is then a splitting morphism.

(2) $\Rightarrow$ (3) : Let X be an R-module and let $\alpha : I \to X$ be an R-monomorphism. By the hypothesis, the short exact sequence

$$0 \longrightarrow I \stackrel{\alpha}{\longrightarrow} X \stackrel{\natural}{\longrightarrow} X/\operatorname{Im} \alpha \longrightarrow 0$$

splits. It follows by Theorem 6.11 that $\operatorname{Im} \alpha$ is a direct summand of X.

(3) $\Rightarrow$ (1) : By Theorem 12.4 there is an injective module Q and a monomorphism $\alpha : I \to Q$. By the hypothesis, $\operatorname{Im} \alpha$ is a direct summand of Q, say $Q = \operatorname{Im} \alpha \oplus J$. We show that $\operatorname{Im} \alpha$ is injective. For this purpose, consider the diagram

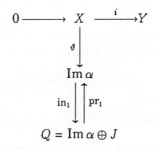

Since Q is injective, there is an R-morphism $f : Y \to Q$ such that $f \circ i = \text{in}_1 \circ \vartheta$. Then $\text{pr}_1 \circ f : Y \to \operatorname{Im} \alpha$ is such that

$$\text{pr}_1 \circ f \circ i = \text{pr}_1 \circ \text{in}_1 \circ \vartheta = \text{id}_I \circ \vartheta = \vartheta.$$

This we see that $\operatorname{Im}\alpha$ is injective. It now follows by Theorem 12.1 and the fact that $\operatorname{Im}\alpha \simeq I$ that I is also injective. ◊

We end our discussion of injective modules with the following connection between character modules and short exact sequences.

Theorem 12.6 *A sequence of R-modules and R-morphisms*

$$0 \longrightarrow A \xrightarrow{\alpha} B \xrightarrow{\beta} C \longrightarrow 0$$

is short exact if and only if the corresponding sequence of character modules

$$0 \longleftarrow A^+ \xleftarrow{\alpha^*} B^+ \xleftarrow{\beta^*} C^+ \longleftarrow 0$$

is short exact.

Proof The necessity follows from the fact that $\mathbb{Q}/\mathbb{Z}$ is injective. As for sufficiency, it is enough to show that the sequence

$$M \xrightarrow{f} N \xrightarrow{g} P$$

is exact whenever the corresponding sequence of character modules

$$M^+ \xleftarrow{f^*} N^+ \xleftarrow{g^*} P^+$$

is exact. Suppose then that $\operatorname{Im} g^* = \operatorname{Ker} f^*$. Given $m \in M$, suppose that $f(m) \notin \operatorname{Ker} g$, so that $(g \circ f)(m) \neq 0$. By the observation (β) following Theorem 12.3, there exists $\alpha \in P^+$ such that $\alpha[(g \circ f)(m)] \neq 0$. But we have $\alpha \circ g \circ f = f^*[g^*(\alpha)] = 0$ since the character sequence is exact. This contradiction shows that $\operatorname{Im} f \subseteq \operatorname{Ker} g$. To obtain the reverse inclusion suppose, again by way of obtaining a contradiction, that there exists $n \in N$ such that $n \in \operatorname{Ker} g$ and $n \notin \operatorname{Im} f$. By the observation (γ) following Theorem 12.3, there exists $\beta \in N^+$ such that $\beta^\rightarrow(\operatorname{Im} f) = \{0\}$ and $\beta(n) \neq 0$. Now $\beta^\rightarrow(\operatorname{Im} f) = \operatorname{Im}(\beta \circ f) = \operatorname{Im} f^*(\beta)$, so we obtain $f^*(\beta) = 0$ whence $\beta \in \operatorname{Ker} f^* = \operatorname{Im} g^*$ and consequently $\beta = g^*(t) = t \circ g$ for some $t \in P^+$. In particular, $\beta(n) = t[g(n)] = 0$ since $n \in \operatorname{Ker} g$. This contradiction yields the reverse inclusion $\operatorname{Ker} g \subseteq \operatorname{Im} f$. ◊

EXERCISES

12.1 Consider the diagram of R-modules and R-morphisms

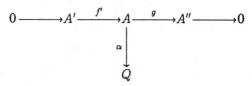

in which the row is exact and α is a monomorphism. Show that this diagram can be extended to a commutative diagram

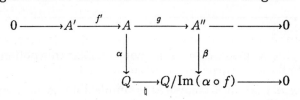

Use this result to prove that an R-module P is projective if and only if, given a diagram

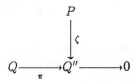

in which Q is injective and the row is exact, there is an R-morphism $\vartheta : P \to Q$ such that $\pi \circ \vartheta = \zeta$.

12.2 If $(M_i)_{i \in I}$ is a family of injective modules prove that $\underset{i \in I}{\times} M_i$ is injective.

12.3 Let R be a commutative integral domain and let Q be its field of quotients. Prove that Q is an injective R-module.

[*Hint*. If L is an ideal of R let $f : L \to Q$ be an R-morphism. Show that if $a, b \in L \setminus \{0\}$ then $f(a)/a = f(b)/b$. Deduce that the assignment $r \mapsto rf(a)/a$ extends f.]

12.4 An R-module N is an *extension* of an R-module M if there is a monomorphism $f : M \to N$. Such an extension is an *essential* extension if every non-zero submodule of N has a non-zero intersection with M. If N is an essential extension of M and if I is an injective module containing M, prove that id_M can be extended to a monomorphism from N to I.

12.5 Prove that an R-module M is injective if and only if it has no essential extension.

[*Hint.* $\Rightarrow$: Suppose that N is an essential extension of M. Then, by Theorem 12.5(3), M is a direct summand of N.

$\Leftarrow$: Let I be an injective module that contains M and let M' be a submodule of I that is maximal with respect to $M \cap M' = \{0\}$. Prove that I/M' is an essential extension of $(M \oplus M')/M \simeq M$.]

13 Simple and semisimple modules

In Chapter 5, in connection with Jordan-Hölder towers, we met the notion of a *simple* module. We shall now consider this notion more closely. As we shall see, it turns out to be one of the basic 'building blocks'.

We recall that an R-module M is simple if it has no submodules other than M itself and the zero submodule. Equivalently, by Theorem 5.5, M is simple if and only if is generated by each of its non-zero elements.

Example 13.1 Consider a unitary ring R as an R-module. Then every minimal left ideal of R, when such exists, is a simple submodule of R.

- Note that not all modules contain simple submodules. For example, the $\mathbb{Z}$-module $\mathbb{Z}$ has no simple submodules since the ring $\mathbb{Z}$ has no minimal (left) ideals.

Example 13.2 The R-module R is simple if and only if R is a division ring. In fact, if R is simple then $R = Rx$ for every $x \neq 0$. In particular, $1_R = x'x$ for some x', i.e. x has a left inverse x'. Similarly, x' has a left inverse x''. Then $x'' = x''1_R = x''x'x = 1_Rx = x$ and consequently x' is a two-sided inverse of x. Thus R is a division ring. Conversely, if R is a division ring and L is a left ideal of R then for every $x \in L \setminus \{0\}$ we have $1_R = x^{-1}x \in L$ whence $L = R$. Thus the R-module R is simple.

Example 13.3 If L is a left ideal of R then R/L is simple if and only if L is maximal. Indeed, all simple modules arise in this way. For, if M is simple then $M = Rx$ for every $x \neq 0$ in M, and $f : R \to M$ given by $f(r) = rx$ is an epimorphism, whence $M \simeq R/\text{Ker} f$. The fact that M is simple then implies, by the correspondence theorem, that $\text{Ker} f$ is a maximal left ideal of R.

A further example is provided by the following result.

Theorem 13.1 *Let V a non-zero finite-dimensional module over a division ring D. Then V is a simple $\text{End}_D V$-module.*

Proof Define an action $\text{End}_D V \times V \to V$ by $(f, v) \mapsto f \cdot v = f(v)$. Then it is readily seen that V is an $\text{End}_D V$-module.

Now let x be any non-zero element of V. Since D is a division ring the set $\{x\}$ is linearly independent and so can be extended to a basis $\{x, x_1, \ldots, x_{n-1}\}$ of V. Given any $y \in V$, consider the mapping $f_y : V \to V$ given by
$$f(d_0 x + d_1 x_1 + \ldots + d_{n-1} x_{n-1}) = d_0 y.$$

Simple and semisimple modules

Clearly, f_y is a D-morphism. Moreover, $y = f_y(x) = f_y \cdot x$ and so it follows that $V = \mathrm{End}_D V \cdot x$ and therefore, by Theorem 5.5, V is a simple $\mathrm{End}_D V$-module. ◊

By way of considering simple modules as building blocks, we now consider modules that are sums of simple modules.

Theorem 13.2 *For an R-module M the following are equivalent* :
  (1) *M is a sum of a family of simple modules;*
  (2) *M is a direct sum of a family of simple modules;*
  (3) *every submodule of M is a direct summand.*

Proof (1) $\Rightarrow$ (2),(3) : We shall prove that if $(M_i)_{i \in I}$ is a family of simple submodules of M such that $M = \sum_{i \in I} M_i$ then for every submodule N of M there is a subset J of I such that $M = N \oplus \bigoplus_{i \in J} M_i$.

If $N = M$ then clearly $J = \emptyset$ suffices. Suppose then that $N \subset M$, so that for some $k \in I$ we have $M_k \not\subset N$. Since M_k is simple we deduce that $N \cap M_k = \{0\}$ and so the sum $N + M_k$ is direct. Now let C be the set of those subsets H of I for which the sum $N + \sum_{i \in H} M_i$ is direct. We have just shown that $C \neq \emptyset$. We now show that C is inductively ordered. For this purpose let T be a totally ordered subset of C and let $K^\star = \bigcup\{K \ ; \ K \in T\}$; we claim that $K^\star \in C$. To see this, we observe that if $x \in \sum_{i \in K^\star} M_i$ then $x = m_{i_1} + \ldots + m_{i_n}$. Since each i_j belongs to some subset I_J of T, and since T is totally ordered, all the sets $I_1, \ldots, I_n$ are contained in one of them, say I_p. Then, since $I_p \in C$,

$$N \cap \sum_{i \in K^\star} M_i \subseteq N \cap \sum_{i \in I_p} M_i = \{0\}$$

whence the sum $N + \sum_{i \in K^\star} M_i$ is direct. This shows that $K^\star \in C$, and so C is inductively ordered. It follows by Zorn's axiom that C has maximal elements. Let J be a maximal element of C. We show that $N \oplus \bigoplus_{i \in J} M_i = M$. Suppose, by way of obtaining a contradiction, that $N \oplus \bigoplus_{i \in J} M_i \subset M$. Then for some $j \in I$ we have $M_j \not\subset N \oplus \bigoplus_{i \in J} M_i$ and, since M_i is simple, we deduce that $M_j \cap \left(N \oplus \bigoplus_{i \in J} M_i\right) = \{0\}$, whence the sum $M_j + N \oplus \bigoplus_{i \in J} M_i$ is direct. But then $J \cup \{j\}$ belongs to C, and this contradicts the maximality of J. Hence we have $M = N \oplus \bigoplus_{i \in J} M_i$.

(2) $\Rightarrow$ (1) : This is clear.

(3) ⇒ (1) : Suppose now that (3) holds. Then we observe that

(a) *if N is a submodule of M then every submodule of N is a direct summand of N.*

In fact, let P be a submodule of N. Since P is also a submodule of M there is, by (3), a submodule Q of M such that $M = P \oplus Q$. Then

$$N = N \cap M = N \cap (P \oplus Q) = P \oplus (N \cap Q),$$

the last equality following by the modular law.

(b) *every non-zero submodule of M has a simple submodule.*

In fact, if N is a non-zero submodule of M and x is a non-zero element of N, consider the submodule Rx. We show that Rx has a simple submodule, whence so does N. For this purpose, let $\mathcal{E}$ be the set of submodules of Rx that do not contain x. Clearly, the zero submodule belongs to $\mathcal{E}$ and so $\mathcal{E} \neq \emptyset$. Moreover, $\mathcal{E}$ is inductively ordered; for if $\mathcal{F}$ is a totally ordered subset of $\mathcal{E}$ then the sum of all the submodules of Rx contained in $\mathcal{F}$ is also a submodule which clearly does not contain x. We can therefore apply Zorn's axiom to obtain a maximal element L of $\mathcal{E}$. By (a) we deduce the existence of a submodule P of Rx such that $Rx = L \oplus P$. We claim that P is simple. In fact, L is a maximal submodule of Rx and so Rx/L is simple; and $Rx/L = (L \oplus P)/L \simeq P$.

To conclude the proof of (3) ⇒ (1), let Σ be the sum of all the simple submodules of M. Then $M = \Sigma \oplus Q$ for some submodule Q of M. If $Q \neq \{0\}$ then by (b) it contains simple submodules; and this is impossible since all the simple submodules of M are contained in Σ with $\Sigma \cap Q = \{0\}$. We deduce that $Q = \{0\}$ whence $M = \Sigma$. ◇

Definition An R-module M that satisfies any of the equivalent properties of Theorem 13.2 is said to be *semisimple*.

Theorem 13.3 *Every submodule and every quotient module of a semisimple module is semisimple.*

Proof Let M be semisimple, let N be a submodule of M and let P be a submodule of N. By Theorem 13.2 there is a submodule Q of M such that $M = P \oplus Q$. Then $N = N \cap M = N \cap (P \oplus Q) = P \oplus (N \cap Q)$ and so P is a direct summand of N. By Theorem 13.2 again, N is therefore semisimple.

Suppose now that M is the sum of the family $(M_i)_{i \in I}$ of simple submodules and let N be a submodule of M. Consider the natural epimorphism $\natural_N : M \to M/N$. By Theorem 5.6(1), for each index i, the submodule $\vec{\natural_N}(M_i)$ of M/N is either zero or is simple. Let J be the set of indices

Simple and semisimple modules   175

i for which $\natural_{\vec{N}}(M_i)$ is non-zero. Then

$$M/N = \operatorname{Im} \natural_N = \natural_{\vec{N}}\left(\sum_{i \in I} M_i\right) = \sum_{i \in J} \natural_{\vec{N}}(M_i)$$

and so M/N is semisimple by Theorem 13.2. ◊

Definition A unitary ring R will be called *semisimple* if R, considered as an R-module, is semisimple.

- The reader should note that the term 'semisimple ring' is used in differing senses in the literature.

- Note that here we do not use the term *left semisimple* for a ring that is a semisimple module. The reason for this is that a left semisimple ring is also right semisimple; see Exercise 13.12.

We are now in a position to answer the question concerning which rings R are such that every R-module is projective. In fact, we obtain a lot more :

Theorem 13.4 *For a unitary ring R the following are equivalent* :

(·1) *R is semisimple;*
(2) *every non-zero R-module is semisimple;*
(3) *every R-module is projective;*
(4) *every R-module is injective.*

Proof (1) $\Rightarrow$ (2) : If R is semisimple then it is clear that every free R-module, being a direct sum of copies of R (Corollary to Theorem 7.6), is semisimple. Since every module is isomorphic to a quotient module of a free module (Theorem 8.7), it follows from Theorem 13.3 that every non-zero R-module M is also semisimple.

(2) $\Rightarrow$ (1) : This is clear.

(2) $\Rightarrow$ (3) : Let M be an R-module and consider a short exact sequence

$$0 \longrightarrow A \longrightarrow B \longrightarrow M \longrightarrow 0.$$

By (2), B is semisimple and so every submodule of B is a direct summand. The above sequence therefore splits and consequently M is projective.

(3) $\Rightarrow$ (2) : Let N be a submodule of M. By (3), M/N is projective and so the short exact sequence

$$0 \longrightarrow N \longrightarrow M \longrightarrow M/N \longrightarrow 0$$

splits. Then N is a direct summand of M and consequently M is semi-simple.

(2) $\Leftrightarrow$ (4) : This is dual to the proof of (2) $\Leftrightarrow$ (3) by virtue of Theorem 12.5. $\Diamond$

As to the structure of semisimple rings, we shall use the following result.

Theorem 13.5 *If M is a semisimple module then the following statements are equivalent*:

(1) M is both noetherian and artinian;
(2) M is finitely generated.

Proof (1) $\Rightarrow$ (2) : If (1) holds then, by Theorem 5.9, M has a Jordan-Hölder tower of submodules. If $M = \bigoplus_{i \in I} N_i$ where each N_i is simple then necessarily I is finite, say $I = \{1, \ldots, n\}$. Since $N_i = Rx_i$ for some non-zero x_i it follows that $M = \bigoplus_{i=1}^{n} Rx_i$ whence (2) follows.

(2) $\Rightarrow$ (1) : Let $M = \bigoplus_{i \in I} N_i$ where each N_i is simple and suppose that $M = \sum_{j=1}^{n} Rx_j$. For $j = 1, \ldots, n$ there is a finite subset I_j of I such that $x \in \bigoplus_{i \in I_j} N_i$. Thus if $I^{\star} = \bigcup_{j=1}^{n} I_j$ we have $M \subseteq \bigoplus_{i \in I^{\star}} N_i \subseteq M$ and from the resulting equality we see that M is a finite direct sum of simple submodules, say $M = \bigoplus_{i=1}^{k} N_i$. Thus M has a Jordan-Hölder tower

$$\{0\} \subset N_1 \subset N_1 \oplus N_2 \subset \ldots \subset N_1 \oplus \cdots \oplus N_k = M,$$

whence (1) follows. $\Diamond$

Corollary *A unitary ring R is semisimple if and only if it is a finite direct sum of minimal left ideals.*

Proof It suffices to note that R is generated by $\{1_R\}$. $\Diamond$

We thus see that if R is semisimple then $R = \bigoplus_{i=1}^{n} L_i$ where each L_i is a minimal left ideal. Let us now group together those minimal left ideals that are isomorphic. To be more precise, suppose that $L_1, \ldots, L_m$ are representatives of the isomorphism classes of the minimal left ideals L_i, in the sense that each L_i is isomorphic to one and only one of $L_1, \ldots, L_m$. For $t = 1, \ldots, m$ define

$$R_t = \bigoplus_{L_j \simeq L_t} L_j.$$

Then we can write
$$R = \bigoplus_{i=1}^{n} L_i = \bigoplus_{t=1}^{m} R_t.$$
In order to investigate these R_t we use the following observation.

Theorem 13.6 *If L is a minimal left ideal of R and M is a simple R-module then either $L \simeq M$ or $LM = \{0\}$.*

Proof LM is a submodule of M so either $LM = M$ or $LM = \{0\}$. If $LM = M$ let $x \in M$ be such that $Lx \neq \{0\}$. Then since M is simple we have $Lx = M$ and the mapping $\vartheta : L \to M$ given by $\vartheta(a) = ax$ is an isomorphism by Theorem 5.6. ◊

Corollary *If L_i, L_j are minimal left ideals of R that are not isomorphic then $L_i L_j = \{0\}$.* ◊

It follows immediately from this that, with the above notation, we have
$$R_i R_j = \{0\} \quad (i \neq j).$$
Consequently, each R_j being a left ideal, we see that
$$R_j = R_j 1_R \subseteq R_j R = R_j R_j \subseteq R_j$$
whence $R_j R = R_j$ and therefore each R_j is also a right ideal of R.

Now if R^+ denotes the additive group of R, and R_i^+ that of R_i, we have $R^+ = \bigoplus_{t=1}^{m} R_i^+$. Since every $x \in R^+$ can be written uniquely in the form $x = \sum_{t=1}^{m} x_t$ where $x_t \in R_t^+$ we deduce from
$$xy = \left(\sum_{t=1}^{m} x_i\right)\left(\sum_{t=1}^{m} y_i\right) = \sum_{t=1}^{m} x_t y_t + \sum_{i \neq j} x_i y_j = \sum_{t=1}^{m} x_t y_t$$
that the mapping $f : \bigoplus_{t=1}^{m} R_i \to R$ given by
$$f(x_1, \ldots, x_m) = \sum_{t=1}^{m} x_i$$
is a ring isomorphism. Thus we have established the following result.

Theorem 13.7 *If R is a semisimple ring then R has only finitely many minimal two-sided ideals $R_1, \ldots, R_m$ and $R = \bigoplus_{t=1}^{m} R_t$.* ◊

Corollary *The identity elements $e_1, \ldots, e_m$ of $R_1, \ldots, R_m$ form an orthogonal set of idempotents in the sense that $e_1 + \ldots + e_m = 1_R$ and $e_j e_j = 0$ for $i \neq j$.*

Proof We can write $1_R = e_1 + \ldots + e_m$ where $e_i \in R_i$ for $i = 1, \ldots, m$. For any $x \in R$ write $x = \sum_{i=1}^{m} x_i$ where $x_i \in R_i$ for $i = 1, \ldots, m$. Then for each j we have, since $R_j R_i = \{0\}$ for $j \neq i$,

$$x_j = x_j 1_R = x_j e_1 + \ldots + x_j e_m = x_j e_j$$

and similarly $e_j x_j = x_j$. Thus e_j is the identity element of R_j. ◊

The above results concerning the ideals R_i lead us to consider the following notion.

Definition A ring R is said to be *simple* if it is semisimple and the only two-sided ideals of R are R itself and the zero ideal.

With this terminology we can rephrase Theorem 13.7 as

Theorem 13.7 *Every semisimple ring is a finite direct sum of simple rings.* ◊

Our interest in simple rings is highlighted by the following result.

Theorem 13.8 *Let V be a finite-dimensional module over a division ring D. Then the ring $\mathrm{End}_D V$ is simple.*

Proof We have shown in Theorem 13.1 that V is a simple $\mathrm{End}_D V$-module. Let $\{b_1, \ldots, b_n\}$ be a basis for V over D and consider the mapping $f : \mathrm{End}_D V \to V^n$ given by

$$f(\alpha) = \bigl(\alpha(b_1), \ldots, \alpha(b_n)\bigr).$$

Clearly, f is an $\mathrm{End}_D V$-morphism. Now $\alpha \in \mathrm{Ker}\, f$ if and only if $\alpha(b_i) = 0$ for each basis vector b_i. Hence $\mathrm{Ker}\, f = \{0\}$ and f is injective. That f is also surjective follows from the fact that for any $z = (z_1, \ldots, z_n) \in V^n$ the mapping $\beta_z : V \to V$ given by

$$\beta_z \left(\sum_{i=1}^{n} x_i b_i \right) = \sum_{i=1}^{n} x_i z_i$$

is a D-endomorphism on V with

$$f(\beta_z) = \bigl(\beta_z(b_1), \ldots, \beta_z(b_n)\bigr) = (z_1, \ldots, z_n).$$

Thus f is an $\mathrm{End}_D V$-isomorphism, and consequently $\mathrm{End}_D V$ is a semisimple $\mathrm{End}_D V$-module, i.e. the ring $\mathrm{End}_D V$ is semisimple. To show that

it is simple, it therefore suffices to show that it has no two-sided ideals other than itself and the zero ideal. This we shall do using matrices, using the fact that if the dimension of V is n then the rings $\text{End}_D V$ and $M = \text{Mat}_{n \times n}(D)$ are isomorphic. Suppose that J is a two-sided ideal of A with $J \neq \{0\}$. Let $A = [a_{ij}] \in J$ be such that $a_{rs} \neq 0$. Let $B = \text{diag}\{a_{rs}^{-1}, \ldots, a_{rs}^{-1}\}$ and let E_{ij} denote the matrix which has 1 in the (i,j)-th position and 0 elsewhere. Then since J is a two-sided ideal we have $E_{rr} B A E_{ss} \in J$ where BA has a 1 in the (r,s)-th position. Observing that $E_{rr} X$ retains the r-th row of X and reduces all other rows to zero, and that $X E_{ss}$ retains the s-th column of X and reduces all other columns to zero, we deduce that $E_{rr} B A E_{ss} = E_{rs}$ and hence that $E_{rs} \in J$ for all r, s. Consequently,
$$I_n = E_{11} + \ldots + E_{nn} \in J$$
and hence J is the whole of M. $\Diamond$

Our objective now is to prove that every simple ring arises as the endomorphism ring of a finite-dimensional module over a division ring. For this purpose, we consider the following notion.

Definition If M is an R-module and S is a subset of $\text{End}_\mathbb{Z} M$ then the *centraliser* of S is the set $C(S)$ of those $\mathbb{Z}$-endomorphisms on M that commute with every $\mathbb{Z}$-endomorphism in S.

It is readily verified that $C(S)$ is a subring of $\text{End}_\mathbb{Z} M$.

Example 13.4 Let M be an R-module and for each $a \in R$ consider the *homothety* $h_a \in \text{End}_\mathbb{Z} M$ given by $h_a(m) = am$. Let $H_R(M))$ be the ring of homotheties on M. Then we have $\varphi \in C(H_R(M))$ if and only if
$$(\forall a \in R)(\forall x \in M) \ \varphi(ax) = (\varphi \circ h_a)(x) = (h_a \circ \varphi)(x) = a\varphi(x),$$
which is the case if and only if $\varphi \in \text{End}_R M$. Thus we have $C(H_R(M)) = \text{End}_R M$.

Definition The *bicentraliser* of S is defined to be $B(S) = C(C(S))$.

For each S we clearly have $S \subseteq B(S)$.

By abuse of language, we now define the centraliser of an R-module M to be $C(M) = C(H_R(M)) = \text{End}_R M$; then $B(M) = B(H_R(M)) = C(\text{End}_R M)$.

Theorem 13.9 $B(R) = H_R(R) \simeq R$.

Proof On the one hand we have $H_R(R) \subseteq B(H_R(R)) = B(R)$.
Suppose now that $\varphi \in B(R) = C(\text{End}_R(R))$. For each $a \in R$ the right translation $\rho_a : R \to R$ defined by $\rho_a(x) = xa$ is in $\text{End}_R(R)$ and so

commutes with φ, so that $\varphi(xa) = \varphi(x)a$. For every $t \in R$ we then have $\varphi(t) = \varphi(1_R t) = \varphi(1_R)t$ and consequently $\varphi = h_{\varphi(1_R)} \in H_R(R)$. Hence $H_R(R) = B(R)$.

That $H_R(R) \simeq R$ follows from the fact that the mapping $h : R \to H_R(R)$ given by $h(r) = h_r$ is a ring isomorphism. $\diamond$

Theorem 13.10 *Let $(M_i)_{i \in I}$ be a family of R-modules each of which is isomorphic to a given R-module M. Then*

$$B\left(\bigoplus_{i \in I} M_i\right) \simeq B(M).$$

Proof For each $k \in I$ let $\alpha_k : M \to M_k$ be an isomorphism and define $\eta_k = \mathrm{in}_k \circ \alpha_k$ and $\varphi_k = \alpha_k^{-1} \circ \mathrm{pr}_k$. Note that we then have $\varphi_k \circ \eta_k = \mathrm{id}$ and that $\sum_{k \in I}(\eta_k \circ \varphi_k) = \sum_{k \in I}(\mathrm{in}_k \circ \mathrm{pr}_k) = \mathrm{id}$.

Let k be a fixed index in I and define $f : B\left(\bigoplus_{i \in I} M_i\right) \to \mathrm{End}_{\mathbf{Z}} M$ by $f(\beta) = \varphi_k \circ \beta \circ \eta_k$. It is clear that f is a ring morphism. Now if $\vartheta \in \mathrm{End}_R M$ we have $\eta_k \circ \vartheta \circ \varphi_k \in \mathrm{End}_R \bigoplus_{i \in I} M_i$ and so commutes with β. Consequently,

$$\begin{aligned} f(\beta) \circ \vartheta &= \varphi_k \circ \beta \circ \eta_k \circ \vartheta \\ &= \varphi_k \circ \beta \circ \eta_k \circ \vartheta \circ \varphi_k \circ \eta_k \\ &= \varphi_k \circ \eta_k \circ \vartheta \circ \varphi_k \circ \beta \circ \eta_k \\ &= \vartheta \circ \varphi_k \circ \beta \circ \eta_k \\ &= \vartheta \circ f(\beta) \end{aligned}$$

and so we have that $f(\beta) \in C(\mathrm{End}_R M) = B(M)$. Thus we see that $\mathrm{Im}\, f \subseteq B(M)$.

In fact, as we shall now show, $\mathrm{Im}\, f = B(M)$. For this purpose, let $\delta \in B(M)$ and consider the mapping $g_\delta : \bigoplus_{i \in I} M_i \to \bigoplus_{i \in I} M_i$ given by

$$g_\delta(x) = \sum_{i \in I}(\eta_i \circ \delta \circ \varphi_i)(x).$$

It is clear that $g_\delta \in \mathrm{End}_{\mathbf{Z}} \bigoplus_{i \in I} M_i$. In fact we claim that $g_\delta \in B\left(\bigoplus_{i \in I} M_i\right)$. To see this, let $\gamma \in \mathrm{End}_R \bigoplus_{i \in I} M_i$. Then for all $i, j \in I$ we have $\varphi_i \circ \gamma \circ \eta_j \in \mathrm{End}_R \bigoplus_{i \in I} M_i$ and so commutes with δ. Thus, for all $x \in M$,

$$\begin{aligned} (g_\delta \circ \gamma \circ \eta_j)(x) &= \sum_{i \in I}(\eta_i \circ \delta \circ \varphi_i \circ \gamma \circ \eta_j)(x) \\ &= \sum_{i \in I}(\eta_i \circ \varphi_i \circ \gamma \circ \eta_j \circ \delta)(x) \\ &= (\gamma \circ \eta_j \circ \delta)(x), \end{aligned}$$

Simple and semisimple modules

whereas

$$(\gamma \circ g_\delta \circ \eta_j)(x) = \gamma\left(\sum_{i \in I}(\eta_i \circ \delta \circ \varphi_i \circ \eta_j)(x)\right) = (\gamma \circ \eta_j \circ \delta)(x).$$

We therefore have $g_\delta \circ \gamma \circ \eta_j = \gamma \circ g_\delta \circ \eta_j$. If now $y \in \bigoplus_{i \in I} M_i$ then

$$(g_\delta \circ \gamma)(y) = \sum_{i \in I}(g_\delta \circ \gamma \circ \eta_i \circ \varphi_i)(x) = \sum_{i \in I}(\gamma \circ g_\delta \circ \eta_i \circ \varphi_i)(x) = (\gamma \circ g_\delta)(x),$$

whence $g_\delta \circ \gamma = \gamma \circ g_\delta$ and so $g_\delta \in B\left(\bigoplus_{i \in I} M_i\right)$. That $\operatorname{Im} f = B(M)$ now follows from the observation that $f(g_\delta) = \varphi_k \circ g_\delta \circ \eta_k = \delta$.

To complete the proof, it suffices to show that f is injective. Suppose then that $\beta \in \operatorname{Ker} f$, so that $\varphi_k \circ \beta \circ \eta_k = 0$. Now for each index i we have $\eta_k \circ \varphi_i \in \operatorname{End}_R \bigoplus_{i \in I} M_i$ and so $\eta_k \circ \varphi_i \circ \beta = \beta \circ \eta_k \circ \varphi_i$ whence, composing on the left with φ_k, we obtain $\varphi_i \circ \beta = 0$ for each $i \in I$. Hence $\beta = 0$ and f is injective. ◊

Theorem 13.11 [Wedderburn-Artin] *Every simple ring is isomorphic to the ring of endomorphisms of a finite-dimensional module over a division ring.*

Proof Let R be a simple ring. Then R is semisimple and has no two-sided ideals other than R itself and $\{0\}$. It follows that R is a direct sum of miminal left ideals all of which are isomorphic. Let V be one of these minimal left ideals. Since V is simple as an R-module, $\operatorname{End}_R V$ is a division ring (by the Corollary to Theorem 5.6), and V is a module over this division ring under the action $\operatorname{End}_R V \times V \to V$ given by $(f, v) \mapsto f \cdot v = f(v)$. By Theorems 13.9 and 13.10, we have

$$R \simeq B(R) \simeq B(V).$$

Now $B(V) = C(\operatorname{End}_R V)$ and so $\alpha \in B(V)$ if and only if

$$(\forall f \in \operatorname{End}_R V)\ \alpha \circ f = f \circ \alpha,$$

which is the case if and only if

$$(\forall f \in \operatorname{End}_R V)(\forall v \in V)\ \alpha(f \cdot v) = \alpha[f(v)] = f[\alpha(v)] = f \cdot \alpha(v),$$

i.e. if and only if $\alpha \in \operatorname{End}_{\operatorname{End}_R V} V$. Thus we see that $R \simeq \operatorname{End}_D V$ where $D = \operatorname{End}_R V$. ◊

Corollary *A ring R is semisimple if and only if there are finitely many division rings $\Delta_1, \ldots, \Delta_m$ and positive integers $n_1, \ldots, n_m$ such that*

$$R \simeq \bigoplus_{i=1}^{m} \mathrm{Mat}_{n_i \times n_i} \Delta_i. \quad \Diamond$$

EXERCISES

13.1 Prove that for every R-module M the ring $R/\mathrm{Ann}_R M$ is isomorphic to the ring $H_R(M)$ of homotheties on M.

13.2 Let M be an R-module such that, for every $x \neq 0$ in M, the submodule Rx is simple. Prove that either M is simple or the ring $H_R(M)$ of homotheties on M is a division ring.

[*Hint.* If M is not simple then $Rx \neq M$ for some $x \neq 0$. Prove that if $y \notin Rx$ then $\mathrm{Ann}_R(x+y) = \mathrm{Ann}_R(x) + \mathrm{Ann}_R(y)$ and deduce that $\mathrm{Ann}_R(M) = \mathrm{Ann}_R(x)$. Now use Exercise 13.1.]

13.3 Let R be a unitary ring. Prove that a left ideal L of R is a direct summand of R if and only if L is generated by an idempotent.

[*Hint.* $\Rightarrow$: If $R = I \oplus J$ write $1_R = i + j$ where $i \in I$ and $j \in J$. Show that $I = Ri$ with $i^2 = i$.

$\Leftarrow$: If $I = Rx$ with $x^2 = x$ show that $R = Rx \oplus R(1_r - x)$.]

Deduce that a minimal left ideal I if R is a direct summand of R if and only if $I^2 \neq \{0\}$.

[*Hint.* $\Leftarrow$: Show that there exists $x \in I$ with $Ix \neq \{0\}$ and $R = I \oplus \mathrm{Ann}_R(x)$.]

13.4 If M is an R-module, prove that the following are equivalent :

(1) M is semisimple and of finite length;

(2) M is noetherian and every maximal submodule is a direct summand;

(3) M is noetherian and every simple submodule is a direct summand.

[*Hint.* (2) $\Rightarrow$ (1) : Show that M has simple (= quotient by maximal) submodules and look at the sum of all the simple submodules.

(3) $\Rightarrow$ (1) : If S is a simple submodule and $M = S \oplus T$, show that T is maximal. Now show that the intersection of all the maximal submodules of M is $\{0\}$. Let P be a maximal element in the

set of finite intersections of maximal submodules. Show that P is contained in every maximal submodule N of M (consider $N \cap P$). Deduce that there exist maximal submodules $N_1, \ldots, N_k$ such that $\bigcap_{i=1}^{k} N_i = \{0\}$. Now produce a monomorphism $f : M \to \underset{i=1}{\overset{k}{\times}} M/N_i$.]

13.5 [*Chinese remainder theorem*] Let R be a unitary ring and suppose that $I_1, \ldots, I_k$ are two-sided ideals of R such that, for each j, R is generated by $\{I_j, \bigcap_{t \neq j} I_t\}$. Prove that the mapping $f : R \to \underset{j=1}{\overset{k}{\times}} R/I_j$ given by $f(x) = (x + I_1, \ldots, x + I_k)$ is an R-epimorphism.

[*Hint.* Show that for $k = 2$ the problem reduces to a solution of the simultaneous congruences $x \equiv r_1(I_1), x \equiv r_2(I_2)$. Since R is generated by $\{I_1, I_2\}$ write $r_1 = r_{11} + r_{12}, r_2 = r_{21} + r_{22} (r_{ij} \in I_j)$ and consider $r_{12} + r_{21}$. Now use induction.]

13.6 Let R be a simple ring. Show that R has a finite number of maximal ideals $I_1, \ldots, I_n$. Prove that $R \simeq \underset{j=i}{\overset{n}{\times}} R/I_j$.

[*Hint.* Let $I_1, \ldots, I_k$ be the maximal ideals which are such that $\bigcap_{t \neq j} I_t \not\subseteq I_j$. Construct f as in Exercise 13.5. Show that, for $k + 1 \leq p \leq n$, $\bigcap_{t=1}^{k} I_t \subseteq I_p$ whence $\bigcap_{t=1}^{k} I_t = \{0\}$ and f is injective. Finally, observe that $\underset{t=1}{\overset{k}{\times}} R/I_t$ has precisely k maximal ideals, whence $k = n$.]

13.7 An R-module is said to be *faithful* if $\mathrm{Ann}_R M = \{0\}$. A ring R is said to be *quasi-simple* if its only two-sided ideals are $\{0\}$ and R. Prove that a unitary ring is quasi-simple if and only if every simple R-module is faithful.

[*Hint.* Let I be a maximal ideal and J a maximal left ideal containing I. Show that $\mathrm{Ann}_R R/J$ contains I.]

13.8 Let M be an R-module and let $x \in M$ be such that $\mathrm{Ann}_R(x)$ is the intersection of a finite number of maximal left ideals $I_1, \ldots, I_n$. By showing that Rx is isomorphic to a submodule of $\underset{j=1}{\overset{n}{\times}} R/I_j$, prove that Rx is semisimple.

Deduce that an R-module M is semisimple if and only if, for every $x \neq 0$ in M, $\mathrm{Ann}_R(x)$ is the intersection of a finite number of maximal left ideals.

13.9 An R-module is said to be *isotypic of type T* if it is the direct sum of a family of simple submodules each of which is isomorphic to the simple R-module T. Prove that if M is a semisimple R-module then there is a unique family $(M_i)_{i \in I}$ of submodules (called the *isotypic components* of M) such that, for some family $(T_i)_{i \in I}$ of simple R-modules,

(1) $(\forall i \in I)$ M_i is isotypic of type T_i;

(2) $T_i \not\simeq T_j$ for $i \neq j$;

(3) every simple submodule of M is contained in precisely one M_i;

(4) $M = \bigoplus_{i \in I} M_i$.

13.10 Let N be a submodule of a semisimple R-module M. Prove that N is a sum of isotypic components of M if and only if N is stable under every endomorphism f of M, in the sense that $f^{\to}(N) \subseteq N$.

13.11 Let R be a semisimple ring. Prove that the isotypic components of the R-module R are the minimal ideals of R.

[*Hint.* Use Exercise 13.10 to show that the left ideals that can be expressed as sums of isotypic components are the two-sided ideals.]

13.12 If V is a finite-dimensional vector space over a division ring D, prove that the rings $\mathrm{End}_D V$ and $\mathrm{End}_D V^d$ are isomorphic.

[*Hint.* Associate with each $\alpha \in \mathrm{End}_D V$ the mapping $\vartheta(\alpha) : V^d \to V^d$ given by $[\vartheta(\alpha)](\xi^d) = \xi^d \circ \alpha$.]

Deduce from this that if we had used the term *left semisimple* instead of just semisimple and defined similarly a *right semisimple* ring then a ring is left semisimple if and only if it is right semisimple.

14 The Jacobson radical

Definition If M is an R-module then its *Jacobson radical* $\operatorname{Rad}_J M$ is the intersection of all the maximal submodules of M. If M has no maximal submodules then we define $\operatorname{Rad}_J M = M$.

A useful simple characterisation of the Jacobson radical is the following.

Theorem 14.1 $\operatorname{Rad}_J M = \bigcap_f \operatorname{Ker} f$ *where f is an R-morphism from M to a simple R-module.*

Proof Let $f : M \to S$ be an R-morphism where S is simple. Then by Theorem 5.6(2) either $f = 0$ or f is an epimorphism. In the former case $\operatorname{Ker} f = M$, and in the latter $S = \operatorname{Im} f \simeq M/\operatorname{Ker} f$ in which case the simplicity of S implies that $\operatorname{Ker} f$ is maximal. $\diamondsuit$

Immediate properties of the Jacobson radical are given in the following results.

Theorem 14.2 *If $f \in \operatorname{Mor}_R(M, N)$ then $f^{\to}(\operatorname{Rad}_J M) \subseteq \operatorname{Rad}_J N$.*

Proof If S is a simple R-module and $g \in \operatorname{Mor}_R(N, S)$ then $g \circ f \in \operatorname{Mor}_R(M, S)$ and so, by Theorem 14.1, $(g \circ f)^{\to}(\operatorname{Rad}_J M) = \{0\}$. It follows that $f^{\to}(\operatorname{Rad}_J M) \subseteq \operatorname{Ker} g$ whence, by Theorem 14.1 again, we obtain $f^{\to}(\operatorname{Rad}_J M) \subseteq \operatorname{Rad}_J N$. $\diamondsuit$

Theorem 14.3 *If N is a submodule of M then*
(1) $\operatorname{Rad}_J N \subseteq \operatorname{Rad}_J M$;
(2) $(N + \operatorname{Rad}_J M)/N \subseteq \operatorname{Rad}_J(M/N)$;
(3) *if $N \subseteq \operatorname{Rad}_J M$ then $\operatorname{Rad}_J M/N = \operatorname{Rad}_J(M/N)$.*

Proof (1) Apply Theorem 14.2 to the natural monomorphism $\iota_N : N \to M$.

(2) Apply Theorem 14.2 to the natural epimorphism $\natural_N : M \to M/N$ to obtain

$$(N + \operatorname{Rad}_J M)/N = \natural_N^{\to}(N + \operatorname{Rad}_J M) = \natural_N^{\to}(\operatorname{Rad}_J M) \subseteq \operatorname{Rad}_J(M/N).$$

(3) There is a bijection between the maximal submodules of M/N and the maximal submodules of M that contain N, i.e. the maximal submodules of M since $N \subseteq \operatorname{Rad}_J M$. $\diamondsuit$

Corollary *The Jacobson radical of M is the smallest submodule N of M with the property that $\text{Rad}_J M/N = \{0\}$.*

Proof First we note that, taking $N = \text{Rad}_J M$ in Theorem 14.3(3),

$$\text{Rad}_J(M/\text{Rad}_J M) = \text{Rad}_J M/\text{Rad}_J M = \{0\}.$$

Suppose now that N is a submodule of M such that $\text{Rad}_J M/N = \{0\}$. Then by Theorem 14.3(2) we have $N + \text{Rad}_J M = N$ whence $\text{Rad}_J M \subseteq N$. ◊

Theorem 14.4 *For every family $(M_i)_{i \in I} M_i$ of R-modules*

(1) $\text{Rad}_J \left(\underset{i \in I}{\times} M_i \right) \subseteq \underset{i \in I}{\times} \text{Rad}_J M_i$;

(2) $\text{Rad}_J \left(\underset{i \in I}{\bigoplus} M_i \right) = \underset{i \in I}{\bigoplus} \text{Rad}_J M_i$.

Proof (1) Let N_i be a maximal submodule of M_i. Then $N_i \times \underset{j \neq i}{\times} M_j$ is a maximal submodule of $\underset{i \in I}{\times} M_i$. Hence

$$\text{Rad}_J \underset{i \in I}{\times} M_i \subseteq \underset{i}{\cap} \left(\text{Rad}_J M_i \times \underset{i \in I}{\times} M_i \right) = \text{Rad}_J M_i.$$

(2) In a similar way, we see that if N_i is a maximal submodule of M_i then $N_i \oplus \underset{j \neq i}{\bigoplus} M_j$ is a maximal submodule of $\underset{i \in I}{\bigoplus} M_i$ and

$$\text{Rad}_J \underset{i \in I}{\bigoplus} M_i \subseteq \underset{i \in I}{\bigoplus} \text{Rad}_J M_i.$$

On the other hand, by Theorem 14.3(1), $\text{Rad}_J M_j \subseteq \text{Rad}_J \underset{i \in I}{\bigoplus} M_i$, whence we have equality. ◊

We can use Theorem 14.4 to investigate those R-modules for which the Jacobson radical is as small as possible.

Theorem 14.5 *An R-module M has zero Jacobson radical if and only if M is isomorphic to a submodule of a cartesian product of simple modules.*

Proof If M is simple then clearly $\text{Rad}_J M = \{0\}$. Consequently, for a family $(M_i)_{i \in I}$ of simple modules we have, by Theorem 14.4(1),

$$\text{Rad}_J \underset{i \in I}{\times} M_i \subseteq \underset{i \in I}{\times} \text{Rad}_J M_i = \{0\}.$$

The Jacobson radical

Thus, if M is a submodule of $\underset{i\in I}{\bigtimes} M_i$ it follows by Theorem 14.3(1) that $\operatorname{Rad}_J M = \{0\}$.

Conversely, suppose that $\operatorname{Rad}_J M = \{0\}$. Let $(N_i)_{i\in I}$ be the family of maximal submodules of M. Then each quotient module M/N_i is simple. Define $f : M \to \underset{i\in I}{\bigtimes} M/N_i$ by $f(x) = \big(\natural_i(x)\big)_{i\in I}$. Then f is an R-morphism with

$$\operatorname{Ker} f = \bigcap_{i\in I} \operatorname{Ker} \natural_i = \bigcap_{i\in I} N_i = \operatorname{Rad}_J M = 0.$$

Thus f is a monomorphism and $M \simeq \operatorname{Im} f$. ◊

Corollary *Every simple and every semisimple module has zero Jacobson radical.*

Proof Direct sums are submodules of cartesian products. ◊

If M is a finitely generated non-zero module then every submodule N of M is contained in a maximal submodule. This can readily be seen by showing that the set of submodules that contain N is inductively ordered and so we can apply Zorn's axiom to produce a maximal such submodule. It follows that $\operatorname{Rad}_J M$ is properly contained in M.

Theorem 14.6 [Nakayama] *If M is a finitely generated module and N is a submodule of M such that $N + \operatorname{Rad}_J M = M$ then necessarily $N = M$.*

Proof Suppose, by way of obtaining a contradiction, that $N \subset M$. Then since M/N is non-zero and also finitely generated we have, from the above observation, that $\operatorname{Rad}_J M/N \subset M/N$. But since by hypothesis $N + \operatorname{Rad}_J M = M$ we have, by Theorem 14.3(2), the contradiction $M/N = \operatorname{Rad}_J M/N$. ◊

Corollary *If $M = Rx$ then $y \in \operatorname{Rad}_J M$ if and only if $M = R(x + ry)$ for every $r \in R$.*

Proof $\Rightarrow$: If $y \in \operatorname{Rad}_J M$ then for every $z \in M = Rx$ we have

$$z = \lambda x = \lambda(x + ry) - \lambda ry \in R(x + ry) + \operatorname{Rad}_J M$$

whence $M = R(x + ry) + \operatorname{Rad}_J M$. It now follows by Theorem 14.6 that $R(x + ry) = M$.

$\Leftarrow$: If $y \notin \operatorname{Rad}_J M$ then there is a maximal submodule N such that $y \notin N$. Then $N + Ry = M = Rx$ and so $x + ry \in N$ for every $r \in R$. It follows that $R(x + ry) \ne M$. ◊

We now consider the Jacobson radical of a unitary ring R. This is defined as the Jacobson radical of the R-module R and therefore is the intersection of all the maximal left ideals of R. Since R is finitely generated, namely by $\{1_R\}$, it follows that $\operatorname{Rad}_J R \subset R$.

We can characterise $\operatorname{Rad}_J R$ in a manner similar to Theorem 14.1 :

Theorem 14.7 *If R is a unitary ring then $\operatorname{Rad}_J R$ is the intersection of the annihilators of all the simple R-modules, and therefore is a two-sided ideal of R.*

Proof If S is a simple R-module and x is a non-zero element of S then the mapping $\rho_x : R \to S$ given by $\rho_x(r) = rx$ is a non-zero R-morphism. By Theorem 5.6(2) we have $\operatorname{Im} \rho_x = S$ and so

$$R/\operatorname{Ann}_R(x) = R/\operatorname{Ker} \rho_x \simeq \operatorname{Im} \rho_x = S.$$

Since S is simple, $\operatorname{Ann}_R(x)$ is then a maximal left ideal of R. Thus, on the one hand, $\operatorname{Rad}_J R \subseteq \bigcap_{x \in S} \operatorname{Ann}_R(x) = \operatorname{Ann}_R S$. On the other hand, suppose that $r \in \operatorname{Ann}_R S$ for every simple R-module S and let L be a maximal left ideal of R. Since the R-module R/L is simple we have in particular that $r \in \operatorname{Ann}_R R/I$ and so $r + L = r(1_R + L) = 0 + L$ whence $r \in L$. Thus we conclude that $\operatorname{Rad}_J R = \bigcap_S \operatorname{Ann}_R S$.

Now for every R-module S, if $r \in R$ annihilates S then r annihilates λx for every $x \in S$ and so $r\lambda$ annihilates S. Thus the left ideal $\operatorname{Ann}_R S$ is also a right ideal. It follows that $\operatorname{Rad}_J R$ is a two-sided ideal of R. ◊

Theorem 14.8 *If R is a unitary ring and $x \in R$ then the following are equivalent* :

(1) $x \in \operatorname{Rad}_J R$;
(2) $(\forall r \in R)$ $1_R - rx$ *has a left inverse in R.*

Proof This is immediate from the Corollary to Theorem 14.6 since the R-module R is generated by $\{1_R\}$, and $\{1_R - rx\}$ generates R if and only if $1_R = \lambda(1_R - rx)$ for some $\lambda \in R$. ◊

Corollary 1 *$\operatorname{Rad}_J R$ is the largest two-sided ideal I of R such that $1_R - x$ has a two-sided inverse for all $x \in I$.*

Proof If $x \in \operatorname{Rad}_J R$ then $1_R - x$ has a left inverse by Theorem 14.8. Let $y \in R$ be such that $y(1_R - x) = 1_R$. Then $1_R - y = -yx \in \operatorname{Rad}_J R$ and so, by Theorem 14.8 again, there exists $z \in R$ such that

$$1_R = z[1_R - (1_R - y)] = zy.$$

Hence y has both a left and a right inverse. Now

$$z = z 1_R = zy(1_R - x) = 1_R - x.$$

The Jacobson radical

and so y is also a right inverse of $1_R - x$.

Suppose now that I is a two-sided ideal of R such that $1_R - x$ is invertible for all $x \in I$. Then clearly $1_R - rx$ is invertible for all $r \in R$ and so $x \in \text{Rad}_J R$. ◊

For a unitary ring R we define the *opposite ring* R^{op} to be the ring obtained by defining on the abelian group of R the multiplication $(x, y) \mapsto yx$. We then have

Corollary 2 $\text{Rad}_J R = \text{Rad}_J R^{\text{op}}$.

Proof This is immediate from Corollary 1. ◊

Corollary 3 $\text{Rad}_J R$ *is also the intersection of all the maximal right ideals of* R.

Proof This follows immediately from Corollary 2 since a right ideal of R is a left ideal of R^{op}. ◊

We now consider the relationship between semisimplicity and the Jacobson radical.

Theorem 14.9 *If M is an R-module then the following are equivalent*:
  (1) M *is semisimple and of finite length*;
  (2) M *is artinian and has zero Jacobson radical*.

Proof (1) $\Rightarrow$ (2) : This is immediate from the Corollary to Theorem 14.5.

(2) $\Rightarrow$ (1) : Let P be a minimal element in the set of all finite intersections of maximal submodules of M. For every maximal submodule N of M we have $N \cap P \subseteq P$ whence $N \cap P = P$ and so $P \subseteq N$. Since $\text{Rad}_J M = \{0\}$ it follows that $P = \{0\}$. Thus there is a finite family $(N_i)_{1 \leq i \leq n}$ of maximal submodules of M such that $\bigcap_{i=1}^{n} N_i = \{0\}$. Then $f : M \to \underset{i=1}{\overset{n}{\times}} M/N_i$ given by $f(x) = (\natural_i(x))_i$ is an R-morphism with $\text{Ker } f = \bigcap_{i=1}^{n} N_i = \{0\}$ and so is injective. Since $\underset{i=1}{\overset{n}{\times}} M/N_i$ is semisimple and of finite length, so is every submodule by Theorem 13.3, whence so is M. ◊

Corollary 1 *A unitary ring is semisimple if and only if it is artinian and has zero Jacobson radical*.

Proof If R is semisimple then the R-module R, being generated by $\{1_R\}$, has finite length by Theorem 13.5. ◊

Corollary 2 *The quotient of an artinian module by its Jacobson radical is a semisimple module of finite length.*

Proof The quotient is artinian, and has zero Jacobson radical since, by Theorem 14.3(3),

$$\text{Rad}_J(M/\text{Rad}_J M) = \text{Rad}_J M/\text{Rad}_J M = \{0\}. \quad \Diamond$$

Corollary 3 *The quotient of an artinian ring by its Jacobson radical is a semisimple ring.* $\Diamond$

In order to characterise the Jacobson radical of an artinian ring, we consider the following notions. For this, we recall that the product of two ideals I, J of a ring R is the set IJ of all finite sums of the form $\sum_{<\infty} a_i b_j$ where $a_i \in I, b_j \in J$.

Definition If R is a ring then $a \in R$ is said to be *nilpotent* if $a^n = 0$ for some positive integer n. An ideal I is called a *nil ideal* if every element of I is nilpotent, and a *nilpotent ideal* if $I^n = \{0\}$ for some positive integer n.

Suppose now that I is a nilpotent ideal with, say, $I^n = \{0\}$. Given any $r \in I$ we have $r^n \in I^n = \{0\}$ and so r is nilpotent. Thus we see that *every nilpotent ideal is a nil ideal*.

Theorem 14.10 *Every nil ideal of R is contained in $\text{Rad}_J R$.*

Proof Let N be a nil left ideal and let $x \in N$. For every $r \in R$ we have $rx \in N$ and so rx is nilpotent, say $(rx)^n = 0$. Since then

$$[1_R + rx + (rx)^2 + \ldots + (rx)^{n-1}](1_R - rx) = 1_R,$$

we see that $1_R - rx$ has a left inverse. It follows that $x \in \text{Rad}_J R$. Similarly we can show that every nil right ideal is contained in $\text{Rad}_J R$. $\Diamond$

Theorem 14.11 *The Jacobson radical of an artinian ring R is the greatest nilpotent two-sided ideal of R.*

Proof By Theorem 14.10 and the observation preceding it, $\text{Rad}_J R$ contains all the nilpotent two-sided ideals of R. It therefore suffices to show that $\text{Rad}_J R$ is nilpotent. For this purpose, write $\text{Rad}_J R = I$.

Since I is a two-sided ideal of R we have $I^2 \subseteq RI \subseteq R$, and this gives rise to a descending chain of ideals

$$I \supseteq I^2 \supseteq \ldots \supseteq I^p \supseteq I^{p+1} \supseteq \ldots.$$

The Jacobson radical

Since R is artinian, there is a positive integer n such that $I^n = I^{n+1} = \cdots$. Let $K = I^n$; we shall prove that $K = \{0\}$ whence the result will follow.

Suppose, by way of obtaining a contradiction, that $K \neq \{0\}$. Since $K^2 = K \neq 0$ the set E of left ideals J of R with $J \subseteq K$ and $KJ \neq \{0\}$ is not empty. Since R satisfies the minimum condition on left ideals, E has a minimal element J_0. Then $KJ_0 \neq \{0\}$ and so there exists $x_0 \in J_0$ such that $Kx_0 \neq \{0\}$. Now Kx_0 is also a left ideal of R and is such that $Kx_0 \subseteq J_0$. Since $K^2 x_0 = Kx_0 \neq \{0\}$ we see that $Kx_0 \in E$. Since J_0 is minimal in E, we deduce that $Kx_0 = J_0$ and so, for some $k \in K$, we have $kx_0 = x_0$. But $k \in \text{Rad}_J R$ and so $1_R - k$ is invertible; so from $(1_R - k)x_0 = \{0\}$ we deduce that $x_0 = 0$, which contradicts the fact that $Kx_0 \neq \{0\}$. We therefore deduce that $K = \{0\}$ whence $I = \text{Rad}_J R$ is nilpotent. $\diamond$

Corollary *In an artinian ring every nil ideal is nilpotent.*

Proof If N is nil then by Theorem 14.10 we have $N \subseteq \text{Rad}_J R$. But by Theorem 14.11 we have $(\text{Rad}_J R)^n = \{0\}$ for some positive integer n. Hence $N^n = \{0\}$ and so N is nilpotent. $\diamond$

We end our discussion of artinian rings with the following remarkable result.

Theorem 14.12 [Hopkins] *Every artinian ring is noetherian.*

Proof Suppose that R is artinian and let $\text{Rad}_J R = I$. Since, by Theorem 14.11, R is nilpotent there is a positive integer n such that

$$R \supset I \supset I^2 \supset \cdots \supset I^n = \{0\}.$$

We consider two cases :

(1) $n = 1$, in which case $I = \{0\}$: in this case R is semisimple by Corollary 1 of Theorem 14.9. Thus R is also noetherian, by the Corollary to Theorem 13.5.

(2) $n \neq 1$, in which case $I \neq \{0\}$: in this case we shall establish, for each t, a chain

$$I^t = K_{i_0} \supset \cdots \supset K_{i_{t_i}} = I^{t+1}$$

of left ideals such that $K_{i_j}/K_{i_{j+1}}$ is simple for $j = 0, \ldots, m_t - 1$. It will then follow that R has a Jordan-Hölder tower and so is noetherian.

Now R/I is an artinian ring with zero Jacobson radical, and so is semisimple. Every non-zero R/I-module is therefore semisimple. Now every R-module M can be regarded as an $R/\text{Ann}_R M$-module; to see this, observe that an action $R/\text{Ann}_R M \times M \to M$ can be defined by $(r + \text{Ann}_R M, m) \mapsto rm$. Thus, since each quotient module I^t/I^{t+1} has annihilator I, it can be regarded as an R/I-module, and as such is simple. We therefore have

$$I^t/I^{t+1} = \bigoplus_k J_k/I^{t+1}$$

where each J_k is a left ideal of R such that $I^{t+1} \subseteq J_k \subseteq I^t$, and since R is artinian this direct sum is finite. Hence I^t/I^{t+1} has a Jordan-Hölder tower

$$I^t/I^{t+1} = K_{i_0}/I^{t+1} \supset \cdots \supset K_{i_{t_i}}/I^{t+1} = \{0\}.$$

The left ideals K_{i_j} then give the tower

$$I^t = K_{i_0} \supset \cdots \supset K_{i_{t_i}} = I^{t+1}$$

in which each quotient $K_{i_j}/K_{i_{j+1}}$ is simple. $\Diamond$

EXERCISES

14.1 Let p be a fixed prime and let S be the set of all sequences $(a_n)_{n \geq 1}$ with $a_n \in \mathbb{Z}_{p^n}$ for each n. Show that S is a ring under the laws given by $(a_n) + (b_n) = (a_n + b_n)$ and $(a_n)(b_n) = (a_n b_n)$. Now let R be the subset of S consisting of those sequences which become zero after a certain point. Show that R is a two-sided ideal of S. Let I be the set of sequences in R of the form $(pt_1, \ldots, pt_n, 0, 0, \ldots)$ where each $t_i \in \mathbb{Z}_{p^i}$. Show that I is a two-sided nil ideal of R that is not nilpotent.

14.2 Let R be a principal ideal domain. Prove that

(1) R has zero Jacobson radical if and only if either R is a field or the set of maximal ideals of R is infinite;

(2) a quotient ring R/Rx has zero Jacobson radical if and only if x has no prime factors of multiplicity greater than 1.

14.3 Determine the Jacobson radical of each of the following rings :

(1) $\mathbb{Z}/4\mathbb{Z}$;

(2) $\mathbb{Z}/6\mathbb{Z}$;

(3) $\mathbb{Z}/pq\mathbb{Z}$ (p, q distinct primes);

(4) $\mathbb{Z}/p^\alpha q^\beta \mathbb{Z}$ (p, q distinct primes, $\alpha, \beta \geq 1$).

14.4 Prove that the ring $F[X]$ of polynomials over a field F has zero Jacobson radical.

The Jacobson radical

14.5 Determine the Jacobson radical of the ring of upper triangular $n \times n$ matrices over a field.

14.6 Let R be a unitary ring. Define a law of composition $\oplus$ on R by $(x,y) \mapsto x \oplus y = x + y - xy$. Show that $(R, \oplus)$ is a semigroup with an identity.

We say that $x \in R$ is *left quasi-regular* if x has a left inverse with respect to $\oplus$. A left ideal L of R is called *left quasi-regular* if every element of L is left quasi-regular. Show that x is left quasi-regular if and only if $1_R - x$ has a left inverse in R and deduce that

$$\text{Rad}_J R = \{x \in R \, ; \, (\forall r \in R) \, rx \text{ is left quasi-regular}\}.$$

If P is a left quasi-regular left ideal of R and M is a simple R-module, prove (via a contradiction) that $PM = \{0\}$. Deduce that $\text{Rad}_J R$ is a left quasi-regular left ideal of R that contains every left quasi-regular left ideal of R.

Let S be the ring of rationals of the form m/n with n odd. Determine the (left) quasi-regular elements of S and show that $\text{Rad}_J S = \langle 2 \rangle$.

14.7 Consider the ring of all 2×2 matrices

$$\begin{bmatrix} a & b \\ 0 & c \end{bmatrix}$$

where (1) $a \in \mathbb{Z}$, $b, c \in \mathbb{Q}$; (2) $a \in \mathbb{Q}$, $b, c \in \mathbb{R}$. Show that in case (1) the ring is right noetherian but not left noetherian, and that in case (2) it is right artinian but not left artinian.

15 Tensor products; flat modules; regular rings

We shall now develop more machinery, thereby laying a foundation for a study of what is often called *multilinear algebra*. Whilst this term will take on a more significant meaning in the next chapter, we begin with the following type of mapping which involves a mixture of left and right modules.

Definition Let M be a right R-module and N a left R-module. If G is a $\mathbb{Z}$-module then a mapping $f : M \times N \to G$ is said to be *balanced* if

$(\forall m_1, m_2 \in M)(\forall n \in N) \quad f(m_1 + m_2, n) = f(m_1, n) + f(m_2, n);$
$(\forall m \in M)(\forall n_1, n_2 \in N) \quad f(m, n_1 + n_2) = f(m, n_1) + f(m, n_2);$
$(\forall m \in M)(\forall n \in N)(\forall \lambda \in R) \quad f(m\lambda, n) = f(m, \lambda n).$

Example 15.1 If M is a left R-module then the mapping $f : M^d \times M \to R$ given by $f(m^d, n) = m^d(n) = \langle n, m^d \rangle$ is balanced.

Given a balanced mapping $f : M \times N \to G$, we shall now consider how to construct a $\mathbb{Z}$-module T with the property that, roughly speaking, f can be 'lifted' to a $\mathbb{Z}$-morphism $h : T \to G$. This construction, together with a similar one that we shall meet with later, gives rise to another important way of constructing new modules from old. Such a 'trading in' of a balanced map for a $\mathbb{Z}$-morphism is a useful device.

Definition Let M be a right R-module and N a left R-module. By a *tensor product* of M and N we shall mean a $\mathbb{Z}$-module T together with a balanced mapping $f : M \times N \to T$ such that, for every $\mathbb{Z}$-module G and every balanced mapping $g : M \times N \to G$, there is a unique $\mathbb{Z}$-morphism $h : T \to G$ such that the diagram

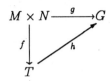

is commutative. We denote such a tensor product by (T, f).

Theorem 15.1 *If (T, f) is a tensor product of the right R-module M and the left R-module N then $\mathrm{Im}\, f$ generates T.*

Tensor products; flat modules; regular rings

Proof This is essentially as that of the second part of Theorem 7.1. ◊

- In contrast to the first part of Theorem 7.1, note that if (T, f) is a tensor product of M and N then f is not injective. In fact, since f is balanced we have

$$(\forall m \in M)(\forall n \in N)(\forall \lambda \in R) \qquad f(m\lambda, n) = f(m, \lambda n)$$

and so, on taking $\lambda = 0$, we have $f(0, n) = f(m, 0)$ for all $m \in M$ and all $n \in N$.

Theorem 15.2 [Uniqueness] *Let (T, f) be a tensor product of M and N. Then (T', f') is also a tensor product of M and N if and only if there is a $\mathbb{Z}$-isomorphism $j : T \to T'$ such that $j \circ f = f'$.*

Proof This is essentially as that of Theorem 7.2. ◊

We shall now settle the question of the existence of tensor products. For this purpose, let M be a right R-module and N a left R-module. Let (F, i) be the free $\mathbb{Z}$-module on $M \times N$ and let H be the subgroup of F that is generated by the elements of the following types :

$$i(m_1 + m_2, n) - i(m_1, n) - i(m_2, n);$$
$$i(m, n_1 + n_2) - i(m, n_1) - i(m, n_2);$$
$$i(m\lambda, n) - i(m, \lambda n).$$

We denote the quotient group F/H by $M \otimes_R N$ and the composite mapping $\natural_H \circ i$ by $\otimes_R$.

Theorem 15.3 [Existence] *If M is a right R-module and N is a left R-module then $(M \otimes_R N, \otimes_R)$ is a tensor product of M and N.*

Proof Let G be a $\mathbb{Z}$-module and $g : M \times N \to G$ a balanced mapping. If (F, i) is the free $\mathbb{Z}$-module on $M \times N$ we shall first show how to obtain $\mathbb{Z}$-morphisms j, h such that the following diagram is commutative :

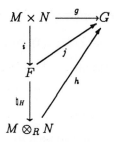

In fact, the existence of a unique $\mathbb{Z}$-morphism $j : F \to G$ such that $j \circ i = g$ results from the fact that (F, i) is free on $M \times N$. It now follows easily from the definition of H and the fact that g is balanced that $H \subseteq \operatorname{Ker} j$. Applying Theorem 3.3, we deduce the existence of a unique $\mathbb{Z}$-morphism $h : F/H \to G$ such that $h \circ \natural_H = j$. The resulting commutative diagram yields

$$h \circ \otimes_H = h \circ \natural_H \circ i = j \circ i = g.$$

We now have to establish the uniqueness of h with respect to this property. For this purpose, suppose that $k : M \otimes_R N \to G$ is also a $\mathbb{Z}$-morphism such that $k \circ \otimes_R = g$. Then we have $k \circ \natural_H \circ i = g$ and so, by the uniqueness of j, we deduce that $k \circ \natural_H = j = h \circ \natural_H$. Since $\natural_H$ is surjective, hence right cancellable, it follows that $k = h$. $\Diamond$

By the above results there is, to within abelian group isomorphism, a unique tensor product of M and N. By *the* tensor product of M and N we shall mean $M \otimes_R N$ as constructed above. The mapping $\otimes_R$ will be called the associated *tensor map*. Given $(m, n) \in M \times N$ we shall write $\otimes_R(m, n)$ as $m \otimes_R n$ and, by abuse of language, call this the *tensor product of the elements m and n*. When no confusion can arise over R, we shall often omit the subscript R in the symbol $\otimes_R$.

- It is clear from the above that in $M \otimes N$ we have the identities

$$(m_1 + m_2) \otimes n = m_1 \otimes n + m_2 \otimes n;$$
$$m \otimes (n_1 + n_2) = n \otimes n_1 + m \otimes n_2;$$
$$m\lambda \otimes n = m \otimes \lambda n.$$

It is immediate from these identities that

$$0 \otimes n = 0 = m \otimes 0$$

and that, by induction for $\mathbb{N}$ then extended to $\mathbb{Z}$,

$$(\forall k \in \mathbb{Z}) \quad km \otimes n = k(m \otimes n) = m \otimes kn.$$

In what follows these identities will be used without reference.

- Note that since $\operatorname{Im} \otimes$ generates $M \otimes N$, every element of $M \otimes N$ can be expressed as a linear combination $\sum_{i=1}^{t} p_i(m_i \otimes n_i)$ where each $p_i \in \mathbb{Z}$. It follows by the preceding remark that every element of $M \otimes N$ can be written in the form $\sum_{i=1}^{t}(a_i \otimes b_i)$ with $a_i \in M$ and $b_i \in N$ for every i. However, it should be noted that, despite the notation, not every element of $M \otimes N$ is of the form $m \otimes n$ with $m \in M$ and $n \in N$.

Tensor products; flat modules; regular rings

It is important to note that $M \otimes_R N$ as defined above is an abelian group and is not in general an R-module. In certain circumstances, however, we can give $M \otimes_R N$ the structure of a left R-module or that of a right R-module. To see this, we require the following notion.

Definition Let R and S be unitary rings. By an (R, S)-*bimodule* we shall mean a module M which is both a left R-module and a right S-module, the actions being linked by the identity

$$(\forall m \in M)(\forall r \in R)(\forall s \in S) \qquad (rm)s = r(ms).$$

Example 15.2 Every unitary ring R is an (R, R)-bimodule.

Example 15.3 If M is a right R-module then M is a $(\mathbb{Z}, R)$-bimodule.

Example 15.4 If R is a commutative unitary ring then every left R-module M can be given the structure of an (R, R)-bimodule. In fact, it is clear that we can define an action $M \times R \to M$ by $(m, s) \mapsto m \cdot s = sm$ and thereby make M into a right R-module. The multiplication in R being commutative, we also have

$$(rm) \cdot s = s(rm) = (sr)m = (rs)m = r(sm) = r(m \cdot s).$$

- When the ground ring R is commutative we shall take it as understood that each left (respectively right) R-module is endowed with the (R, R)-bimodule structure described in Example 15.4.

Theorem 15.4 *If M is an (R, S)-bimodule and N is a left R-module then $\operatorname{Mor}_R(M, N)$ is a left S-module relative to the action $(s, f) \mapsto sf$ where $(sf)(m) = f(ms)$ for all $m \in M$.*

Proof This is simply a routine verification of the identities $s(f + g) = sf + sg, (s + s')f = sf + s'f, s'(sf) = (s's)f$ and $1_S f = f$, each of which follows easily from the definition of sf. $\diamondsuit$

Theorem 15.5 *Let M be an (R, S)-bimodule and N a left S-module. Then $M \otimes_R N$ is a left R-module relative to the action defined by*

$$\left(r, \sum_{i=1}^{t}(m_i \otimes_S n_i)\right) \mapsto \sum_{i=1}^{t}(rm_i \otimes_S n_i).$$

Proof Given $r \in R$ consider the mapping $\vartheta_r : M \times N \to M \otimes_S N$ described by $\vartheta_r(m, n) = rm \otimes_S n$. It is readily verified that ϑ_r is a balanced mapping and so there exists a unique $\mathbb{Z}$-morphism $f_r : M \otimes_S N \to M \otimes_S N$ such that

$$(\forall m \in M)(\forall n \in N) \qquad f_r(m \otimes_S n) = rm \otimes_S n.$$

Since every element of $M \otimes_S N$ can be written as $\sum_{i=1}^{t}(m_i \otimes_S n_i)$ we can define an action $R \times (M \otimes_S N) \to M \otimes_S N$ by the prescription

$$\left(r, \sum_{i=1}^{t}(m_i \otimes_S n_i)\right) \mapsto f_r\left(\sum_{i=1}^{t}(m_i \otimes_S n_i)\right) = \sum_{i=1}^{t}(rm_i \otimes_S n_i).$$

It is now readily verified that $M \otimes_S N$ is a left R-module. $\Diamond$

- There is, of course, a result that is dual to Theorem 15.5, namely that *if M is a right R-module and N is an (R,S)-bimodule then $M \otimes_R N$ can be given the structure of a right R-module.*

By way of applying Theorems 15.4 and 15.5, we now establish the following result which shows how tensor products may be used to simplify certain morphism groups.

Theorem 15.6 *Let M be a left R-module, N an (S,R)-bimodule and P a left S-module. Then there is a $\mathbb{Z}$-isomorphism*

$$\mathrm{Mor}_R(M, \mathrm{Mor}_S(N, P)) \simeq \mathrm{Mor}_S(N \otimes_R M, P).$$

Proof We note first that, by Theorem 15.4, $\mathrm{Mor}_S(N, P)$ is a left R-module and that, by Theorem 15.5, $N \otimes_R M$ is a left S-module.

Given an R-morphism $f : M \to \mathrm{Mor}_S(N, P)$, consider the mapping $\alpha_f : N \times M \to P$ given by $\alpha_f(n, m) = [f(m)](n)$. It is readily verified that α_f is a balanced mapping and so there is a unique $\mathbb{Z}$-morphism $\vartheta_f : N \otimes_R M \to P$ such that

$$(\forall n \in N)(\forall m \in M) \quad \vartheta_f(n \otimes_R m) = [f(m)](n).$$

Now ϑ_f is an S-morphism; for, by the action defined in Theorem 15.5,

$$\vartheta_f[s(n \otimes_R m)] = \vartheta_f(sn \otimes_R m) = [f(m)](sn)$$
$$= s \cdot [f(m)](n) = s\vartheta_f(n \otimes_R m)$$

and $N \otimes_R M$ is generated by the elements of the form $n \otimes_R m$. We can therefore define a mapping

$$\vartheta : \mathrm{Mor}_R(M, \mathrm{Mor}_S(N, P)) \to \mathrm{Mor}_S(N \otimes_R M, P)$$

by the prescription $\vartheta(f) = \vartheta_f$. It is clear that ϑ is a $\mathbb{Z}$-morphism. Our objective is to show that it is a $\mathbb{Z}$-isomorphism.

For this purpose, let $g : N \otimes_R M \to P$ be an S-morphism and define a mapping $\beta_g : M \to \mathrm{Mor}_S(N, P)$ by assigning to every $m \in M$ the S-morphism $\beta_g(m) : N \to P$ given by $[\beta_g(m)](n) = g(n \otimes_R m)$. That each

Tensor products; flat modules; regular rings

$\beta_g(m)$ is an R-morphism follows by Theorem 15.4. We can now define a mapping
$$\beta : \mathrm{Mor}_S(N \otimes_R M, P) \to \mathrm{Mor}_R(M, \mathrm{Mor}_S(N, P))$$
by the prescription $\beta(g) = \beta_g$. Clearly, β is a $\mathbb{Z}$-morphism. We shall show that ϑ and β are mutually inverse $\mathbb{Z}$-isomorphisms whence the result will follow.

Since $(\beta \circ \vartheta)(f) = \beta_{\vartheta_f}$ with $[\beta_{\vartheta_f}(m)](n) = \vartheta_f(n \otimes m) = [f(m)](n)$, we see that $\beta_{\vartheta_f} = f$ and so $\beta \circ \vartheta$ is the appropriate identity map. Likewise, $(\vartheta \circ \beta)(g) = \vartheta_{\beta_g}$ with $\vartheta_{\beta_g}(n \otimes m) = [\beta_g(m)](n) = g(n \otimes m)$ and so ϑ_{β_g} and g coincide on a set of generators of $N \otimes M$. It follows that $\vartheta_{\beta_g} = g$ whence $\vartheta \circ \beta$ is the appropriate identity map. Thus ϑ and β are mutually inverse $\mathbb{Z}$-isomorphisms. ◊

Corollary *If M is a left R-module and N is a right R-module then*
$$\mathrm{Mor}_R(M, N^+) \simeq (N \otimes_R M)^+.$$

Proof Take $S = \mathbb{Z}$ and $P = \mathbb{Q}/\mathbb{Z}$ in the above and use Example 15.3. ◊

A less involved consequence of Theorem 15.5 is the following.

Theorem 15.7 *If M is a left R-module then there is a unique isomorphism*
$$\vartheta : R \otimes M \to M$$
such that $\vartheta(r \otimes m) = rm$.

Proof By Theorem 13.5, $R \otimes M$ is a left R-module. The mapping $f : R \times M \to M$ given by $f(r, m) = rm$ is clearly balanced. There is therefore a unique $\mathbb{Z}$-morphism $\vartheta : R \otimes M \to M$ such that $\vartheta \circ \otimes = f$. This $\mathbb{Z}$-morphism ϑ is in fact an R-morphism. For, given $r, s \in R$ and $m \in M$ we have, relative to the action defined in Theorem 15.5,
$$\vartheta[s(r \otimes m)] = \vartheta(sr \otimes m) = srm = s\vartheta(r \otimes m),$$
from which it follows that $\vartheta(tn) = t\vartheta(n)$ for all $t \in R$ and all $n \in R \otimes M$, since every such n is a linear combination of elements of the form $r \otimes m$.

To show that ϑ is an R-isomorphism, we shall construct an inverse for ϑ. For this purpose, consider the R-morphism $\xi : M \to R \otimes M$ given by $\xi(m) = 1_R \otimes m$. For every $m \in M$ we have
$$(\vartheta \circ \xi)(m) = \xi(1_R \otimes m) = m$$
and so $\vartheta \circ \xi = \mathrm{id}_M$. On the other hand, for every $r \in R$ and every $m \in M$ we have
$$(\xi \circ \vartheta)(r \otimes m) = \xi(rm) = 1_R \otimes rm = r \otimes m$$

and so, since $R \otimes M$ is generated by the elements of the form $r \otimes m$, we see that $\xi \circ \vartheta$ is the identity map on $R \otimes M$. Thus ϑ and ξ are mutually inverse R-isomorphisms. ◊

Corollary $R \otimes R \simeq R$. ◊

There is, of course, a dual result to the above, namely :

Theorem 15.8 *If M is a right R-module then there is a unique isomorphism*
$$\vartheta : M \otimes R \to M$$
such that $\vartheta(m \otimes r) = mr$. ◊

We shall now investigate how tensor products behave with respect to exact sequences. For this purpose, we require the following notions. Given morphisms $f : M_1 \to M_2$ and $g : N_1 \to N_2$, consider the diagram

$$\begin{array}{ccc} M_1 \times N_1 & \xrightarrow{\otimes_1} & M_1 \otimes N_1 \\ {\scriptstyle f \times g} \downarrow & & \\ M_2 \times N_2 & \xrightarrow{\otimes_2} & M_2 \otimes N_2 \end{array}$$

in which $f \times g$ is the *cartesian product morphism* given by

$$(f \times g)(m_1, n_1) = \bigl(f(m_1), g(n_1)\bigr).$$

It is readily seen that $\otimes_2 \circ (f \times g)$ is a balanced mapping and so there is a unique $\mathbb{Z}$-morphism $h : M_1 \otimes N_1 \to M_2 \otimes N_2$ that completes the above diagram in a commutative manner. We call this $\mathbb{Z}$-morphism the *tensor product* of f and g and denote it by $f \otimes g$.

- Although the notation $f \otimes g$ for the tensor product of the morphisms f and g is quite standard, it constitutes an indefensible abuse; for if $f : M \to N$ and $g : P \to Q$ are R-morphisms then, by our previously agreed conventions, $f \otimes g$ ought to denote an element of
$$\mathrm{Mor}_R(M, N) \otimes_R \mathrm{Mor}_R(P, Q).$$

Since we shall rarely require this latter (and proper) interpretation of $f \otimes g$, we shall adhere to the standard practice of using $f \otimes g$ for the tensor product of f and g as defined above. The reader should remain fully aware of this convention.

Tensor products; flat modules; regular rings 201

The principal properties of the tensor product of R-morphisms are summarised in the following result.

Theorem 15.9 *If M is a right R-module and N is a left R-module then*

$$\mathrm{id}_M \otimes \mathrm{id}_N = \mathrm{id}_{M \otimes N}.$$

Moreover, given the diagram $M \xrightarrow{f} M' \xrightarrow{f'} M''$ of right R-modules and the diagram $N \xrightarrow{g} N' \xrightarrow{g'} N''$ of left R-modules, we have

$$(f' \circ f) \otimes (g' \circ g) = (f' \otimes g') \circ (f \otimes g).$$

Proof It suffuces to consider the diagrams

$$\begin{array}{ccc} M \times N & \xrightarrow{\otimes} & M \otimes N \\ {\scriptstyle \mathrm{id}_M \times \mathrm{id}_N} \downarrow & & \downarrow {\scriptstyle \mathrm{id}_{M \otimes N}} \\ M \times N & \xrightarrow{\otimes} & M \otimes N \end{array}$$

$$\begin{array}{ccc} M \times N & \xrightarrow{\otimes} & M \otimes N \\ {\scriptstyle f \times g} \downarrow & & \downarrow {\scriptstyle f \otimes g} \\ M' \times N' & \xrightarrow{\otimes} & M' \otimes N' \\ {\scriptstyle f' \times g'} \downarrow & & \downarrow {\scriptstyle f' \otimes g'} \\ M'' \times N'' & \xrightarrow{\otimes} & M'' \otimes N'' \end{array}$$

and observe that each is commutative. The result therefore follows by the definition of the tensor product of two morphisms. ◊

In what follows we shall make use of the following notation. If $f : A \to B$ is an R-morphism then for any given R-module M the induced $\mathbb{Z}$-morphism $\mathrm{id}_M \otimes f : M \otimes A \to M \otimes B$ will be denoted by $^\otimes f$ and the induced $\mathbb{Z}$-morphism $f \otimes \mathrm{id}_M : A \otimes M \to B \otimes M$ will be denoted by $f^\otimes$. This notation will be used only when there is no confusion over the R-module M.

Theorem 15.10 *Let M be a right R-module and let*

$$A' \xrightarrow{f} A \xrightarrow{g} A'' \to 0$$

be an exact sequence of left R-modules and R-morphisms. Then there is the induced exact sequence of $\mathbb{Z}$-modules and $\mathbb{Z}$-morphisms

$$M \otimes A' \xrightarrow{\otimes f} M \otimes A \xrightarrow{\otimes g} M \otimes A'' \longrightarrow 0.$$

Proof Since $g \circ f = 0$ we deduce from Theorem 15.9 that $^\otimes g \circ {^\otimes f} = 0$ so that $\operatorname{Im} {^\otimes f} \subseteq \operatorname{Ker} {^\otimes g}$. Now let $\natural : M \otimes A \to (M \otimes A)/\operatorname{Im} {^\otimes f}$ be the natural $\mathbb{Z}$-morphism. By Theorem 3.3, there is a unique $\mathbb{Z}$-morphism $\vartheta : (M \otimes A)/\operatorname{Im} {^\otimes f} \to M \otimes A''$ such that $\vartheta \circ \natural = {^\otimes g}$. Again by Theorem 3.3, to show that $\operatorname{Im} {^\otimes f} = \operatorname{Ker} {^\otimes g}$ it suffices to show that ϑ is injective; and for this it suffices to find a $\mathbb{Z}$-morphism $\xi : M \otimes A'' \to (M \otimes A)/\operatorname{Im} {^\otimes f}$ such that $\xi \circ \vartheta$ is the identity map on $(M \otimes A)/\operatorname{Im} {^\otimes f}$.

We construct such a $\mathbb{Z}$-morphism ξ as follows. Given $a, b \in A$ we have

$$\begin{aligned} g(a) = g(b) &\Rightarrow a - b \in \operatorname{Ker} g = \operatorname{Im} f \\ &\Rightarrow (\exists a' \in A')\ a - b = f(a') \\ &\Rightarrow (\forall m \in M)\ m \otimes a - m \otimes b = m \otimes f(a') \in \operatorname{Im} {^\otimes f}. \end{aligned}$$

Since g is surjective, it follows that we can define a mapping

$$\alpha : M \times A'' \to (M \otimes A)/\operatorname{Im} {^\otimes f}$$

by the prescription

$$\alpha(m, a'') = m \otimes a + \operatorname{Im} {^\otimes f}$$

where $a \in A$ is such that $g(a) = a''$. It is readily verified that α is a balanced mapping and so there is a unique $\mathbb{Z}$-morphism

$$\xi : M \otimes A'' \to (M \otimes A)/\operatorname{Im} {^\otimes f}$$

such that $\xi \circ \otimes = \alpha$. Now for every $m \in M$ and every $a \in A$ we have

$$\begin{aligned} (\xi \circ \vartheta)(m \otimes a + \operatorname{Im} {^\otimes f}) &= \xi[{^\otimes g}(m \otimes a)] \\ &= \xi[m \otimes g(a)] \\ &= (\xi \circ \otimes)(m, g(a)) \\ &= \alpha(m, g(a)) \\ &= m \otimes a + \operatorname{Im} {^\otimes f}. \end{aligned}$$

Thus we see that $\xi \circ \vartheta$ coincides with the identity map on a set of generators of $(M \otimes A)/\operatorname{Im} {^\otimes f}$ whence we have that $\xi \circ \vartheta$ is the identity map.

To complete the proof, it remains to show that $^\otimes g$ is surjective. Now, given $a'' \in A''$ there exists $a \in A$ such that $g(a) = a''$ whence, for every $m \in M$,

$$m \otimes a'' = m \otimes g(a) = {^\otimes g}(m \otimes a) \in \operatorname{Im} {^\otimes g}.$$

Since then $\operatorname{Im} {}^\otimes g$ contains a set of generators of $M \otimes A''$, we conclude that $\operatorname{Im} {}^\otimes g = M \otimes A''$ and so ${}^\otimes g$ is surjective. ◊

There is of course a dual to Theorem 15.10 which we state without proof:

Theorem 15.11 *Let M be a left R-module and let*

$$A' \xrightarrow{f} A \xrightarrow{g} A'' \longrightarrow 0$$

be an exact sequence of right R-modules and R-morphisms. Then there is the induced exact sequence of $\mathbb{Z}$-modules and $\mathbb{Z}$-morphisms

$$A' \otimes M \xrightarrow{f^\otimes} A \otimes M \xrightarrow{g^\otimes} A'' \otimes M \longrightarrow 0. \quad ◊$$

- By Theorems 15.10 and 15.11 we see that 'tensoring by M' preserves a certain amount of exactness. We note here, however, that it does not go as far as preserving short exact sequences in general. For example, consider the short exact sequence

$$0 \longrightarrow \mathbb{Z} \xrightarrow{f} \mathbb{Z} \xrightarrow{\natural} \mathbb{Z}/2\mathbb{Z} \longrightarrow 0$$

in which f is given by $f(n) = 2n$. Consider now the induced exact sequence

$$\mathbb{Z}/2\mathbb{Z} \otimes \mathbb{Z} \xrightarrow{{}^\otimes f} \mathbb{Z}/2\mathbb{Z} \otimes \mathbb{Z} \xrightarrow{{}^\otimes \natural} \mathbb{Z}/2\mathbb{Z} \otimes \mathbb{Z}/2\mathbb{Z} \longrightarrow 0.$$

Since

$$\begin{aligned}
{}^\otimes f(n + 2\mathbb{Z} \otimes m) &= n + 2\mathbb{Z} \otimes f(m) \\
&= n + 2\mathbb{Z} \otimes 2m \\
&= 2(n + 2\mathbb{Z}) \otimes m \\
&= 0 + 2\mathbb{Z} \otimes m \\
&= 0,
\end{aligned}$$

it follows that ${}^\otimes f$ is the zero map. Its kernel is therefore $\mathbb{Z}/2\mathbb{Z} \otimes \mathbb{Z}$ which, by Theorem 15.8, is $\mathbb{Z}$-isomorphic to $\mathbb{Z}/2\mathbb{Z}$ and so cannot be a zero module. Thus ${}^\otimes f$ is not a monomorphism and the induced sequence is not short exact. Despite this, however, we do have the following preservation of split exact sequences.

Theorem 15.12 *Let M be a right R-module and let*

$$0 \longrightarrow A' \xrightarrow{f} A \xrightarrow{g} A'' \longrightarrow 0$$

be a split short exact sequence of left R-modules and R-morphisms. Then there is the induced split short exact sequence

$$0 \longrightarrow M \otimes A' \xrightarrow{{}^\otimes f} M \otimes A \xrightarrow{{}^\otimes g} M \otimes A'' \longrightarrow 0$$

of $\mathbb{Z}$-modules and $\mathbb{Z}$-morphisms.

Proof By virtue of Theorem 15.10 it suffices to show, using the fact that the given sequence splits, that $^\otimes f$ is injective. For this purpose, let f^0 be a splitting morphism for f and define a mapping $\alpha : M \times A \to M \otimes A'$ by $\alpha(m, a) = m \otimes f^0(a)$. It is readily verified that α is a balanced mapping and so there is a unique $\mathbb{Z}$-morphism $\vartheta : M \otimes A \to M \otimes A'$ such that $\vartheta(m \otimes a) = m \otimes f^0(a)$ for all $m \in M$ and all $a \in A$. Now given any $m \in M$ and any $a' \in A'$ we have

$$(\vartheta \circ {}^\otimes f)(m \otimes a') = \vartheta(m \otimes f(a')) = m \otimes f^0[f(a')] = m \otimes a'$$

and so, since $M \otimes A'$ is generated by the elements of the form $m \otimes a'$, we deduce that $\vartheta \circ {}^\otimes f$ is the identity map on $M \otimes A'$, whence $^\otimes f$ is injective. ◊

There is, of course, a dual result to Theorem 15.12.

Before proceeding to discuss those R-modules M which, when tensored into a short exact sequence, induce a short exact sequence, we derive some useful consequences of Theorems 15.10 and 15.11.

Theorem 15.13 *If* $E' \xrightarrow{\alpha} E \xrightarrow{\beta} E'' \longrightarrow 0$ *is an exact sequence of right R-modules and* $F' \xrightarrow{\gamma} F \xrightarrow{\delta} F'' \longrightarrow 0$ *is an exact sequence of left R-modules then there is a short exact sequence*

$$0 \longrightarrow \mathrm{Im}(\mathrm{id}_E \otimes \gamma) + \mathrm{Im}(\alpha \otimes \mathrm{id}_F) \xrightarrow{i} E \otimes F \xrightarrow{\beta \otimes \delta} E'' \otimes F'' \longrightarrow 0$$

of $\mathbb{Z}$-modules and $\mathbb{Z}$-morphisms, in which i is the natural inclusion.

Proof Consider the diagram

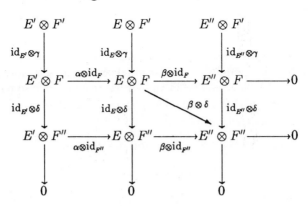

By Theorems 15.10 and 15.11, both rows and all three columns are exact; moreover, by Theorem 15.9, the diagram is commutative. The result will

follow, therefore, if we can show via a diagram chase that
$$\mathrm{Ker}(\beta\otimes\delta) = \mathrm{Im}(\mathrm{id}_E\otimes\gamma) + \mathrm{Im}(\alpha\otimes\mathrm{id}_F).$$
Now by the exactness of the middle column we see that
$$(\beta\otimes\delta)\circ(\mathrm{id}_E\otimes\gamma) = 0;$$
and likewise, by the exactness of the top row, that
$$(\beta\otimes\delta)\circ(\alpha\otimes\mathrm{id}_F) = 0.$$
It follows that
$$\mathrm{Im}(\mathrm{id}_E\otimes\gamma) + \mathrm{Im}(\alpha\otimes\mathrm{id}_F) \subseteq \mathrm{Ker}(\beta\otimes\delta).$$
To obtain the reverse inclusion, let $z \in \mathrm{Ker}(\beta\otimes\delta)$; then
$$(\mathrm{id}_E\otimes\delta)(z) \in \mathrm{Ker}(\beta\otimes\mathrm{id}_{F''}) = \mathrm{Im}(\alpha\otimes\mathrm{id}_{F''})$$
and so there exists $x \in E'\otimes F''$ such that $(\mathrm{id}_E\otimes\delta)(z) = (\alpha\otimes\mathrm{id}_{F''})(x)$. Since $\mathrm{id}_{E'}\otimes\delta$ is surjective there then exists $y \in E'\otimes F$ such that
$$(\mathrm{id}_E\otimes\delta)(z) = (\alpha\otimes\mathrm{id}_{F''})[(\mathrm{id}_{E'}\otimes\delta)(y)].$$
Now let $z' = z - (\alpha\otimes\mathrm{id}_F)(y)$. Then $z' \in \mathrm{Ker}(\mathrm{id}_E\otimes\delta) = \mathrm{Im}(\mathrm{id}_E\otimes\gamma)$ and so
$$z = z' + (\alpha\otimes\mathrm{id}_F)(y) \in \mathrm{Im}(\mathrm{id}_E\otimes\gamma) + \mathrm{Im}(\alpha\otimes\mathrm{id}_F),$$
whence the reverse inclusion follows. ◊

Corollary 1 *If, in the above, E' and F' are submodules of E and F respectively and if $\iota_{E'}, \iota_{F'}$ are the corresponding natural inclusions, then there is a $\mathbb{Z}$-isomorphism*
$$E/E' \otimes F/F' \simeq (E\otimes F)/(\mathrm{Im}(\mathrm{id}_E\otimes\iota_{F'}) + \mathrm{Im}(\iota_{E'}\otimes\mathrm{id}_F)).$$

Proof It suffices to apply the above to the canonical exact sequences in which $E'' = E/E'$ and $F'' = F/F'$. ◊

Corollary 2 *Let I be a right ideal of R and let M be a left R-module. Then there is a $\mathbb{Z}$-isomorphism*
$$R/I \otimes M \simeq M/IM.$$

Proof Taking $E = R, E' = I, F = M, F' = \{0\}$ in Corollary 1, we obtain
$$R/I \otimes M \simeq (R\otimes M)/\mathrm{Im}(\iota_I\otimes\mathrm{id}_M).$$
The result now follows by Theorem 15.7. ◊

- Note that the isomorphism of Corollary 2 can be described by the assignment $(r+I) \otimes M \mapsto rm/IM$.

We now give consideration to those modules which, on tensoring into a short exact sequence, induce a short exact sequence. More explicitly, recalling Theorem 15.10, we introduce the following notion.

Definition A right R-module M is said to be *flat* if, for every monomorphism $f : A \to A'$ of left R-modules, the induced $\mathbb{Z}$-morphism

$$M \otimes A' \xrightarrow{\otimes f} M \otimes A$$

is a monomorphism. Flat left modules are defined similarly.

An immediate example of a flat module is provided by the following.

Theorem 15.14 *The right (respectively left) R-module R is flat.*

Proof From the proof of Theorem 15.7 there is an isomorphism $\xi_A : A \to R \otimes A$ given by $a \mapsto 1_R \otimes a$. For any R-morphism $f : A' \to A$ we therefore have the commutative diagram

$$\begin{array}{ccc} A' & \xrightarrow{f} & A \\ {\scriptstyle \xi_{A'}}\downarrow & & \downarrow{\scriptstyle \xi_A} \\ R \otimes A' & \xrightarrow[\otimes f]{} & R \otimes A \end{array}$$

which yields $\otimes f = \xi_A \circ f \circ \xi_{A'}^{-1}$. Thus we see that $\otimes f$ is a monomorphism whenever f is a monomorphism. Consequently, R is flat. $\diamond$

In order to obtain an abundant supply of flat modules, we shall now investigate how tensor products behave in relation to direct sums.

Theorem 15.15 *If $(M_i)_{i \in I}$ is a family of left R-modules and if M is a right R-module then $\left(M \otimes \bigoplus_{i \in I} M_i, (\mathrm{id}_M \otimes \mathrm{in}_i)_{i \in I}\right)$ is a coproduct of the family $(M \otimes M_i)_{i \in I}$.*

Proof We shall use the fact that the natural epimorphisms $\mathrm{pr}_j^\oplus$ and the natural monomorphisms in_j satisfy the properties given in Theorem 6.7.

For every $j \in I$ consider the $\mathbb{Z}$-morphisms

$$\alpha_j = \mathrm{id}_M \otimes \mathrm{pr}_j^\oplus : M \otimes \bigoplus_{i \in I} M_i \to M \otimes M_j;$$
$$\beta_j = \mathrm{id}_M \otimes \mathrm{in}_j : M \otimes M_j \to M \otimes \bigoplus_{i \in I} M_i.$$

Tensor products; flat modules; regular rings 207

It is immediate from Theorem 15.9 that

$$\alpha_k \circ \beta_j = \begin{cases} \mathrm{id}_{M \otimes M_j} & \text{if } k = j; \\ 0 & \text{if } k \neq j. \end{cases}$$

Now if $m \in M$ and $(m_i)_{i \in I} \in \bigoplus_{i \in I} M_i$ then from

$$\alpha_j(m \otimes (m_i)_{i \in I}) = m \otimes \mathrm{pr}_j^\oplus((m_i)_{i \in I})$$

and the fact that $\mathrm{pr}_j^\oplus((m_i)_{i \in I})$ is zero for all but finitely many $j \in I$, we see that $\alpha_j(m \otimes (m_i)_{i \in I})$ is zero for all but finitely many $j \in I$. Moreover,

$$\sum_{j \in I}(\beta_j \circ \alpha_j)(m \otimes (m_i)_{i \in I}) = m \otimes \sum_{j \in I}(\mathrm{in}_j \circ \mathrm{pr}_j^\oplus)((m_i)_{i \in I}) = m \otimes (m_i)_{i \in I}.$$

The result now follows on appealing to Theorem 6.7. ◊

Corollary *If $(M_i)_{i \in I}$ is a family of left R-modules and M is a right R-module then there is a $\mathbb{Z}$-isomorphism*

$$M \otimes \bigoplus_{i \in I} M_i \simeq \bigoplus_{i \in I}(M \otimes M_i).$$

Proof This is immediate by Theorem 6.5. ◊

Theorem 15.16 *If $(M_i)_{i \in I}$ is a family of left R-modules then $\bigoplus_{i \in I} M_i$ is flat if and only if every M_i is flat.*

Proof Let $f : M' \to M$ be a monomorphism of right R-modules and consider the diagram

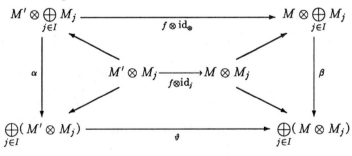

in which $\mathrm{id}_\oplus$ denotes the identity map on $\bigoplus_{j \in I} M_j$, id_j denotes the identity map on M_j, the non-horizontal and non-vertical maps are the obvious monomorphisms, α and β are the unique isomorphisms that make the left and right triangles commutative (Theorems 15.13 and 6.5), and ϑ

is the unique morphism that makes the lower trapezium commutative (definition of coproduct). We leave the reader to check that α is such that
$$\alpha((m' \otimes m_j)_{j \in I}) = m' \otimes (m_j)_{j \in I},$$
that β satisfies a similar identity, that
$$\vartheta((m' \otimes m_j)_{j \in I}) = (m \otimes m_j)_{j \in I},$$
and that the entire diagram is commutative. Now since α and β are isomorphisms we see that $f \otimes \mathrm{id}_\oplus$ is injective if and only if ϑ is injective; and clearly ϑ is injective if and only if every $f \otimes \mathrm{id}_j$ is injective. It therefore follows that $\bigoplus_{i \in I} M_i$ is flat if and only if every M_i is flat. $\diamond$

Corollary *Every projective module is flat.*

Proof Suppose that P is a projective R-module. By Theorem 8.8, P is a direct summand of a free R-module F. By the Corollary to Theorem 7.6, F is isomorphic to a direct sum of copies of R and so, by Theorems 15.14 and 15.16, F is flat. By Theorem 15.16 again, it follows that P is flat. $\diamond$

We now consider the following interesting connection between flat modules and injective modules.

Theorem 15.17 *A right R-module M is flat if and only if its character module M^+ is injective.*

Proof $\Rightarrow$: Suppose that M is flat and let $f : A' \to A$ be a monomorphism of left R-modules. Then we have the exact sequence
$$0 \longrightarrow M \otimes A' \xrightarrow{\otimes f} M \otimes A.$$
Since the $\mathbb{Z}$-module $\mathbb{Q}/\mathbb{Z}$ is injective, we can construct the diagram

$$\begin{array}{ccccc}
0 & \longleftarrow & (M \otimes A')^+ & \xleftarrow{(\otimes f)^*} & (M \otimes A)^+ \\
& & \vartheta_1 \uparrow & & \uparrow \vartheta_2 \\
& & \mathrm{Mor}_R(A', M^+) & \xleftarrow{f^*} & \mathrm{Mor}_R(A, M^+)
\end{array}$$

in which the top row is exact and ϑ_1, ϑ_2 are the isomorphisms of the Corollary to Theorem 15.6. We note in fact from the proof of Theorem 15.6 that, for example, ϑ_1 is such that
$$[\vartheta_1(f)](m \otimes a') = [f(a')](m).$$

We now verify that this diagram is commutative. If $\alpha \in \mathrm{Mor}_R(A, M^+)$ then we have on the one hand

$$(\vartheta_1 \circ f^\star)(\alpha) = \vartheta_1[f^\star(\alpha)] = \vartheta_1(\alpha \circ f^\star)$$

where

$(\star)$ $\quad [\vartheta_1(\alpha \circ f)](m \otimes a') = [(\alpha \circ f)(a')](m).$

On the other hand,

$$[(^\otimes f)^\star \circ \vartheta_2](\alpha) = (^\otimes f)^\star[\vartheta_2(\alpha)]$$

where

$(\star\star)$ $\quad (^\otimes f)^\star[\vartheta_2(\alpha)](m \otimes a') = \vartheta_2(\alpha)[m \otimes f(a')] = [(\alpha \circ f)(a')](m).$

The commutativity now follows from $(\star)$ and $(\star\star)$. It follows by this commutativity that $f^\star$ is an epimorphism, whence we see that M^+ is injective.

$\Leftarrow$: Conversely, suppose that M^+ is injective. Then we can construct a diagram similar to the above in which $f^\star$ is an epimorphism. The commutativity of such a diagram shows that $(^\otimes f)^\star$ is also an epimorphism whence, by Theorem 12.6, we have the exact sequence

$$0 \longrightarrow M \otimes A' \xrightarrow{^\otimes f} M \otimes A.$$

Consequently, M is flat. $\diamond$

The above result gives the following criterion for flatness.

Theorem 15.18 *A right R-module is flat if and only if, for every left ideal I of R, the induced sequence*

$$0 \longrightarrow M \otimes I \xrightarrow{^\otimes \iota} M \otimes R$$

is exact, where $\iota : I \to R$ is the natural inclusion.

Proof Since the necessity is clear from the definition of flatness, we need establish only sufficiency.

Suppose then that every induced sequence of the above form is exact. Then, just as in the proof of the necessity in Theorem 15.17, we can show that every sequence

$$0 \longleftarrow \mathrm{Mor}_R(I, M^+) \xleftarrow{\iota^\star} \mathrm{Mor}_R(R, M^+)$$

is exact. By the Corollary to Theorem 12.2, it follows that M^+ is injective and so, by Theorem 15.17, M is flat. $\diamond$

If I is a left ideal of R then for every right R-module M the mapping $\alpha : M \times I \to MI$ given by $\alpha(m, r) = mr$ is clearly balanced and so there is a unique $\mathbb{Z}$-morphism $\vartheta_I : M \otimes I \to MI$ such that $\vartheta_I(m \otimes r) = mr$. It is clear that ϑ_I is an epimorphism. The above result therefore yields the following

Corollary *A right R-module M is flat if and only if, for every left ideal I of R, the map ϑ_I is a $\mathbb{Z}$-isomorphism.*

Proof For every left ideal I and natural inclusion $\iota_I : I \to R$ consider the diagram

$$\begin{array}{ccc} M \otimes I & \xrightarrow{\otimes \iota} & M \otimes R \\ {\scriptstyle \vartheta_I} \downarrow & & \downarrow {\scriptstyle \vartheta} \\ MI & \xrightarrow{j} & MR \end{array}$$

in which ϑ is the isomorphism of Theorem 15.8 and j is the natural inclusion. This diagram is clearly commutative and so $\otimes \iota = \vartheta^{-1} \circ j \circ \vartheta_I$. Since ϑ_I is an epimorphism and $\vartheta^{-1} \circ j$ is a monomorphism, we deduce from Theorem 3.4 that Ker $\vartheta_I =$ Ker $\otimes \iota$. It follows that every ϑ_I is a $\mathbb{Z}$-isomorphism if and only if every $\otimes \iota$ is a $\mathbb{Z}$-monomorphism, which is the case if and only if M is flat. $\Diamond$

- Note that in the above Corollary we can restrict I to be a *finitely generated* left ideal. In fact, if for every such ideal I the morphism ϑ_I is injective then so is the corresponding morphism ϑ_J for every left ideal J of R. To see this, suppose that $\sum_{i=1}^{n}(m_i \otimes a_i) \in$ Ker ϑ_J, i.e. that $\sum_{i=1}^{n} m_i a_i = 0$. Then $\sum_{i=1}^{n}(m_1 \otimes a_i) \in$ Ker ϑ_I where $I = \sum_{i=1}^{n} Ra_i$ is finitely generated. By hypothesis, $\sum_{i=1}^{n}(m_i \otimes a_i) = 0$ as an element of $M \otimes I$. If $i_I : I \to J$ is the inclusion map then $\sum_{i=1}^{n}(m_i \otimes a_i)$ is the zero of $M \otimes J$.

Yet another useful criterion for flatness is given in the following result.

Theorem 15.19 *Let M be a right R-module, F a free right R-module, and $\pi : F \to M$ an epimorphism. Then M is flat if and only if, for every (finitely generated) left ideal I of R,*

$$FI \cap \operatorname{Ker} \pi = (\operatorname{Ker} \pi) I.$$

Tensor products; flat modules; regular rings

Proof We have the exact sequence

$$0 \longrightarrow \operatorname{Ker} \pi \xrightarrow{\iota} F \xrightarrow{\pi} M \longrightarrow 0$$

in which F, being free and therefore projective, is flat by the Corollary to Theorem 15.6. For every left ideal I of R we then have the exact sequence

$$\operatorname{Ker} \pi \otimes I \xrightarrow{\otimes \iota} F \otimes I \xrightarrow{\otimes \pi} M \otimes I \longrightarrow 0.$$

Since F is flat, we have a $\mathbb{Z}$-isomorphism $F \otimes I \simeq FI$ by the Corollary to Theorem 15.18; and under this isomorphism $\operatorname{Im} \otimes \iota$ corresponds to $(\operatorname{Ker} \pi)I$. We therefore have $M \otimes I = \operatorname{Im} \otimes \pi \simeq FI/(\operatorname{Ker} \pi)I$. By the Corollary to Theorem 15.18 again, we see that M is flat if and only if $MI \simeq FI/(\operatorname{Ker} \pi)I$ for every left ideal I. Now MI consists of all finite sums of the form

$$\sum_{i=1}^{n} \pi(f_i) r_i = \sum_{i=1}^{n} \pi(f_i r_i) = \pi\left(\sum_{i=1}^{n} f_i r_i\right)$$

where $f_i \in F$ and $r_i \in R$. Thus $MI = \pi^{\rightarrow}(FI) \simeq FI/(FI \cap \operatorname{Ker} \pi)$ by the first isomorphism theorem. We deduce, therefore, that M is flat if and only if

$$FI/(FI \cap \operatorname{Ker} \pi) \simeq FI/(\operatorname{Ker} \pi)I$$

for all left ideals I. Now clearly $(\operatorname{Ker} \pi)I \subseteq FI \cap \operatorname{Ker} \pi$ and so we conclude from the correspondence theorem that M is flat if and only if $FI \cap \operatorname{Ker} \pi = (\operatorname{Ker} \pi)I$. $\Diamond$

We can apply the above result to determine which $\mathbb{Z}$-modules are flat. For this purpose, we require the following notion.

Definition A $\mathbb{Z}$-module M is said to be *torsion-free* if

$$(\forall x \in M)(\forall n \in \mathbb{Z}) \qquad nx = 0 \Rightarrow x = 0.$$

Example 15.5 The $\mathbb{Z}$-module $\mathbb{Z}$ is torsion-free; but when $n \neq 0$ the $\mathbb{Z}$-modules $\mathbb{Z}/n\mathbb{Z}$ are not torsion-free.

Theorem 15.20 *A $\mathbb{Z}$-module is flat if and only if it is torsion-free.*

Proof Let M be a $\mathbb{Z}$-module and $\pi : F \to M$ an epimorphism with F a free $\mathbb{Z}$-module. By Theorem 15.19 and the fact that every ideal of $\mathbb{Z}$ is of the form $n\mathbb{Z}$ for some $n \in \mathbb{Z}$, we see that M is flat if and only if

$$(\forall n \in \mathbb{Z}) \qquad F(n\mathbb{Z}) \cap \operatorname{Ker} \pi \subseteq (\operatorname{Ker} \pi)n\mathbb{Z};$$

in other words, if and only if

$$(\forall n \in \mathbb{Z})(\forall f \in F) \qquad nf \in \operatorname{Ker} \pi \Rightarrow f \in \operatorname{Ker} \pi.$$

On passing to quotients, we see that this is equivalent to

$$(\forall n \in \mathbb{Z})(\forall m \in M \simeq F/\operatorname{Ker} \pi) \qquad nm = 0 \Rightarrow n = 0,$$

which is precisely the condition that M be torsion free. $\diamond$

- We shall consider later the notion of a torsion-free module over a more general ring, when we shall generalise Theorem 15.20.
- By Theorem 15.20 we can assert that the $\mathbb{Z}$-module $\mathbb{Q}$, for example, is flat. Although $\mathbb{Q}$ is an injective $\mathbb{Z}$-module (see Exercise 12.3 for the details), it is not a projective $\mathbb{Z}$-module (this will be established later). Thus the class of flat modules is wider than that of projective modules.
- Later, we shall be concerned with a particular type of module for which the notions of free, projective, and flat coincide.

Do there exist rings R such that every R-module is flat? We shall answer this question in the affirmative, and for this we consider the following type of ring.

Definition A unitary ring R is said to be *regular* (or to be a *von Neumann ring*) if for every $a \in R$ there exists $x \in R$ such that $axa = a$.

Example 15.6 If R is a unitary ring then the set of idempotents of R forms a regular ring under the multiplication of R and the addition $x \oplus y = x + y - xy$.

Theorem 15.21 *If R is a unitary ring then the following statements are equivalent* :

(1) *R is a regular ring;*
(2) *every principal left ideal of R is generated by an idempotent;*
(3) *for every principal left ideal Ra of R there exists $b \in R$ such that $R = Ra \oplus Rb$;*
(4) *every principal left ideal of R is a direct summand of R;*
(5) *similar statements concerning right ideals.*

Proof We give a proof for left ideals; the symmetry of (1) shows that we can replace 'left' by 'right' throughout.

(1) $\Rightarrow$ (2) : Given $a \in R$ let $x \in R$ be such that $a = axa$. Then $xa = xaxa = (xa)^2$. Moreover, it is clear that $Rxa \subseteq Ra$ and $Ra = Raxa \subseteq Rxa$, so that $Ra = Raxa$ from which (2) follows.

(2) $\Rightarrow$ (3) : Let $e \in R$ be an idempotent such that $Ra = Re$. Since $1_R = e + (1_R - e)$ we see that $R = Re + R(1_R - e)$. Moreover, if $x \in Re \cap R(1_R - e)$ then $x = ye = z(1_R - e)$ for some $y, z \in R$ whence

$$x = ye = ye^2 = (z(1_R - e))e = z(e - e^2) = z0 = 0.$$

Thus we see that $R = Re \oplus R(1_R - e)$.

(3) $\Rightarrow$ (4) : This is trivial.

(4) $\Rightarrow$ (1) : Given any $a \in R$, there exists by the hypothesis a left ideal J such that $R = Ra \oplus J$. Then $1_R = xa + b$ where $x \in R$ and $b \in J$, and consequently $a = axa + ab$. Since then

$$ab = a - axa \in Ra \cap J = \{0\},$$

we conclude that $a = axa$. $\Diamond$

Corollary 1 *If R is a regular ring then every finitely generated left or right ideal of R is principal.*

Proof Suppose that the left ideal I is generated by $\{x_1, \ldots, x_n\}$. Then by Theorem 15.21(2) there exist idempotents $e_1, \ldots, e_n$ such that

$$I = Rx_1 + \ldots + Rx_n = Re_1 + \ldots + Re_n.$$

We show by induction that $Re_1 + \ldots + Re_n$ is principal. The result is trivial for $n = 1$. It holds for $n = 2$ as follows. First we observe that

$$a_1 e_1 + a_2 e_2 = (a_1 + a_2 e_2)e_1 + a_2(e_2 - e_2 e_1),$$

from which we deduce that $Re_1 + Re_2 = Re_1 + R(e_2 - e_2 e_1)$. Now since R is regular there exists $x \in R$ such that $(e_2 - e_2 e_1)x(e_2 - e_2 e_1) = e_2 - e_2 e_1$, and $e_2^\star = x(e_2 - e_2 e_1)$ is idempotent. Moreover, we see that $Re_1 + Re_2 = Re_1 + Re_2^\star$ with $e_2^\star e_1 = x(e_2 - e_2 e_1)e_1 = 0$. But since

$$a_1 e_1 + a_2 e_2^\star = (a_1 e_1 + a_2 e_2^\star)(e_1 + e_2^\star - e_1 e_2^\star),$$

we also have $Re_1 + Re_2^\star = R(e_1 + e_2^\star - e_1 e_2^\star)$ where $e_1 + e_2^\star - e_1 e_2^\star$ is idempotent. Hence we have $Re_1 + Re_2 = Re_3$ where $e_3 = e_2 + e_2^\star - e_1 e_2^\star$. The inductive step readily follows from this observation. $\Diamond$

Corollary 2 *A unitary ring is regular if and only if every finitely generated left/right ideal of R is a direct summand of R.* $\Diamond$

Example 15.7 If D is a division ring and if V is a module over D then the ring $\text{End}_D V$ is regular. To see this, it suffices to show that every principal right ideal of $\text{End}_D V$ is generated by an idempotent. Given $f \in \text{End}_D V$, let p project onto $\text{Im}\, f$. Then p is an idempotent and $V = \text{Im}\, p \oplus \text{Ker}\, p$ with $\text{Im}\, p = \text{Im}\, f$. Let $(f(v_i))_{i \in I}$ be a basis of $\text{Im}\, p$ and define $g \in \text{End}_D V$ by

$$\begin{cases} g(x) = 0 & \text{if } x \in \text{Ker}\, p = \text{Ker}\, f; \\ g[f(v_i)] = v_i & \text{for all } i \in I. \end{cases}$$

Then for $x \in \text{Ker}\, p$ we have $(f \circ g)(x) = 0 = p(x)$; and for all $i \in I$,

$$(f \circ g)[f(v_i)] = f(v_i) = (p \circ f)(v_i).$$

Thus we have $f \circ g = p$. But the restriction of p to $\text{Im}\, p = \text{Im}\, f$ is the identity map, and therefore we have $f = p \circ f$. These observations show that the right ideals of $\text{End}_D V$ generated by f and by p coincide. Consequently, the ring $\text{End}_D V$ is regular by Theorem 15.21.

Theorem 15.22 *If R is a unitary ring then the following statements are equivalent* :

(1) *R is regular*;
(2) *every left R-module is flat*;
(3) *every right R-module is flat*.

Proof Since the concept of regularity is symmetric, it suffices to establish the equivalence of (1) and (3).

(1) $\Rightarrow$ (3) : Suppose that R is regular and let M be a right R-module. By Corollary 1 of Theorem 15.21, every finitely generated right ideal of R is principal and so, in the notation of Theorem 15.19, it suffices to show that for every $r \in R$ we have $Fr \cap \text{Ker}\, \pi \subseteq (\text{Ker}\, \pi)r$. But if $x \in Fr \cap \text{Ker}\, \pi$ then we have $x = fr$ where $f \in \text{Ker}\, \pi$ and so, by the regularity of R,

$$x = fr = frr'r = xr'r \in (\text{Ker}\, \pi)r.$$

(3) $\Rightarrow$ (1) : If $r \in R$ then by (3) the right R-module R/rR is flat. By Theorem 15.19 with $F = R$ and $\pi = \natural_{rR}$, we see that for every left ideal A of R we have $A \cap rR = rA$. In particular, taking $A = Rr$ we obtain $r \in Rr \cap rR = rRr$ whence $r = rr'r$ for some $r' \in R$. $\Diamond$

Finally, we have the following connection with semisimplicity.

Theorem 15.23 *If R is a unitary ring then the following statements are equivalent* :

(1) *R is semisimple*;

Tensor products; flat modules; regular rings

(2) R is noetherian and regular.

Proof If R is semisimple then it is clearly noetherian. Since every left ideal of R is a direct summand, so in particular is every finitely generated left ideal whence R is also regular.

Conversely, if R is noetherian then every left ideal is finitely generated, and if R is regular then every left ideal is principal and a direct summand of R. Hence R is semisimple. ◊

EXERCISES

15.1 Let R be a unitary ring. If I is a right ideal of R and if M is a left R-module prove that there is a $\mathbb{Z}$-isomorphism

$$f : M \otimes R/I \to M/IM$$

such that $f(m \otimes (r+I)) = rm + IM$. Deduce that if L is a left ideal of R then, as abelian groups,

$$R/L \otimes R/I \simeq R/(I+L).$$

[*Hint.* Use Corollary 1 of Theorem 15.13.]

15.2 Let G be an additive abelian group. For every positive integer n let $nG = \{ng \; ; \; g \in G\}$. Establish a $\mathbb{Z}$-module isomorphism

$$\mathbb{Z}/n\mathbb{Z} \otimes_{\mathbb{Z}} G \simeq G/nG.$$

[*Hint.* Use Corollary 2 of Theorem 15.13.]

15.3 Prove that $\mathbb{Z}/n\mathbb{Z} \otimes_{\mathbb{Z}} \mathbb{Q} = \{0\}$.

15.4 If $(M_i)_{i \in I}$ is a family of right R-modules and if $(N_j)_{j \in J}$ is a family of left R-modules, establish a $\mathbb{Z}$-isomorphism

$$\bigoplus_{i \in I} M_i \otimes \bigoplus_{j \in J} N_j \simeq \bigoplus_{(i,j)} (M_i \otimes N_j).$$

15.5 Given a short exact sequence

$$0 \longrightarrow M' \longrightarrow M \longrightarrow M'' \longrightarrow 0$$

of R-modules and R-morphisms in which M' and M'' are flat, prove that M is flat.

[*Hint.* Use the Corollary to Theorem 15.18.]

15.6 Prove that a ring R is regular if and only if $IJ = I \cap J$ for every left ideal I and every right ideal J of R. Deduce that if R is commutative then R is regular if and only if $I^2 = I$ for every ideal I of R.

15.7 Prove that a regular ring has zero Jacobson radical.

15.8 Prove that the centre of a regular ring is regular.

15.9 Prove that a regular ring with no zero divisors is a division ring.

15.10 Suppose that R is a unitary ring with no non-zero nilpotent elements. Prove that every idempotent of R is in the centre of R.

[*Hint.* If $e^2 = e$ consider $[(1_R - e)ae]^2$ and $[ea(1_R - e)]^2$.]

If R is regular, deduce that every left ideal of R is two-sided.

15.11 Prove that in a regular ring the intersection of two principal left ideals is a principal left ideal.

[*Hint.* Observe that if e is an idempotent then so is $1_R - e$, and that $Re = \text{Ann}_R(1_R - e)R$.]

16 Tensor products; tensor algebras

We shall now concentrate on tensor products in the case where the ground ring R is commutative. In this case all the morphism groups in question may be regarded as R-modules and we can generalise the notion of tensor product to an arbitrary family of R-modules. The motivation for this is as follows.

Definition Let M, N, P be R-modules. A mapping $f : M \times N \to P$ is said to be *bilinear* if the following identities hold :

$$f(m + m', n) = f(m, n) + f(m', n);$$
$$f(m, n + n') = f(m, n) + f(m, n');$$
$$f(\lambda m, n) = \lambda f(m, n) = f(m, \lambda n).$$

Example 14.1 If R is a commutative unitary ring and M is an R-module then the mapping $f : M^d \times M \to R$ given by $f(x^d, x) = x^d(x) = \langle x, x^d \rangle$ is bilinear. This follows from identities (α) to (δ) of Chapter 9 and the fact that R is commutative.

Theorem 16.1 *If R is a commutative unitary ring and M, N are R-modules then $M \otimes_R N$ is an R-module in which $\lambda m \otimes_R n = \lambda(m \otimes_R n) = m \otimes_R \lambda n$, and $\otimes_R$ is bilinear.*

Proof Both M and N are (R, R)-bimodules and so $M \otimes_R N$ is an R-module by Theorem 15.5 and the subsequent remark; in fact the action in question is given by

$$\lambda \left(\sum_{i=1}^{t} x_i \otimes_R y_i \right) = \sum_{i=1}^{t} \lambda x_i \otimes_R y_i = \sum_{i=1}^{t} x_i \otimes_R \lambda y_i.$$

It is readily verified that $\otimes_R$ is bilinear. $\diamond$

Theorem 16.2 *Let R be a commutative unitary ring and let M and N be R-modules. Then the R-module $M \otimes_R N$ satisfies the following property : if P is an R-module and if $g : M \times N \to P$ is a bilinear mapping there is a unique R-morphism $h : M \otimes_R N \to P$ such that the diagram*

$$\begin{array}{ccc} M \times N & \xrightarrow{g} & P \\ {\scriptstyle \otimes_R} \downarrow & \nearrow {\scriptstyle h} & \\ M \otimes_R N & & \end{array}$$

is commutative.

Proof We know that there is a unique $\mathbb{Z}$-morphism $h : M \otimes_R N \to P$ such that $h \circ \otimes_R = g$. It therefore suffices to show that h is an R-morphism; and this follows from the equalities

$$h[\lambda(x \otimes y)] = h(\lambda x \otimes y) = g(\lambda x, y) = \lambda g(x,y) = \lambda h(x \otimes y)$$

and the fact that $M \otimes_R N$ is generated by $\text{Im} \otimes_R$. ◊

- As we shall see, when R is commmutative the R-module $M \otimes_R N$ is characterised by the property given in Theorem 16.2. In fact, the results of Theorems 16.1 and 16.2 give rise to the following general situation.

Definition Let R be a commutative unitary ring and let $(M_i)_{i \in I}$ be a family of R-modules. If N is an R-module then a mapping $f : \underset{i \in I}{\text{\Large$\times$}} M_i \to N$ is said to be *multilinear* if, whenever $(x_i)_{i \in I}, (y_i)_{i \in I}, (z_i)_{i \in I} \in \underset{i \in I}{\text{\Large$\times$}} M_i$ are such that, for some k, $z_k = x_k + y_k$ and $z_i = x_i = y_i$ for $i \neq k$, then

$$f((z_i)_{i \in I}) = f((x_i)_{i \in I}) + f((y_i)_{i \in I});$$

and whenever $(x_i)_{i \in I}, (y_i)_{i \in I} \in \underset{i \in I}{\text{\Large$\times$}} M_i$ are such that, for some $k \in I$ $y_k = \lambda x_k$ and $y_i = x_i$ for $i \neq k$, then

$$f((y_i)_{i \in I}) = \lambda f((x_i)_{i \in I}).$$

- In the case where $I = \{1, \ldots, n\}$ we shall use the term *n-linear* instead of multilinear; and in particular when $n = 2, 3$ we shall use the terms *bilinear*, *trilinear*.

Motivated by the desire to trade in multilinear mappings for morphisms, we now introduce the following concept.

Definition Let R be a commutative unitary ring and let $(M_i)_{i \in I}$ be a family of R-modules. Then by a *tensor product* of the family $(M_i)_{i \in I}$ we shall mean an R-module T together with a multilinear mapping $f : \underset{i \in I}{\text{\Large$\times$}} M_i \to T$ such that, for every R-module N and every multilinear mapping $g : \underset{i \in I}{\text{\Large$\times$}} M_i \to N$, there is a unique R-morphism $h : T \to N$ such that the diagram

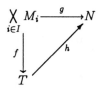

is commutative. We denote such a tensor product by (T, f).

Theorem 16.3 *If (T, f) is a tensor product of the family $(M_i)_{i \in I}$ of R-modules then $\operatorname{Im} f$ generates T.*

Proof This is as in the second part of Theorem 7.1. ◊

Theorem 16.4 [Uniqueness] *Let (T, f) be a tensor product of the family $(M_I)_{i \in I}$ of R-modules. Then (T', f') is also a tensor product of this family if and only if there is a unique R-isomorphism $j : T \to T'$ such that $j \circ f = f'$.*

Proof This is similar to that of Theorem 7.2. ◊

- Because of this uniqueness up to isomorphism, we see that when R is commutative $M \otimes_R N$ is characterised by the property stated in Theorem 16.2. In other words, we can assert by these results the existence of tensor products when the cardinality of I is 2.

To settle the question concerning the existence of tensor products for an arbitrary family $(M_i)_{i \in I}$ of R-modules, let (F, i) be the free R-module on $\bigtimes_{i \in I} M_i$ and let G be the submodule of F generated by the elements of either of the forms

(1) $i(x_j)_{j \in I} + i(y_j)_{j \in I} - i(z_j)_{j \in I}$ where, for some $k \in I$, $z_k = x_k + y_k$ and, for $j \neq k$, $z_j = x_j = y_j$;

(2) $i(x_j)_{j \in I} - \lambda i(y_j)_{j \in I}$ where, for some $k \in I$, $x_k = \lambda y_k$ and, for $j \neq k$, $x_j = y_j$.

We denote the quotient module F/G by $\bigotimes_{i \in I} M_i$ and the composite map $\natural_G \circ i$ by $\otimes_R$ (or simply $\otimes$ if no confusion can arise).

Theorem 16.5 [Existence] *If R is a commutative unitary ring and $(M_i)_{i \in I}$ is a family of R-modules then $\left(\bigotimes_{i \in I} M_i, \otimes_R\right)$ is a tensor product of the family.*

Proof This follows the same lines as that of Theorem 15.3 with $\mathbb{Z}$-morphisms becoming R-morphisms; we leave the details to the reader. ◊

- By the above results, tensor products exist and are unique up to isomorphism. By *the* tensor product of the family $(M_i)_{i\in I}$ we shall mean the R-module $\bigotimes_{i\in I} M_i$ constructed above, together with the mapping $\otimes$ which we shall call the associated *tensor map*. As noted previously, $\otimes$ is not injective.

- In the case where $I = \{1,\ldots,n\}$ we shall use the notation $\bigotimes_{i=1}^{n} M_i$ or, equivalently, $M_1 \otimes \ldots \otimes M_n$. Also, given $(m_1,\ldots,m_n) \in \underset{i=1}{\overset{n}{\times}} M_i$ we shall write $\otimes(m_1,\ldots,m_n)$ in the form $m_1 \otimes \ldots \otimes m_n$ and, by abuse of language, call this the *tensor product of the elements* $m_1,\ldots,m_n$.

- Care should be taken to note that $\bigotimes_{i=1}^{n} M_i$ is generated by the elements of the form $m_1 \otimes \ldots \otimes m_n$ and, despite the notation, *not every element of $M_1 \otimes \ldots \otimes M_n$ is of this form.*

We shall now establish some properties of tensor products in the case where R is commutative.

Theorem 16.6 [Commutativity of $\otimes$] *If R is a commutative unitary ring and M, N are R-modules then there is a unique R-isomorphism $\vartheta : M \otimes N \to N \otimes M$ such that*

$$(\forall m \in M)(\forall n \in N) \qquad \vartheta(m \otimes n) = n \otimes m.$$

Proof The mapping $f : M \times N \to N \otimes M$ given by $f(m,n) = n \otimes m$ is clearly bilinear. There is therefore a unique R-morphism $\vartheta : M \otimes N \to N \otimes M$ such that $\vartheta \circ \otimes = f$, i.e. such that $\vartheta(m \otimes n) = n \otimes m$. In a similar way we have a unique R-morphism $\zeta : N \otimes M \to M \otimes N$ such that $\zeta(n \otimes m) = m \otimes n$. We thus have

$$(\zeta \circ \vartheta)(m \otimes n) = m \otimes n, \qquad (\vartheta \circ \zeta)(n \otimes m) = n \otimes m.$$

Since $M \otimes N$ is generated by the set of elements of the form $m \otimes n$, we deduce that $\zeta \circ \vartheta = \text{id}_{M \otimes N}$, and likewise that $\vartheta \circ \zeta = \text{id}_{N \otimes M}$. It follows that ϑ, ζ are mutually inverse R-isomorphisms. $\Diamond$

Theorem 16.7 [Associativity of $\otimes$] *If R is a commutative unitary ring and M, N, P are R-modules then there is a unique R-isomorphism $\vartheta : M \otimes N \otimes P \to (M \otimes N) \otimes P$ such that*

$$(\forall m \in M)(\forall n \in N)(\forall p \in P) \qquad \vartheta(m \otimes n \otimes p) = (m \otimes n) \otimes p.$$

Tensor products; tensor algebras

Likewise, there is a unique R-isomorphism $\zeta : M \otimes N \otimes P \to M \otimes (N \otimes P)$ *such that*

$$(\forall m \in M)(\forall n \in N)(\forall p \in P) \qquad \zeta(m \otimes n \otimes p) = m \otimes (n \otimes p).$$

Proof The mapping $f : M \times N \times P \to (M \otimes N) \otimes P$ given by

$$f(m, n, p) = (m \otimes n) \otimes p$$

is trilinear and so there is a unique R-morphism $\vartheta : M \otimes N \otimes P \to (M \otimes N) \otimes P$ such that $\vartheta(m \otimes n \otimes p) = (m \otimes n) \otimes p$. We show that ϑ is an R-isomorphism by producing an inverse for ϑ.

For every $p \in P$ the mapping $f_p : M \times N \to M \otimes N \otimes P$ given by

$$f_p(m, n) = m \otimes n \otimes p$$

is bilinear and so there is a unique R-morphism $g_p : M \otimes N \to M \otimes N \otimes P$ such that $g_p(m \otimes n) = m \otimes n \otimes p$. The mapping $g : (M \otimes N) \times P \to M \otimes N \otimes P$ given by $g(t, p) = g_p(t)$ is now bilinear and so there is a unique R-morphism $h : (M \otimes N) \otimes P \to M \otimes N \otimes P$ such that $h(t \otimes p) = g(t, p)$. We deduce that

$$h[(m \otimes n) \otimes p] = g(m \otimes n, p) = g_o(m \otimes n) = m \otimes n \otimes p.$$

Since then

$$(h \circ \vartheta)(m \otimes n \otimes p) = m \otimes n \otimes p, \qquad (\vartheta \circ h)[(m \otimes n) \otimes p] = (m \otimes n) \otimes p,$$

it follows that $h \circ \vartheta$ and $\text{id}_{M \otimes N \otimes P}$ coincide on a set of generators of $M \otimes N \otimes P$, and likewise that $\vartheta \circ h$ and $\text{id}_{(M \otimes N) \otimes P}$ coincide on a set of generators of $(M \otimes N) \otimes P$. It follows that ϑ and h are mutually inverse R-isomorphisms.

The second part of the theorem is established similarly. ◇

- We have already used several times the fact that $\text{Im} \otimes$ generates $M \otimes N$, so that every $t \in M \otimes N$ can be written in the form $t = \sum_{i=1}^{k} \lambda_i(M_i \otimes N_i)$ where $m_i \in M, n_i \in N$ and $\lambda_i \in R$ for $i = 1, \ldots, k$. Since $\otimes$ is bilinear, it follows that every $t \in M \otimes N$ can be written in the form $t = \sum_{i=1}^{k} a_i \otimes b_i$ where $a_i \in M, b_i \in N$.

- It should be noted that some collapsing can occur with tensor products. For example, $\mathbb{Z}/2\mathbb{Z} \otimes \mathbb{Z}/3\mathbb{Z}$ is generated by the elements of the form $x \otimes y$ where $x \in \mathbb{Z}/2\mathbb{Z}$ and $y \in \mathbb{Z}/3\mathbb{Z}$. But

$$x \otimes y = 3(x \otimes y) - 2(x \otimes y) = (x \otimes 3y) - (2x \otimes y) = (x \otimes 0) - (0 \otimes y) = 0.$$

Thus we see that $\mathbb{Z}/2\mathbb{Z} \otimes \mathbb{Z}/3\mathbb{Z} = \{0\}$.

The reader will have no difficulty in extending the $\mathbb{Z}$-isomorphism of Theorem 15.15 to an R-isomorphism in the case where R is commutative. Using this, we can now establish the following result.

Theorem 16.8 *If M and N are free R-modules over a commutative unitary ring R then so also is $M \otimes N$. If $\{m_i \ ; \ i \in I\}$ is a basis of M and $\{n_j \ ; \ j \in J\}$ is a basis of N then $\{m_i \otimes n_j \ ; \ (i,j) \in I \times J\}$ is a basis of $M \otimes N$.*

Proof Given $j \in J$ consider the map $f : M \times Rn_j \to M$ described by $f(m, rn_j) = rm$. Clearly, f is bilinear and so there is a unique R-morphism $\vartheta : M \otimes Rn_j \to M$ such that $\vartheta \circ \otimes = f$. Consider now the R-morphism $\zeta : M \to M \otimes Rn_j$ given by $\zeta(m) = m \otimes n_j$. We have

$$(\zeta \circ \vartheta)(m \otimes rn_j) = \zeta(rm) = rm \otimes n_j = m \otimes rn_j$$

and so, since $M \otimes Rn_j$ is generated by the elements of the form $m \otimes rn_j$, we deduce that $\zeta \circ \vartheta$ is the identity map on $M \otimes Rn_j$. On the other hand, we also have

$$(\vartheta \circ \zeta)(m) = \vartheta(m \otimes n_j) = m$$

and so $\vartheta \circ \zeta = \mathrm{id}_M$. Thus ϑ, ζ are mutually inverse R-isomorphisms. Since the R-isomorphism ζ carries bases to bases, we deduce that $\{m_i \otimes n_j \ ; \ i \in I\}$ is a basis of $M \otimes Rn_j$. Now by the analogue of Theorem 15.15 and the fact that we are dealing with internal direct sums we have

$$M \otimes N = M \otimes \bigoplus_{j \in J} Rn_j = \bigoplus_{j \in J}(M \otimes Rn_j).$$

The result now follows by Theorem 7.8. $\Diamond$

It is clear that, the ground ring R being commutative, if M, N, P are R-modules then the set $\mathrm{Bil}_R(M \times N, P)$ of bilinear mappings $f : M \times N \to P$ forms an R-module. The most basic property that relates R-morphisms, bilinear mappings, and tensor products is then the following.

Theorem 16.9 *If R is a commutative unitary ring and M, N, P are R-modules then there are R-isomorphisms*

$$\mathrm{Mor}_R(M, \mathrm{Mor}_R(N, P)) \simeq \mathrm{Bil}_R(M \times N, P) \simeq \mathrm{Mor}_R(M \otimes N, P).$$

Proof For every $\alpha \in \mathrm{Bil}_R(M \times N, P)$ and every $m \in M$ let $\alpha_m : N \to P$ be given by $\alpha_m(n) = \alpha(m, n)$. It is clear that $\alpha \in \mathrm{Mor}_R(N, P)$. Now let $\vartheta_\alpha : M \to \mathrm{Mor}_R(N, P)$ be given by $\vartheta_\alpha(m) = \alpha_m$. Then it is clear that

$\vartheta_\alpha \in \text{Mor}_R(M, \text{Mor}_R(N, P))$ and that the assignment $\alpha \mapsto \vartheta_\alpha$ yields an R-morphism

$$\vartheta : \text{Bil}_R(M \times N, P) \to \text{Mor}_R(M, \text{Mor}_R(N, P)).$$

Now let $f \in \text{Mor}_R(M, \text{Mor}_R(N, P))$ and let $\zeta_f : M \times N \to P$ be the bilinear mapping given by $\zeta_f(m, n) = [f(m)](n)$. Then the assignment $f \mapsto \zeta_f$ yields an R-morphism

$$\zeta : \text{Mor}_R(M, \text{Mor}_R(N, P)) \to \text{Bil}_R(M \times N, P).$$

We leave to the reader the easy task of showing that ϑ and ζ are mutually inverse R-isomorphisms.

Consider now the mapping

$$\xi : \text{Bil}_R(M \times N, P) \to \text{Mor}_R(M \otimes N, P)$$

given by $\xi(f) = f_*$ where $f_* : M \otimes N \to P$ is the unique R-morphism such that $f_* \circ \otimes = f$. It is readily verified that ξ is an R-morphism, and that $f_* = g_*$ gives $f = g$, so that ξ is injective. To see that ξ is also surjective, it suffices to observe that if $g \in \text{Mor}_R(M \otimes N, P)$ then $g \circ \otimes \in \text{Bil}_R(M \times N, P)$ with $(g \circ \otimes)_* = g$. $\diamond$

We shall now investigate how tensor products behave in relation to duality. As we are restricting our attention to a commutative ground ring R, we may take $P = R$ in Theorem 16.9 and obtain R-isomorphisms

$$\text{Mor}_R(M, N^d) \simeq \text{Bil}_R(M \times N, R) \simeq (M \otimes N)^d.$$

It is clear that we can interchange M and N in this to obtain an R-isomorphism

$$\text{Mor}_R(N, M^d) \simeq (M \otimes N)^d.$$

We also have the following result.

Theorem 16.10 *Let R be a commutative unitary ring. If M and N are R-modules then there is a unique R-morphism*

$$\vartheta_{M,N} : M^d \otimes N \to \text{Mor}_R(M, N)$$

such that $\vartheta_{M,N}(m^d \otimes n) : x \mapsto \langle x, m^d \rangle n$.

Moreover, $\vartheta_{M,N}$ is a monomorphism whenever N is projective, and is an isomorphism whenever M or N is projective and finitely generated.

Proof Given $m^d \in M^d$ and $n \in N$, let $f_{m^d,n} : M \to N$ be the R-morphism given by $f_{m^d,n}(x) = \langle x, m^d \rangle n$. Then it is readily seen that the mapping

$f : M^d \times N \to \mathrm{Mor}_R(M, N)$ given by $f(m^d, n) = f_{m^d, n}$ is bilinear. The first statement now follows by the definition of tensor product.

Suppose now that N is projective. Then by Theorem 8.8 there is a free R-module F of which N is a direct summand, say $F = N \oplus P$. Since P is then also projective, the canonical short exact sequence

$$0 \longrightarrow N \xrightarrow{i} F \xrightarrow{\pi} P \longrightarrow 0$$

splits. Given any R-module M we can now construct the diagram

$$\begin{array}{ccc} M^d \otimes N & \xrightarrow{\otimes i} & M^d \otimes F \\ {\scriptstyle \vartheta_{M,N}} \downarrow & & \downarrow {\scriptstyle \vartheta_{M,F}} \\ \mathrm{Mor}_R(M, N) & \xrightarrow{i_*} & \mathrm{Mor}_R(M, F) \end{array}$$

in which, by Theorem 8.4 and the analogue of Theorem 15.12, $\otimes i$ and i_* are injective. Moreover, this diagram is commutative; for, as is readily seen, we have

$$f_{m^d, i(n)} = i \circ f_{m^d, n}.$$

We show first that $\vartheta_{M,F}$ is a monomorphism (so that the result holds for free R-modules); that $\vartheta_{M,N}$ is a monomorphism will then follow from the commutativity of the diagram. Suppose then that $\{b_i \ ; \ i \in I\}$ is a basis of F. Then every $t \in M^d \otimes F$ can be written as a finite sum

$$t = \sum_{i \in I} (m_i^d \otimes b_i).$$

Now $\vartheta_{M,F}(t)$ is the morphism from M to F described by

$$x \mapsto \sum_{i \in I} \langle x, m_i^d \rangle b_i.$$

Thus if $\vartheta_{M,F}(t)$ is the zero morphism then $\langle x, m_i^d \rangle = 0$ for all $x \in M$ and all $m_i^d \in M^d$, whence $m_i^d = 0$ for all $m_i^d \in M^d$, and hence $t = 0$. Consequently, $\vartheta_{M,F}$ is injective.

Suppose now that N is projective and finitely generated. Then by the remark following Theorem 8.6 there is a finite-dimensional R-module F and an epimorphism $\pi : F \to N$. We therefore have the commutative diagram

$$\begin{array}{ccc} M^d \otimes F & \xrightarrow{\otimes \pi} & M^d \otimes N \\ {\scriptstyle \vartheta_{M,F}} \downarrow & & \downarrow {\scriptstyle \vartheta_{M,N}} \\ \mathrm{Mor}_R(M, F) & \xrightarrow{\pi_*} & \mathrm{Mor}_R(M, N) \end{array}$$

in which both horizontal maps are epimorphisms. It is clear from this that $\vartheta_{M,N}$ is surjective whenever $\vartheta_{M,F}$ is surjective. It therefore suffices to prove that $\vartheta_{M,F}$ is an isomorphism; in other words, that the result holds for finite-dimensional R-modules. We achieve this by induction on the dimension of F. Suppose first that $\dim F = 1$. Then there is an R-isomorphism $\alpha : R \to F$, namely that given by $r \mapsto rb$ where $\{b\}$ is a basis of F, and we have the commutative diagram

$$\begin{array}{ccc} M^d \otimes R & \xrightarrow{\otimes \alpha} & M^d \otimes F \\ \vartheta_{M,R} \downarrow & & \downarrow \vartheta_{M,F} \\ M^d = \mathrm{Mor}_R(M, R) & \xrightarrow{\alpha_*} & \mathrm{Mor}_R(M, F) \end{array}$$

in which $\otimes \alpha$ and α_* are R-isomorphisms. Now it is readily seen that $\vartheta_{M,R}$ coincides with the R-isomorphism $m^d \otimes r \mapsto m^d r$ of Theorem 15.8. Consequently, we see that $\vartheta_{M,F}$ is also an R-isomorphism. Suppose now, by way of induction, that the result holds for all free R-modules of dimension at most $n-1$ where $n > 2$ and let F be a free R-module of dimension n. If $\{b_1, \ldots, b_n\}$ is a basis of F, so that $F = \bigoplus_{i=1}^{n} Rb_i$, consider the submodules $A = \bigoplus_{i=2}^{n} Rb_i$ and $B = Rb_1$. Since there is an R-isomorphism $\vartheta : F/A \to B$, we have the split short exact sequence

$$0 \longrightarrow A \xrightarrow{f} F \xrightarrow{g} B \longrightarrow 0$$

where f is the natural inclusion and $g = \vartheta \circ \natural_A$. Moreover, A and B are free, of dimensions $n-1$ and 1 respectively. The commutative diagram

$$\begin{array}{ccccccccc} 0 & \longrightarrow & M^d \otimes A & \xrightarrow{\otimes f} & M^d \otimes F & \xrightarrow{\otimes g} & M^d \otimes B & \longrightarrow & 0 \\ & & \vartheta_{M,A} \downarrow & & \vartheta_{M,F} \downarrow & & \vartheta_{M,B} \downarrow & & \\ 0 & \longrightarrow & \mathrm{Mor}_R(M, A) & \xrightarrow{f_*} & \mathrm{Mor}_R(M, F) & \xrightarrow{g_*} & \mathrm{Mor}_R(M, B) & \longrightarrow & 0 \end{array}$$

then has exact rows so, by the induction hypothesis, $\vartheta_{M,A}, \vartheta_{M,B}$ are isomorphisms. It follows by the Corollary to Theorem 3.10 that $\vartheta_{M,F}$ is also an isomorphism.

Suppose now that M is projective and finitely generated. Then on the one hand there is a free R-module F, of which M is a direct summand,

and a commutative diagram

$$\begin{array}{ccc} F^d \otimes N & \xrightarrow{(f^t)^\otimes} & M^d \otimes N \\ \vartheta_{F,N} \downarrow & & \downarrow \vartheta_{M,N} \\ \mathrm{Mor}_R(F,N) & \xrightarrow{f^*} & \mathrm{Mor}_R(M,N) \end{array}$$

in which both horizontal maps are epimorphisms.

On the other hand, there is a finite-dimensional R-module (which we shall also denote by F without confusion) and an epimorphism $\pi : F \to M$ that gives rise to a commutative diagram

$$\begin{array}{ccc} M^d \otimes N & \xrightarrow{(\pi^t)^\otimes} & F^d \otimes N \\ \vartheta_{M,N} \downarrow & & \downarrow \vartheta_{F,N} \\ \mathrm{Mor}_R(M,N) & \xrightarrow{\pi^*} & \mathrm{Mor}_R(F,N) \end{array}$$

in which both horizontal maps are monomorphisms.

It follows immediately from these diagrams that if $\vartheta_{F,N}$ is an isomorphism then so is $\vartheta_{M,N}$. It therefore suffices to show that $\vartheta_{F,N}$ is an isomorphism; in other words, that the result holds for free R-modules of finite dimension. Suppose then that $\{b_1, \ldots, b_t\}$ is a basis of F and let $\{b_1^d, \ldots, b_t^d\}$ be the corresponding dual basis. To show that $\vartheta_{F,N} : F^d \otimes N \to \mathrm{Mor}_R(F,N)$ is an R-isomorphism, it suffices to produce an inverse for it. For this purpose, consider the R-morphism $\gamma : \mathrm{Mor}_R(F,N) \to F^d \otimes N$ given by

$$\gamma(f) = \sum_{i=1}^t (b_i^d \otimes f(b_i)).$$

Given $x = \sum_{j=1}^t x_j b_j \in F$, we have

$$\begin{aligned} \left[\vartheta_{F,N}(b_i^d \otimes f(b_i))\right](x) &= \langle x, b_i^d \rangle f(b_i) \\ &= \left\langle \sum_{j=1}^t x_j b_j, b_i^d \right\rangle f(b_i) \\ &= \sum_{j=1}^t x_j \langle b_j, b_i^d \rangle f(b_i) \\ &= x_i f(b_i) \\ &= f(x_i b_i) \end{aligned}$$

Tensor products; tensor algebras

and so
$$\sum_{i=1}^{t}\left[\vartheta_{F,N}(b_i^d \otimes f(b_i))\right](x) = \sum_{i=1}^{t} f(x_i b_i) = f(x).$$

Consequently, we have
$$(\vartheta_{F,N} \circ \gamma)(f) = \sum_{i=1}^{t} \vartheta_{F,N}(b_i^d \otimes f(b_i)) = f$$

and so $\vartheta_{F,N} \circ \gamma$ is the identity map on $\text{Mor}_R(F, N)$. Also, given any $n \in N$ and $m^d \in \sum_{j=1}^{t} \lambda_j b_j^d \in F^d$, we have

$$\begin{aligned}
(\gamma \circ \vartheta_{F,N})(m^d \otimes n) = \gamma(f_{m^d, n}) &= \sum_{i=1}^{t} (b_i^d \otimes \langle b_i, m^d \rangle n) \\
&= \sum_{i=1}^{t} \left(b_i^d \otimes \sum_{j=1}^{t} \lambda_j \langle b_i, b_j^d \rangle n \right) \\
&= \sum_{i=1}^{t} (b_i^d \otimes \lambda_i n) \\
&= \left(\sum_{i=1}^{t} \lambda_i b_i^d \right) \otimes n \\
&= m^d \otimes n.
\end{aligned}$$

Since $f^d \otimes N$ is generated by the elements of the form $m^d \otimes n$, we deduce that $\gamma \circ \vartheta_{F,N}$ is the identity map on $M^d \otimes N$. Thus $\vartheta_{F,N}$ and γ are mutually inverse R-isomorphisms. $\Diamond$

Corollary *Let R be a commutative unitary ring. If M and N are R-modules, at least one of which is projective and finitely generated, then*
$$(M \otimes N)^d \simeq M^d \otimes N^d.$$

Proof This is immediate from Theorem 16.10 and the isomorphisms that immediately precede it. $\Diamond$

We shall now describe an important application of Theorem 16.10. Let R be a commutative unitary ring and let F be a free R-module of finite dimension. By Theorem 16.10, there is an R-isomorphism
$$\vartheta_{F,F} : F^d \otimes F \to \text{Mor}_R(F, F).$$

In the proof of that theorem we constructed an inverse for $\vartheta_{F,F}$ in terms of a given basis of F. The uniqueness of $\vartheta_{F,F}$ implies that the isomorphism $\vartheta_{F,F}^{-1}$ is independent of this choice of basis. We can therefore call $\vartheta_{F,F}$ the *canonical* R-isomorphism from $F^d \otimes F$ to $\text{Mor}_R(F, F)$.

Observing that the mapping from $F^d \otimes F$ to R described by $(x^d, x) \mapsto \langle x, x^d \rangle$ is bilinear, and therefore induces a unique R-morphism $\alpha_F : F^d \otimes F \to R$ such that $\alpha_F(x^d \otimes x) = \langle x, x^d \rangle$, we consider the following notion.

Definition If F is a free R-module of finite dimension then

$$\alpha_F \circ \vartheta_{F,F}^{-1} : \mathrm{Mor}_R(F, F) \to R$$

is called the *trace form* on $\mathrm{Mor}_R(F, F)$ and is denoted by tr_F (or simply tr if no confusion can arise). For every $f \in \mathrm{Mor}_R(F, F)$ we call $\mathrm{tr}\, f$ the *trace* of f.

For a given $f \in \mathrm{Mor}_R(F, F)$ the trace of f has a convenient interpretation in terms of matrices. We shall now describe this. For this purpose, suppose that $\{b_1, \ldots, b_n\}$ is a basis of F and that $\{b_1^d, \ldots, b_n^d\}$ is the corresponding dual basis. Suppose also that

$$\mathrm{Mat}[\,f, (b_i)_n, (b_i)_n\,] = [\,a_{ij}\,]_{n \times n}.$$

Then, using the formula for $\gamma = \vartheta_{F,F}^{-1}$ given in the proof of Theorem 16.10, we have

$$\begin{aligned}
\mathrm{tr}\, f = (\alpha_F \circ \vartheta_{F,F}^{-1})(f) &= \sum_{i=1}^{n} \alpha\bigl(b_i^d \otimes f(b_i)\bigr) \\
&= \sum_{i=1}^{n} \langle f(b_i), b_i^d \rangle \\
&= \sum_{i=1}^{n} \sum_{j=1}^{n} a_{ji} \langle b_j, b_i^d \rangle \\
&= \sum_{i=1}^{n} a_{ii}.
\end{aligned}$$

Because of this, we define the *trace* of an $n \times n$ matrix A over a commutative unitary ring R to be the sum of the diagonal entries of A. Bearing in mind that $\vartheta_{F,F}^{-1}$ is independent of the choice of basis, we therefore deduce immediately the folowing facts :

(1) *the trace of an R-morphism f is the trace of any matrix that represents f;*

(2) *similar matrices have the same trace.*

- The converse of (2) is not true. For example, in $\mathrm{Mat}_{2 \times 2}(\mathbb{R})$ the zero matrix and the matrix

$$\begin{bmatrix} 1 & 0 \\ 0 & -1 \end{bmatrix}$$

have the same trace but are not similar.

We also have the following result which gives a characterisation of trace forms.

Theorem 16.11 *Let R be a commutative unitary ring and let F be a free R-module of finite dimension. Then*

$$(\forall f, g \in \mathrm{Mor}_R(F,F)) \qquad \mathrm{tr}(f \circ g) = \mathrm{tr}(g \circ f).$$

Moreover, if τ is any linear form on $\mathrm{Mor}_R(F,F)$ such that $\tau(f \circ g) = \tau(g \circ f)$ for all $f, g \in \mathrm{Mor}_R(F,F)$ then $\tau = \lambda \mathrm{tr}$ for a unique $\lambda \in R$.

Proof To show that $\mathrm{tr}(f \circ g) = \mathrm{tr}(g \circ f)$ it suffices to show that if A, B are matrices that represent f, g with respect to some fixed ordered bases then AB and BA have the same trace; and, R being commutative, this follows from

$$\mathrm{tr}(AB) = \sum_{i=1}^{n}\left(\sum_{j=1}^{n} a_{ij} b_{ji}\right) = \sum_{j=1}^{n}\left(\sum_{i=1}^{n} b_{ji} a_{ij}\right) = \mathrm{tr}(BA).$$

As for the second statement, it again suffices to establish a similar result for matrices. Suppose then that $\tau : \mathrm{Mat}_{n \times n}(R) \to R$ is such that $\tau(AB) = \tau(BA)$ for all $A, B \in \mathrm{Mat}_{n \times n}(R)$. The result is trivial for $n = 1$ so we shall assume that $n \geq 2$. Let E_{ij} denote the $n \times n$ matrix that has 1_R in the (i,j)-th position and 0 elsewhere. Then it is readily seen that

$$E_{ij} E_{pq} = \begin{cases} E_{iq} & \text{if } p = j; \\ 0 & \text{if } p \neq j. \end{cases}$$

Taking $A = E_{ij}, B = E_{jk}$ with $i \neq k$ we obtain $\tau(E_{ik}) = 0$; and taking $A = E_{ij}, B = E_{ji}$ with $i \neq j$ we obtain $\tau(E_{ii}) = \tau(E_{jj})$. Observing that $\{E_{ij} \;;\; i, j = 1, \ldots, n\}$ is a basis for $\mathrm{Mat}_{n \times n}(R)$, we have

$$\tau(A) = \tau\left(\sum_{i,j} a_{ij} E_{ij}\right) = \sum_{i,j} a_{ij} \tau(E_{ij}) = \lambda \sum_{i=1}^{n} A_{ii}$$

where $\lambda = \tau(E_{ii}) = \tau(E_{jj})$ for all i and j. ◇

We end the present section by considering the concept of the *tensor algebra* of an R-module. Here we shall be primarily interested in existence and uniqueness; the importance of such a concept will emerge in the discussion in the next chapter.

Definition Let R be a commutative unitary ring. If M is an R-module then by a *tensor algebra* over M we shall mean an associative unitary R-algebra T together with an R-morphism $f : M \to T$ such that, for every associative unitary R-algebra X and every R-morphism $g : M \to X$, there is a unique 1-preserving R-algebra morphism $h : T \to X$ such that the diagram

is commutative. We denote such a tensor algebra by (T, f).

We recall from Chapter 4 that a subalgebra of an R-algebra A is a submodule of A that is also an R-algebra with respect to the multiplication of A. We say that an R-algebra A is *generated* by a subset S of A if A is the smallest subalgebra of A that contains S, i.e. the intersection of all the subalgebras that contain S.

Theorem 16.12 *Let R be a commutative unitary ring. If M is an R-module and (T, f) is a tensor algebra over M then $\operatorname{Im} f \cup \{1_T\}$ generates T.*

Proof This is similar to that of Theorem 7.1. ◊

Theorem 16.13 [Uniqueness] *Let (T, f) be a tensor algebra over the R-module M. Then (T', f') is also a tensor algebra over M if and only if there is an R-algebra isomorphism $j : T \to T'$ such that $j \circ f = f'$.*

Proof This is similar to that of Theorem 7.2. ◊

We shall now settle the question concerning the existence of tensor algebras. For this purpose, let R be a commutative unitary ring and M an R-module. Suppose that $(M_i)_{1 \leq i \leq n}$ is a finite family of R-modules each of which is isomorphic to M. Then we shall call $\bigotimes_{i=1}^{n} M_i$ the *n-th tensor power* of M and denote it henceforth by $\bigotimes^n M$. For convenience, we also define $\bigotimes^0 M$ to be R.

With this notation, consider the R-module

$$\bigotimes M = \bigoplus_{n \in \mathbb{N}} (\bigotimes^n M).$$

For every $j \in \mathbb{N}$ we shall identify $\bigotimes^j M$ with the submodule $\operatorname{in}_j^\to(\bigotimes^j M)$ of $\bigotimes M$, thereby regarding the above direct sum as an internal direct sum of submodules.

Tensor products; tensor algebras

We shall now define a multiplication on $\bigotimes M$ such that it becomes an associative unitary R-algebra. For this purpose, we note that the R-module $\bigotimes M$ is generated by the set consisting of $1_R \in \bigotimes^0 M$ and the elements $x_1 \otimes \ldots \otimes x_n \in \bigotimes^n M$ for each $n \geq 1$. We can define products for these generators by setting

$$\begin{cases} 1_R(x_1 \otimes \ldots \otimes x_n) = x_1 \otimes \ldots \otimes x_n = (x_1 \otimes \ldots \otimes x_n)1_R; \\ (x_1 \otimes \ldots \otimes x_n)(y_1 \otimes \ldots \otimes y_m) = x_1 \otimes \ldots \otimes x_n \otimes y_1 \otimes \ldots \otimes y_m. \end{cases}$$

Now, given $x \in \bigotimes^n M$ and $y \in \bigotimes^m M$ expressed as linear combinations of generators, say $x = \sum_i \lambda_i x_i$ and $y = \sum_j \mu_j y_j$ where each $x_i \in \bigotimes^n M$ and each $y_j \in \bigotimes^m M$, we define the product xy by

$$xy = \sum_{i,j} \lambda_i \mu_j x_i y_j.$$

It is clear that this definition of xy is, by the above, independent of the linear combinations representing x and y. Now every $z \in \bigotimes M$ can be expressed uniquely in the form $z = \sum_i m_i$ where $m_i \in \bigotimes^i M$ for each i with all but finitely many m_i equal to zero. Given in this way $z = \sum_i m_i$ and $z' = \sum_j m'_j$, we now define the product zz' by

$$zz' = \sum_{i,j} m_i m'_j.$$

It is now readily verified that the multiplication so defined makes $\bigotimes M$ into an associative algebra over R with identity element 1_R.

- It is important to note from the above definition of multiplication in $\bigotimes M$ that for all $i, j \in \mathbb{N}$ we have

$$\bigotimes^i M \cdot \bigotimes^j M \subseteq \bigotimes^{i+j} M.$$

In particular, if $m_1, \ldots, m_n \in M = \bigotimes^1 M$ then $\prod_{j=1}^n m_j \in \bigotimes^n M$.

Theorem 16.14 [Existence] *If R is a commutative unitary ring and if M is an R-module then $(\bigotimes M, \iota_M)$ is a tensor algebra over M.*

Proof Let N be an associative unitary algebra and $g : M \to N$ an R-morphism. Define a family $(g_i)_{i \in I}$ of R-morphisms $g_i : \bigotimes^i M \to N$ as follows. For $i = 0$ let $g_0 : R \to N$ be the unique R-morphism such that $g_0(1_R) = 1_N$ (recall that $\{1_R\}$ is a basis for R). For $i \geq 1$ let $\bigtimes^i M$ denote the cartesian product of i copies of the R-module M and let $g'_i : \bigtimes^i M \to N$ be the mapping described by

$$g'_i(m_1, \ldots, m_i) = \prod_{j=1}^{i} g(m_j).$$

It is clear that each g'_i is i-linear and so yields a unique R-morphism $g_i : \bigotimes^i M \to N$ such that

$$g(m_1 \otimes \ldots \otimes m_i) = \prod_{j=1}^{i} g(m_j).$$

By the definition of $\bigotimes M$ there is a unique R-morphism $h : \bigotimes M \to N$ such that every diagram

$$\begin{array}{ccc} \bigotimes^i M & \xrightarrow{g_i} & N \\ \text{in}_i \downarrow & \nearrow h & \\ \bigotimes M = \bigoplus_{i \in \mathbb{N}} (\bigotimes^i M) & & \end{array}$$

is commutative. The diagram corresponding to $i = 0$ yields $h(1_R) = 1_N$, so that h is 1-preserving; and that corresponding to $i = 1$ yields $h \circ \iota_M = g_1 = g$. That h is an R-algebra morphism follows immediately from the definition of multiplication in $\bigotimes M$, the definition of g_i, and the commutativity of each of the above diagrams.

Suppose now that $t : \bigotimes M \to N$ is a 1-preserving R-algebra morphism such that $t \circ \iota_M = g$. We show that $t = h$ by showing that $t \circ \text{in}_i = g_i$ for every $i \in \mathbb{N}$ and appealing to the uniqueness of h with respect to this property. For this purpose, we note that if $m_1, \ldots, m_i \in M$ then in $\bigotimes M$ we have the equality

$$\prod_{j=1}^{i} \iota_M(m_j) = \text{in}_i \left(\bigotimes_{j=1}^{i} m_j \right).$$

In fact, the k-th component of the right-hand side is given by

$$\left(\text{in}_i \left(\bigotimes_{j=1}^{i} m_j \right) \right)_k = \begin{cases} \bigotimes_{j=1}^{i} m_j & \text{if } k = i; \\ 0 & \text{if } k \neq i, \end{cases}$$

that of $\iota_M(m_j)$ is given by

$$(\iota_M(m_j))_k = \begin{cases} m_j & \text{if } k = 1; \\ 0 & \text{if } k \neq 1, \end{cases}$$

Tensor products; tensor algebras

and the equality follows from the remark preceding the theorem and the definition of multiplication in $\bigotimes M$. Using this equality, we see that

$$g_i(m_1 \otimes \ldots \otimes m_i) = \prod_{j=1}^{i} g(m_j) = \prod_{j=1}^{i}(t \circ \iota_M)(m_j)$$
$$= t\left(\prod_{j=1}^{i} \iota_M(m_j)\right)$$
$$= (t \circ \mathrm{in}_i)(m_1 \otimes \ldots \otimes m_i).$$

Consequently $t \circ \mathrm{in}_i$ and g_i coincide on a set of generators of $\bigotimes^i M$, whence it follows that $t \circ \mathrm{in}_i = g_i$. Since this holds for every $i \geq 1$, it remains to show that $t \circ \mathrm{in}_0 = g_0$; and this follows from the fact that t is 1-preserving, so that these morphisms coincide on the basis $\{1_R\}$ of R. ◊

The above results establish the existence, and uniqueness up to isomorphism, of a tensor algebra over a given R-module M. By *the* tensor algebra over M we shall mean that constructed in Theorem 16.14.

EXERCISES

16.1 Deduce from Theorem 16.10 the existence of a natural R-morphism

$$\alpha : M \otimes N \to \mathrm{Mor}_R(M^d, N)$$

that is injective whenever M and N are projective.

[*Hint.* Use Theorem 9.5.]

16.2 Let R be a commutative unitary ring and let M_1, M_2 be R-modules of finite dimension. Given $m_1^d \in M_1^d$ and $m_2^d \in M_2^d$, show that there is a unique R-morphism $f : M_1 \otimes M_2 \to R$ such that $f(m_1, m_2) = m_1^d(m_1) m_2^d(m_2)$. Hence show that the assignment $m_1^d \otimes m_2^d \mapsto f$ yields an R-isomorphism

$$M_1^d \otimes M_2^d \simeq (M_1 \otimes M_2)^d.$$

16.3 Identifying $M_1^d \otimes M_2^d$ and $(M_1 \otimes M_2)^d$ under the isomorphism of the previous exercise, establish the identity

$$\langle m_1 \otimes m_2, m_1^d \otimes m_2^d \rangle = \langle m_1, m_1^d \rangle \langle m_2, m_2^d \rangle.$$

16.4 Let M_1, M_2 be finite-dimensional R-modules where R is a commutative unitary ring. If $f_1 \in \mathrm{Mor}_R(M_1, M_1)$ and $f_2 \in \mathrm{Mor}_R(M_2, M_2)$ prove that
$$\mathrm{tr}(f_1 \otimes f_2) = \mathrm{tr}\, f_1 \, \mathrm{tr}\, f_2.$$

16.5 Let M_1, M_2, N_1, N_2 be finite-dimensional R-modules where R is a commutative unitary ring. Prove that if $f_1 \in \text{Mor}_R(M_1, N_1)$ and $f_2 \in \text{Mor}_R(M_2, N_2)$ then

$$(f_1 \otimes f_2)^t = f_1^t \otimes f_2^t.$$

16.6 Let R be a commutative unitary ring and let M be an R-module of finite dimension n. Show that the trace form $\text{tr} : \text{Mor}_R(M, M) \to R$ is surjective, and that Ker tr is a direct summand of $\text{Mor}_R(M, M)$ of dimension $n^2 - 1$.

[*Hint.* Use matrices; consider the submodule of $\text{Mat}_{n \times n}(R)$ that is generated by $\{E_{ij} \; ; \; i \neq j\} \cup \{E_{ii} - E_{11} \; ; \; 2 \leq i \leq n\}$.]

16.7 Let R be a commutative unitary ring. If M and N are projective R-modules prove that so also is $M \otimes N$.

[*Hint.* Use Theorems 8.8, 15.9 and 16.8.]

16.8 Let R be a commutative integral domain. If M and N are free R-modules and if $m \in M, n \in N$ are such that $m \otimes n = 0$, prove that either $m = 0$ or $n = 0$.

[*Hint.* Use Theorem 16.8.]

16.9 Let R be a commutative unitary ring and let M be an R-module of finite dimension. If $\{m_i \; ; \; i \in I\}$ is a basis of M and if, for all $i, i' \in I$, $f_{i,i'} : M \to M$ is the R-morphism such that

$$f_{i,i'}(m_t) = \begin{cases} m_{i'} & \text{if } t = i; \\ 0 & \text{if } t \neq i, \end{cases}$$

prove that $\{f_{i,i'} \; ; \; i, i' \in I\}$ is a basis of $\text{End}_R M$.

Deduce that there is a unique R-isomorphism

$$\vartheta : \text{End}_R M \otimes \text{End}_R N \to \text{End}_R(M \otimes N)$$

such that $\vartheta(f \otimes g) = f \otimes_R g$.

[*Note* : here $f \otimes g$ denotes a generator of $\text{End}_R M \otimes \text{End}_R N$!]

Identifying $\text{End}_R M \otimes \text{End}_R N$ and $\text{End}_R(M \otimes N)$ for all such R-modules M and N, deduce that

$$\bigotimes \text{End}_R M = \bigoplus_{n \in \mathbb{N}} \text{End}_R(\bigotimes^n M).$$

17 Exterior algebras; determinants

We shall now turn our attention to that part of linear algebra that is often called *exterior algebra*. Since we shall be dealing with expressions of the form
$$x_{\sigma(1)} \otimes \ldots \otimes x_{\sigma(n)}$$
where σ is a permutation (=bijection) on $\{1,\ldots,n\}$, we begin by mentioning some properties of permutations. We shall denote by P_n the group of permutations on $\{1,\ldots,n\}$. The basic properties that we shall require are:

(1) *If $n \geq 2$ then every $\sigma \in P_n$ is a composite of transpositions;*
(2) $(\forall \sigma, \vartheta \in P_n)\ \epsilon_{\sigma \circ \vartheta} = \epsilon_\sigma \epsilon_\vartheta$ *where ϵ_σ denotes the signum of σ.*

- For the convenience of the reader, we give brief proofs of these results.

(1) $f \in P_n$ is called a *transposition* of there exist $i,j \in \{1,\ldots,n\}$ such that $i \neq j, f(i) = j, f(j) = i$ and $f(x) = x$ for all $x \neq i,j$. Roughly speaking then, a transposition swaps two elements and leaves the others fixed. We establish the first result by induction on the number t of elements that are *not* fixed by σ. Clearly, the result holds when $t = 2$. Suppose, by way of induction, that the result holds for $t = m - 1$ with $2 < m \leq n$. Let A be the subset of $\{1,\ldots,n\}$ consisting of those elements that are *not* fixed by σ and let $|A| = m$. Then given $i \in A$ we have $\sigma(i) \in A$; for otherwise $\sigma[\sigma(i)] = \sigma(i)$ and consequently $\sigma(i) = i$, a contradiction. Now let τ be the transposition such that $\tau(i) = \sigma(i)$ and $\tau[\sigma(i)] = i$. Then the set of elements *not* fixed under $\tau \circ \sigma$ is $A \setminus \{i\}$ which is of cardinal $m - 1$. Thus, by the induction hypothesis, $\tau \circ \sigma$ is a composite of transpositions, say $\tau \circ \sigma = \tau_1 \circ \ldots \circ \tau_k$. It follows that so also is σ, for then $\sigma = \tau^{-1} \circ \tau_1 \circ \ldots \circ \tau_k$.

(2) For every $\sigma \in P_n$ the *signum* of σ is defined to be
$$\epsilon_\sigma = \prod_{i<j}[\sigma(j) - \sigma(i)] \Big/ \prod_{i<j}(j-i).$$

It is readily seen that if $\tau \in P_n$ is a transposition then $\epsilon_\tau = -1$. Suppose now that τ swaps a and b with, say, $a < b$. Then clearly if $i < j$ we have
$$\tau(j) < \tau(i) \iff i = a, j = b.$$

Equivalently, if $i < j$ we have
$$\tau(i) < \tau(j) \iff (i \neq a \text{ or } j \neq b).$$

Consequently, for every $\sigma \in P_n$,

$$\prod_{i<j}[(\sigma\circ\tau)(j) - (\sigma\circ\tau)(i)]$$
$$= [(\sigma\circ\tau)(b) - (\sigma\circ\tau)(a)] \prod_{\substack{i<j \\ \tau(i)<\tau(j)}} [(\sigma\circ\tau)(j) - (\sigma\circ\tau)(i)]$$
$$= [\sigma(a) - \sigma(b)] \prod_{\substack{\tau(i)<\tau(j) \\ i<j}} [\sigma(j) - \sigma(i)]$$
$$= -\prod_{i<j}[\sigma(j) - \sigma(i)]$$

and so it follows that $\epsilon_{\sigma\circ\tau} = -\epsilon_\sigma = \epsilon_\sigma \epsilon_\tau$. A simple induction now shows that if $\tau_1, \ldots, \tau_k$ are transpositions then

$$\epsilon_{\sigma\circ\tau_1\circ\ldots\circ\tau_k} = \epsilon_\sigma \epsilon_{\tau_1} \cdots \epsilon_{\tau_k}.$$

Using (1) we now see that $\epsilon_{\sigma\circ\vartheta} = \epsilon_\sigma \epsilon_\vartheta$ for all $\sigma, \vartheta \in P_n$.

- Note that, by (1) and (2), ϵ_σ is either 1 or -1 for every $\sigma \in P_n$. We say that σ is an *even* permutation if $\epsilon_\sigma = 1$ and an *odd* permutation if $\epsilon_\sigma = -1$. It is clear from (2) that the even permutations form a subgroup A_n of P_n. Since, by (2), $\sigma \mapsto \epsilon_\sigma$ describes a morphism with kernel A_n, we see that A_n is a normal subgroup of P_n. This is called the *alternating subgroup* of P_n; it can be shown that $|A_n| = \frac{1}{2}n!$.

If M and N are R-modules then for every $\sigma \in P_n$ and every n-linear mapping $f : \bigtimes^n M \to N$ we shall denote by $\sigma f : \bigtimes^n M \to N$ the n-linear mapping given by

$$(\sigma f)(x_1, \ldots, x_n) = f(x_{\sigma(1)}, \ldots, x_{\sigma(n)}).$$

If $g : \bigotimes^n M \to N$ is the unique R-morphism such that $g \circ \otimes = f$, we shall denote by $\sigma g : \bigotimes^n M \to N$ the unique R-morphism such that $\sigma g \circ \otimes = \sigma f$. We therefore have

(1) $\quad (\sigma g)(x_1 \otimes \ldots \otimes x_n) = g(x_{\sigma(1)} \otimes \ldots \otimes x_{\sigma(n)}).$

Given $\sigma \in P_n$, consider now the mapping from $\bigtimes^n M$ to $\bigotimes^n M$ described by

$$(x_1, \ldots, x_n) \mapsto x_{\sigma^{-1}(1)} \otimes \ldots \otimes x_{\sigma^{-1}(n)}.$$

Exterior algebras; determinants

This mapping is clearly n-linear and so there is a unique R-morphism $\sigma^\star : \bigotimes^n M \to \bigotimes^n M$ such that

(2) $\qquad \sigma^\star(x_1 \otimes \ldots \otimes x_n) = x_{\sigma^{-1}(1)} \otimes \ldots \otimes x_{\sigma^{-1}(n)}$.

- The reason for σ^{-1} appearing here will become clear in a moment. Note in particular that if τ is the transposition described by $i \leftrightarrow j$ then

$$\tau^\star(x_1 \otimes \ldots \otimes x_i \otimes \ldots \otimes x_j \otimes \ldots \otimes x_n) = x_1 \otimes \ldots \otimes x_j \otimes \ldots \otimes x_i \otimes \ldots \otimes x_n.$$

With the above notation, it is clear from (1) and (2) that

$$\sigma^{-1}g = g \circ \sigma^\star.$$

Theorem 17.1 $(\forall \vartheta, \sigma \in P_n)\ (\vartheta \circ \sigma)^\star = \vartheta^\star \circ \sigma^\star$.

Proof This follows from the equalities

$$\begin{aligned}
(\vartheta^\star \circ \sigma^\star)(x_1 \otimes \ldots \otimes x_n) &= \vartheta^\star\big(x_{\sigma^{-1}(1)} \otimes \ldots \otimes x_{\sigma^{-1}(n)}\big) \\
&= \vartheta^\star(y_1 \otimes \ldots \otimes y_n) \quad \text{where } y_i = x_{\sigma^{-1}(n)} \\
&= y_{\vartheta^{-1}(1)} \otimes \ldots \otimes y_{\vartheta^{-1}(n)} \\
&= x_{\sigma^{-1}[\vartheta^{-1}(1)]} \otimes \ldots \otimes x_{\sigma^{-1}[\vartheta^{-1}(n)]} \\
&= x_{(\vartheta \circ \sigma)^{-1}(1)} \otimes \ldots \otimes x_{(\vartheta \circ \sigma)^{-1}(n)} \\
&= (\vartheta \circ \sigma)^\star(x_1 \otimes \ldots \otimes x_n). \quad \diamond
\end{aligned}$$

Let us now consider the submodule $A(\bigotimes^n M)$ of $\bigotimes^n M$ that is generated by the elements of the form $m_1 \otimes \ldots \otimes m_n$ where $m_i = m_j$ for some i,j with $i \neq j$; in other words, the submodule generated by the elements x such that $\tau^\star(x) = x$ for some transposition τ. An important property of this submodule is the following.

Theorem 17.2 *For every $\sigma \in P_n$ and every $x \in \bigotimes^n M$,*

$$x - \epsilon_\sigma \sigma^\star(x) \in A(\bigotimes^n M).$$

Proof Suppose first that τ is a transposition. To show that the result holds for τ we have to show that, for every $x \in \bigotimes^n M$,

$$x + \tau^\star(x) \in A(\bigotimes^n M).$$

Suppose then that τ swaps i and j with $i < j$. We have

$x + \tau^\star(x)$
$= (x_1 \otimes \ldots \otimes x_i \otimes \ldots \otimes x_j \otimes \ldots \otimes x_n) + (x_1 \otimes \ldots \otimes x_j \otimes \ldots \otimes x_i \otimes \ldots \otimes x_n)$
$= (x_1 \otimes \ldots \otimes (x_i + x_j) \otimes \ldots \otimes (x_i + x_j) \otimes \ldots \otimes x_n) -$
$\quad (x_1 \otimes \ldots \otimes x_i \otimes \ldots \otimes x_i \otimes \ldots \otimes x_n) - (x_1 \otimes \ldots \otimes x_j \otimes \ldots \otimes x_j \otimes \ldots \otimes x_n)$
$\in A(\bigotimes^n M)$

Now since every $\sigma \in P_n$ is a composite of transpositions, we can proceed to establish the result by induction. Suppose then that the result holds for all composites ϑ of m transpositions where $m \geq 2$. We shall show that it is also true for $\sigma = \tau \circ \vartheta$ where τ is a transposition. By the hypothesis we have, for every $x \in \bigotimes^n M$,

$$x - \epsilon_\vartheta \vartheta^*(x) \in A(\bigotimes^n M).$$

Since $A(\bigotimes^n M)$ is clearly stable under τ^* we deduce that

$$\tau^*(x) - \epsilon_\vartheta (\tau^* \circ \vartheta^*)(x) \in A(\bigotimes^n M).$$

Now by the first part of the proof we have

$$x - \epsilon_\tau \tau^*(x) \in A(\bigotimes^n M).$$

Since $A(\bigotimes^n M)$ is a submodule, it follows that

$$x - \epsilon_\sigma \sigma^*(x) = x - \epsilon_{\tau \circ \vartheta}(\tau \circ \vartheta)^*(x) \in A(\bigotimes^n M). \quad \diamond$$

Definition An n-linear mapping $f : \times^n M \to N$ is said to be *alternating* if $f(x_1, \ldots, x_n) = 0$ whenever $x_i = x_j$ for some i, j with $i \neq j$. An R-morphism $g : \bigotimes^n M \to N$ is called *alternating* if the corresponding n-linear mapping

$$(x_1, \ldots, x_n) \mapsto g(x_1 \otimes \ldots \otimes x_n)$$

is alternating.

Example 17.1 The maping $f : \mathbb{Z}^2 \times \mathbb{Z}^2 \to \mathbb{Z}$ given by

$$f((m, n), (p, q)) = mq - np$$

is bilinear and alternating.

Example 17.2 It is clear that an R-morphism $g : \bigotimes^n M \to N$ is alternating if and only if $A(\bigotimes^n M) \subseteq \text{Ker } g$. In particular, therefore, the natural map

$$\natural : \bigotimes M \to \bigotimes M / A(\bigotimes^n M)$$

is alternating.

Theorem 17.3 *If $f : \bigotimes^n M \to N$ is an alternating R-morphism then*

$$(\forall \sigma \in P_n) \qquad \sigma f = \epsilon_\sigma f.$$

Proof For every $x \in \bigotimes^n M$ and every $\sigma \in P_n$ we have, by Theorem 17.2,

$$f(x - \epsilon_\sigma \sigma^*(x)) = 0.$$

The result now follows from the fact that

$$(\sigma f)(x_1 \otimes \ldots \otimes x_n) = f(x_{\sigma(1)} \otimes \ldots \otimes x_{\sigma(n)})$$
$$= f[\epsilon_\sigma \sigma^\star(x_{\sigma(1)} \otimes \ldots \otimes x_{\sigma(n)})]$$
$$= f[\epsilon_\sigma(x_1 \otimes \ldots \otimes x_n)]$$
$$= (\epsilon f)(x_1 \otimes \ldots \otimes x_n). \quad \Diamond$$

We shall see the significance of Theorem 17.3 in due course. For the present, we proceed to consider the following notion.

Definition Let R be a commutative unitary ring, let M be an R-module, and let n be an integer with $n \geq 2$. By an n-th *exterior power* of M we shall mean an R-module P together with an n-linear alternating mapping $f : \mathsf{X}^n M \to P$ such that, for every R-module X and every n-linear alternating mapping $g : \mathsf{X}^n M \to N$, there is a unique R-morphism $h : P \to N$ such that the diagram

is commutative.

The following two results are immediate.

Theorem 17.4 *If (P, f) is an n-th exterior power of M then $\operatorname{Im} f$ generates P.* $\Diamond$

Theorem 17.5 [Uniqueness] *Let (P, f) b an n-th exterior power of M. Then (P', f') is also an n-th exterior power of M if and only if there is a unique R-isomorphism $j : P \to P'$ such that $j \circ f = f'$.* $\Diamond$

As to the existence of n-th exterior powers, consider the quotient module

$$\bigwedge^n M = \bigotimes^n M / A(\bigotimes^n M).$$

We denote the composite R-morphism

$$\mathsf{X}^n M \xrightarrow{\otimes} \bigotimes^n M \xrightarrow{\natural} \bigwedge^n M$$

by $\wedge$. It is clear that $\wedge$ is n-linear and alternating, and that the R-morphism $\natural$ is alternating (Example 17.2).

Theorem 17.6 [Existence] *Let R be a commutative unitary ring, let M be an R-module, and let n be an integer with $n \geq 2$. Then $(\bigwedge^n M, \wedge)$ is an n-th exterior power of M.*

Proof Consider the diagram

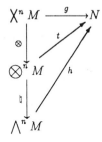

in which N is an arbitrary R-module and g is an n-linear alternating map. By the definition of $\bigotimes^n M$ there is a unique morphism $t : \bigotimes^n M \to N$ such that $t \circ \otimes = g$, Since g is alternating, it is clear that

$$\operatorname{Ker} t \supseteq A(\bigotimes^n M) = \operatorname{Ker} \natural$$

and so, by Theorem 3.4, there is a unique R-morphism $h : \bigwedge^n M \to N$ such that $h \circ \natural = t$. It follows that

$$h \circ \wedge = h \circ \natural \circ \otimes = t \circ \otimes = g.$$

To show that h is unique with respect to this property, suppose that $k : \bigwedge^n M \to N$ is also an R-morphism such that $k \circ \wedge = g$. Then $k \circ \natural \circ \otimes = g$ and so, by the uniqueness of t, we have $k \circ \natural = t = h \circ \natural$. Since $\natural$ is surjective, hence right cancellable, we deduce that $k = h$. ◇

- The above results show that there is, to within R-isomorphism, a unique n-th exterior power of M. We shall call $\bigwedge^n M$ *the n-th exterior power of M*. For every $(x_1, \ldots, x_n) \in \bigtimes^n M$ we write $\wedge(x_1, \ldots, x_n)$ as $x_1 \wedge \ldots \wedge x_n$ and call this the *exterior product of the elements* $x_1, \ldots, x_n$.

- Note that $\operatorname{Im} \wedge$ generates $\bigwedge^n M$ and that, despite the notation, *not every element of $\bigwedge^n M$ is of the form $x_1 \wedge \ldots \wedge x_n$.*

We shall now use exterior powers to construct the *exterior algebra* of a given R-module M. For this purpose, we shall agree to define $\bigwedge^0 M = R$ and $\bigwedge^1 M = M$.

Exterior algebras; determinants

Definition Let R be a commutative unitary ring and M an R-module. By an *exterior algebra* of M we mean an associative unitary R-algebra A together with an R-morphism $f : M \to A$ such that

(1) $(\forall x \in M)\ [f(x)]^2 = 0$;

(2) for every associative unitary R-algebra X and every R-morphism $g : M \to X$ such that $[g(x)]^2 = 0$ for every $x \in M$, there is a unique 1-preserving R-algebra morphism $h : X \to A$ such that the diagram

is commutative.

The following results are immediate.

Theorem 17.7 *If (A, f) is an exterior algebra of the R-module M then $\text{Im } f \cup \{1_A\}$ generates A.* ◊

Theorem 17.8 [Uniqueness] *Let (A, f) be an exterior algebra of the R-module M. Then (A', f') is also an exterior algebra of M if and only if there is a unique R-algebra isomorphism $j : A \to A'$ such that $j \circ f = f'$.* ◊

In order to construct an exterior algebra of M we shall use the tensor algebra $\bigotimes M$. We recall that

$$\bigotimes M = \bigoplus_{i \in \mathbb{N}} (\bigotimes{}^i M).$$

Now $\bigoplus_{i \in \mathbb{N}} A(\bigotimes^i M)$ is a submodule of $\bigotimes M$. It is in fact an ideal of the R-algebra $\bigotimes M$; this follows from the definition of multiplication in $\bigotimes M$ and the fact that if $x_1 \otimes \ldots \otimes x_r \in A(\bigotimes^r M)$ with say $x_j = x_k$ for some $j \neq k$, then for all $y_1 \otimes \ldots \otimes y_p \in \bigotimes^p M$ we have both

$$x_1 \otimes \ldots x_r \otimes y_1 \otimes \ldots \otimes y_p \in A(\bigotimes{}^{r+p} M)$$

and

$$y_1 \otimes \ldots y_p \otimes x_1 \otimes \ldots \otimes x_r \in A(\bigotimes{}^{r+p} M).$$

We can therefore form the quotient algebra

$$\bigotimes M \Big/ \bigoplus_{i \in \mathbb{N}} A(\bigotimes{}^i M).$$

Consider now the R-morphism

$$\otimes M = \bigoplus_{i\in\mathbb{N}}(\otimes^i M) \xrightarrow{\vartheta} \bigoplus_{i\in\mathbb{N}}\left(\otimes^i M / A(\otimes^i M)\right) = \bigoplus_{i\in\mathbb{N}} \bigwedge^i M$$

described by

$$(x_i)_{i\in\mathbb{N}} \mapsto \left(\natural_i(x_i)\right)_{i\in\mathbb{N}},$$

where $\natural_i : \otimes^i M \to \bigwedge^i M$ is the natural epimorphism. It is clear that ϑ is an R-epimorphism with $\operatorname{Ker}\vartheta = \bigotimes_{i\in\mathbb{N}} A(\otimes^i M)$. We therefore have an R-isomorphism

$$\otimes M \Big/ \bigoplus_{i\in\mathbb{N}} A(\otimes^i M) \xrightarrow{f} \bigoplus_{i\in\mathbb{N}} \bigwedge^i M.$$

Appealing to Theorem 4.8, we see that there is then a unique multiplication on $\bigoplus_{i\in\mathbb{N}} \bigwedge^i M$ such that $\bigoplus_{i\in\mathbb{N}} \bigwedge^i M$ is an R-algebra (which is then associative and unitary) with f an R-algebra isomorphism.

For peace of mind, we shall make the following identifications and conventions. We shall agree to identify the algebras $\otimes M \Big/ \bigoplus_{i\in\mathbb{N}} A(\otimes^i M)$ and $\bigoplus_{i\in\mathbb{N}} \bigwedge^i M$ and we shall denote each by simple $\bigwedge M$. We shall also identify each $\bigwedge^j M$ with the submodule $\operatorname{in}_j^{\to}(\bigwedge^j M)$ of $\bigwedge M$, thereby regarding $\bigwedge M$ as an internal direct sum of the submodules $\bigwedge^j M$. In this way, $\bigwedge M$ is generated by $1_R \in \bigwedge^0 M$ and the elements $x_1 \wedge \ldots \wedge x_r \in \bigwedge^r M$. Now since in this R-algebra $x_1 \wedge \ldots \wedge x_r$ is the equivalence class of $x_1 \otimes \ldots \otimes x_r$ modulo the ideal $\bigoplus_{i\in\mathbb{N}} A(\otimes^i M)$, we see that multiplication in $\bigwedge M$ is inherited from that of $\otimes M$ in such a way that

$$\begin{cases} (x_1 \wedge \ldots \wedge x_r)(y_1 \wedge \ldots \wedge y_t) = x_1 \wedge \ldots \wedge x_r \wedge y_1 \wedge \ldots \wedge y_t; \\ 1_R(x_1 \wedge \ldots \wedge x_r) = x_1 \wedge \ldots \wedge x_r = (x_1 \wedge \ldots \wedge x_r)1_R. \end{cases}$$

- Note from the above that $\bigwedge^p M \cdot \bigwedge^q M \subseteq \bigwedge^{p+q} M$.

Theorem 17.9 [Existence] *Let R be a commutative unitary ring and let M be an R-module. Then $(\bigwedge M, \iota_M)$ is an exterior algebra of M.*

Proof Given $x \in M = \bigwedge^1 M$ we have

$$x \otimes x \in A(\otimes^2 M) \subseteq \bigoplus_{i\in\mathbb{N}} A(\otimes^i M)$$

Exterior algebras; determinants 243

and so, on passing to quotients, we obtain $x^2 = x \wedge x = 0$ for all $x \in M$; in other words, $[\iota_M(x)]^2 = 0$ in the R-algebra $\bigwedge M$.

Suppose now that N is an associative unitary R-algebra and that $g : M \to N$ is an R-morphism such that $[g(x)]^2 = 0$ for every $x \in M$. Define a family $(g_i)_{i \in I}$ of R-morphisms $g_i : \bigwedge^i M \to N$ as follows. For $i \geq 1$ let $g'_i : \mathcal{X}^i M \to N$ be the mapping described by

$$(m_1, \ldots, m_i) \mapsto \prod_{j=1}^{i} g(m_j).$$

Since $[g(m_j)]^2 = 0$ for every m_j, we see that each g'_i is i-linear and alternating. For every $i \geq 1$ there is therefore a unique R-morphism $g_i : \bigwedge^i M \to N$ such that

$$g_i(m_1 \wedge \ldots \wedge m_i) = \prod_{j=1}^{i} g(m_j).$$

Now by the definition of $\bigwedge M$ there is a unique R-morphism $h : \bigwedge M \to N$ such that the diagram

$$\begin{array}{ccc} \bigwedge^i M & \xrightarrow{g_i} & N \\ \text{in}_i \downarrow & \nearrow h & \\ \bigwedge M = \bigoplus_{i \in \mathbb{N}} \bigwedge^i M & & \end{array}$$

is commutative. The rest of the proof is now an exact replica of the corresponding part of the proof of Theorem 16.14 with $\otimes$ replaced at each stage by $\wedge$. We leave the details to the reader. $\diamond$

The above results show that to within isomorphism there is a unique exterior algebra over a given R-module M. We shall refer to that constructed in Theorem 17.9 as *the* exterior algebra over M.

We have seen in the above that $x \wedge x = 0$ for all $x \in M = \bigwedge^1 M$. We also have the following property in $\bigwedge M$.

Theorem 17.10 *For every $x_1 \wedge \ldots \wedge x_n \in \bigwedge^n M$ and every $\sigma \in P_n$,*

$$x_{\sigma(1)} \wedge \ldots \wedge x_{\sigma(n)} = \epsilon_\sigma(x_1 \wedge \ldots \wedge x_n).$$

Proof This is immediate from Theorem 17.3 since $\natural : \bigotimes^n M \to \bigwedge^n M$ is an alternating R-morphism (see Example 17.2). $\diamond$

Corollary 1 *If $x \in \bigwedge^p M$ and $y \in \bigwedge^q M$ then $y \wedge x = (-1)^{pq}(x \wedge y)$.*

Proof Let $x = x_1 \wedge \ldots \wedge x_p$ and $y = y_1 \wedge \ldots \wedge y_q$. For convenience, we shall write y_i as x_{p+i} for $i = 1, \ldots, q$. Also, without loss of generality, we shall assume that $q \leq p$. Consider the permutation on $\{1,\ldots,p+q\}$ described by

$$
\begin{array}{ccccccccc}
1 & 2 & \cdots & q & q+1 & \cdots & p & p+1 & \cdots & p+q \\
\downarrow & \downarrow & & \downarrow & \downarrow & & \downarrow & \downarrow & & \downarrow \\
p+1 & p+2 & \cdots & p+q & 1 & \cdots & p-q & p-q+1 & \cdots & p
\end{array}
$$

Alternatively, σ can be described by the prescription

$$\sigma(n) = \begin{cases} n+p & \text{if } n \leq q; \\ n-q & \text{if } n > q. \end{cases}$$

For $i = 1, \ldots, p$ consider also the transpositions described by

$$\begin{aligned} \tau_{i,1} &: p+q \leftrightarrow i \\ \tau_{i,2} &: p+q-1 \leftrightarrow i \\ &\vdots \\ \tau_{i,q} &: p+1 \leftrightarrow i. \end{aligned}$$

It is readily seen that

$$\tau_{p,q} \circ \cdots \circ \tau_{p,1} \circ \tau_{p-1,q} \circ \cdots \circ \tau_{p-1,1} \circ \cdots \circ \tau_{1,q} \circ \cdots \circ \tau_{1,1} \circ \sigma = \mathrm{id},$$

from which we deduce that $(-1)^{pq}\epsilon_\sigma = 1$ and consequently that $\epsilon_\sigma = (-1)^{pq}$. By Theorem 17.10 (with $n = p+q$) we then have

$$\begin{aligned} y \wedge x &= x_{p+1} \wedge \ldots \wedge x_{p+q} \wedge x_1 \wedge \ldots \wedge x_p \\ &= x_{\sigma(1)} \wedge \ldots \wedge x_{\sigma(q)} \wedge x_{\sigma(q+1)} \wedge \ldots \wedge x_{\sigma(p+q)} \\ &= \epsilon_\sigma(x_1 \wedge \ldots \wedge x_q \wedge x_{q+1} \wedge \ldots \wedge x_{p+q}) \\ &= (-1)^{pq}(x \wedge y). \quad \Diamond \end{aligned}$$

Corollary 2 *If $x, y \in \bigwedge^p M$ then $y \wedge x = (-1)^p(x \wedge y)$.*

Proof Take $q = p$ in Corollary 1 and observe that p^2 has the same parity as p. $\Diamond$

We shall now apply the above results to a study of $\bigwedge^r M$ in the case where M is free and of finite dimension.

Theorem 17.11 *Let R be a commutative unitary ring and let M be a free R-module of finite dimension. Then for $r = 0, \ldots, n$ the r-th exterior power $\bigwedge^r M$ is free with*

$$\dim \bigwedge^r M = \binom{n}{r}.$$

Exterior algebras; determinants 245

Moreover, for $r > n$, $\bigwedge^r M = 0$.

Proof If $r = 0$ then $\bigwedge^0 M = R$ is free of dimension 1. Thus the result holds for $r = 0$. We shall now show that it holds for $r = n$. For this purpose, let $\{b_1, \ldots, b_n\}$ be a basis of M. Then by Theorem 16.8 and induction, $\bigotimes^n M$ is of dimension n^n with basis

$$\{b_{i_1} \otimes \ldots \otimes b_{i_n} \,;\, i_j \in \{1, \ldots, n\}\}.$$

Now all of these basis vectors belong to $A(\bigotimes^n M)$ except those of the form $b_{\sigma(1)} \otimes \ldots \otimes b_{\sigma(n)}$ for every $\sigma \in P_n$. Moreover, by Theorem 17.10,

$$b_{\sigma(1)} \wedge \ldots \wedge b_{\sigma(n)} = \epsilon_\sigma (b_1 \wedge \ldots \wedge b_n).$$

It follows, therefore, that the singleton $\{b_1 \wedge \ldots \wedge b_n\}$ is a basis of $\bigwedge^n M$ and so dim $\bigwedge^n M = 1$.

Suppose now that $1 < r < n$. Since $\bigwedge^r M$ is generated by exterior products of r elements of M, it is generated by exterior products of r of the basis elements. We thus see by Theorem 17.10 that $\bigwedge^r M$ is generated by the set of elements of the form

$$b_{i_1} \wedge \ldots \wedge b_{i_r} \quad \text{where} \quad 1 \leq i_1 < \ldots < i_r \leq n.$$

For convenience, we shall denote a typical element of this form by $b_\mathbf{i}$ where $\mathbf{i} : \{1, \ldots, r\} \to \{1, \ldots, n\}$ is an increasing mapping, in the sense that if $j < k$ then $i_j = \mathbf{i}(j) < \mathbf{i}(k) = i_k$. Our aim is to show that these elements constitute a linearly independent set.

For this purpose, suppose that $\sum \lambda_\mathbf{i} b_\mathbf{i} = 0$ in $\bigwedge^r M$ where the sum is over all such increasing sequences $\mathbf{i}$ of r elements of $\{1, \ldots, n\}$, and suppose that $\lambda_\mathbf{k} \neq 0$. The $n - r$ indices that do not belong to Im $\mathbf{k}$ may be arranged in increasing order, say

$$k'_1 < \ldots < k'_{n-r}.$$

Let $b_{\mathbf{k}'}$ denote the corresponding element

$$b_{k'_1} \wedge \ldots \wedge b_{k'_{n-r}}$$

of $\bigwedge^{n-r} M$. It is clear by Theorem 17.10 that

$$b_{\mathbf{k}'} b_\mathbf{k} = b_{\mathbf{k}'} \wedge b_\mathbf{k} = (\pm)(b_1 \wedge \ldots \wedge b_n).$$

On the other hand, if $\mathbf{i} \neq \mathbf{k}'$ then $b_\mathbf{i}$ and $b_{\mathbf{k}'}$ have at least one x_j in common, whence

$$b_{\mathbf{k}'} b_\mathbf{i} = b_{\mathbf{k}'} \wedge b_\mathbf{i} = 0,$$

by Theorem 17.10 and the fact that $x_j \wedge x_j = 0$. We therefore deduce that, in $\bigwedge M$,

$$0 = b_{k'} \wedge 0 = b_{k'} \wedge \sum \lambda_k (b_{k'} \wedge b_k) = \lambda_k(\pm)(b_1 \wedge \ldots \wedge b_n).$$

Since $\{b_1 \wedge \ldots \wedge b_n\}$ is a basis of $\bigwedge^n M$, we have $b_1 \wedge \ldots \wedge b_n \neq 0$, and consequently we have the contradiction $\lambda_k = 0$.

We thus see that the elements b_i constitute a linearly independent set. Since this is also a generating set for $\bigwedge^r M$, it is therefore a basis. The result now follows from the fact that there are $\binom{n}{r}$ subsets of $\{1, \ldots, n\}$ that consist of precisely r elements.

Consider now the case where $r > n$. Since $\bigwedge^r M$ is generated by products of r of the n basis elements, it follows from the fact that $r > n$ that every such product must contain a repeated factor, whence it must be zero since $x \wedge x = 0$ for every $x \in M$. Thus we see that when $r > n = \dim M$ we have $\bigwedge^r M = \{0\}$. $\diamond$

Corollary 1 *If R is a commutative unitary ring and if M is a free R-module of dimension n then $\bigwedge M$ is free and of dimension 2^n.*

Proof The result is immediate from Theorem 7.8 and the fact that

$$\sum_{r=0}^{n} \binom{n}{r} = 2^n. \quad \diamond$$

Corollary 2 *If M is of dimension n then for every r such that $0 \leq r \leq n$ the R-modules $\bigwedge^r M$ and $\bigwedge^{n-r} M$ are isomorphic.* $\diamond$

Our next task will be to illustrate the importance of Theorem 17.11. In fact, we shall show that it surprisingly leads in a very natural way to the notion of the determinant of an R-morphism (and hence that of a square matrix). For this purpose, we require the following notion.

Let R be a commutative unitary ring and let M, N be R-modules. If $f : M \to N$ is an R-morphism then the assignment

$$(x_1, \ldots, x_p) \mapsto f(x_1) \wedge \ldots \wedge f(x_p)$$

yields a p-linear alternating mapping from $\bigtimes^p M$ to $\bigwedge^p N$. There is therefore a unique R-morphism, which we shall denote by

$$\bigwedge^p f : \bigwedge^p M \to \bigwedge^p N$$

such that

$$(\bigwedge^p f)(x_1 \wedge \ldots \wedge x_p) = f(x_1) \wedge \ldots \wedge f(x_p).$$

We call $\bigwedge^p f$ the p-th *exterior power of the R-morphism f.*

Exterior algebras; determinants

- Note that here we perpetrate an abuse of notation that is similar to that in the use of $f \otimes g$.

Theorem 17.12 *If $f : M \to N$ and $g : N \to P$ are R-morphisms then, for every positive integer p,*

$$\bigwedge^p (g \circ f) = (\bigwedge^p g) \circ (\bigwedge^p f).$$

Proof It is clear that

$$(\bigwedge^p g) \circ (\bigwedge^p f) : \bigwedge^p M \to \bigwedge^p P$$

is an R-morphism. Since

$$[(\bigwedge^p g) \circ (\bigwedge^p f)](x_1 \wedge \ldots \wedge x_p) = (\bigwedge^p g)[f(x_1) \wedge \ldots \wedge f(x_p)]$$
$$= (g \circ f)(x_1) \wedge \ldots \wedge (g \circ f)(x_p),$$

it then follows by the uniqueness that $(\bigwedge^p g) \circ (\bigwedge^p f)$ coincides with $\bigwedge^p (g \circ f)$. $\diamond$

Definition If M is an R-module and $f : M \to M$ is an endomorphism on M then we say that f is a *homothety* if there exists $r \in R$ such that $f(x) = rx$ for every $x \in M$.

Example 17.3 If R is commutative then every R-endomorphism $f : R \to R$ is a homothety. In fact, we have

$$(\forall r \in R) \qquad f(r) = f(1_R r) = f(1_R) r.$$

Suppose now that, R being commutative, M is a free R-module of dimension n. Then, by Theorem 17.11, $\bigwedge^n M$ is free and of dimension 1. It is therefore R-isomorphic to R. If $f : M \to M$ is an R-endomorphism, it follows by this isomorphism and Example 17.3 that the R-endomorphism

$$\bigwedge^n f : \bigwedge^n M \to \bigwedge^n M$$

is a homothety. Describing this homothety by the assignment $x \mapsto \lambda x$, we define the *determinant* of f to be the scalar λ. We shall write the determinant of f as $\det f$.

It is clear from the definition of $\bigwedge^n f$ that, for all $x_1, \ldots, x_n \in M$,

$$f(x_1) \wedge \ldots \wedge f(x_n) = (\det f)(x_1 \wedge \ldots \wedge x_n).$$

It is also clear that $\det \mathrm{id}_M = 1_R$. We also have the following result.

Theorem 17.13 *If $f, g : M \to M$ are R-morphisms where M is free and of finite dimension, then*

$$\det(g \circ f) = (\det g)(\det f).$$

Proof Let $\dim M = n$; then the result is an immediate consequence of Theorem 17.12 (with $p = n$). ◊

To see that the above notion of a determinant yields (in a simple way, moreover) the corresponding familiar notion and properties of determinants, suppose that R is a commutative unitary ring and let A be an $n \times n$ matrix over R. Let $\{e_i\ ;\ 1 \leq i \leq n\}$ be the natural ordered basis of the R-module $\text{Mat}_{n \times 1}(R)$ and let

$$f_A : \text{Mat}_{n \times 1}(R) \to \text{Mat}_{n \times 1}(R)$$

be the R-morphism such that A is the matrix of f_A relative to the natural ordered basis. Then we define the *determinant of the $n \times n$ matrix A* to be $\det f_A$. We thus have

$$f_A(e_1) \wedge \ldots \wedge f_A(e_n) = (\det A)(e_1 \wedge \ldots \wedge e_n).$$

Example 17.4 Consider the case $n = 2$. Given the matrix

$$A = \begin{bmatrix} a_{11} & a_{12} \\ a_{21} & a_{22} \end{bmatrix}$$

the associated R-morphism f_A is given by

$$f_A(e_1) = \begin{bmatrix} a_{11} \\ a_{21} \end{bmatrix} = a_{11}e_1 + a_{21}e_2;$$

$$f_A(e_2) = \begin{bmatrix} a_{12} \\ a_{22} \end{bmatrix} = a_{12}e_1 + a_{22}e_2.$$

Since $e_1 \wedge e_1 = 0 = e_2 \wedge e_2$ and since, by Theorem 17.10, $e_2 \wedge e_1 = -(e_1 \wedge e_2)$, we deduce that

$$\begin{aligned} f_A(e_1) \wedge f_A(e_2) &= (a_{11}e_1 + a_{21}e_2) \wedge (a_{12}e_1 + a_{22}e_2) \\ &= (a_{11}a_{22} - a_{21}a_{12})(e_1 \wedge e_2), \end{aligned}$$

whence we see that $\det A = a_{11}a_{22} - a_{21}a_{12}$.

Theorem 17.14 *If A and B are $n \times n$ matrices over the commutative unitary ring R then*

$$\det(AB) = \det A \det B.$$

Proof This is immediate from the definition of the determinant of a matrix and Theorem 17.13. ◊

Corollary 1 *If A is an invertible $n \times n$ matrix over a commutative unitary ring R then $\det A$ is an invertible element of R, and*

$$(\det A)^{-1} = \det A^{-1}.$$

Exterior algebras; determinants

Proof Simply take $B = A^{-1}$ in Corollary 1 and use the fact that the determinant of the $n \times n$ identity matrix is 1_R. ◊

Corollary 2 *Similar matrices have the same determinant.*

Proof If A and B are similar $n \times n$ matrices then there is an invertible $n \times n$ matrix P such that $B = PAP^{-1}$. By Theorem 17.14 and the fact that R is commutative, we see that $\det B = \det A$. ◊

In Example 17.4 above we obtained a formula for $\det A$ when A is of size 2×2. In the general $n \times n$ case we have

$$(\det A)(\mathbf{e}_1 \wedge \ldots \wedge \mathbf{e}_n) = f_A(\mathbf{e}_1) \wedge \ldots \wedge f_A(\mathbf{e}_n)$$

$$= \begin{bmatrix} a_{11} \\ \vdots \\ a_{n1} \end{bmatrix} \wedge \ldots \wedge \begin{bmatrix} a_{1n} \\ \vdots \\ a_{nn} \end{bmatrix}$$

$$= \sum_{i=1}^{n} a_{i1}\mathbf{e}_i \wedge \ldots \wedge \sum_{i=1}^{n} a_{in}\mathbf{e}_i$$

$$= \sum_{\sigma \in P_n} a_{\sigma(1),1} \cdots a_{\sigma(n),n} (\mathbf{e}_{\sigma(1)} \wedge \ldots \wedge \mathbf{e}_{\sigma(n)})$$

$$= \sum_{\sigma \in P_n} \epsilon_\sigma a_{\sigma(1),1} \cdots a_{\sigma(n),n} (\mathbf{e}_1 \wedge \ldots \wedge \mathbf{e}_n)$$

whence we have the general formula

$$\det A = \sum_{\sigma \in P_n} \epsilon_\sigma a_{\sigma(1),1} \cdots a_{\sigma(n),n}.$$

We also note from the above string of equalities that *if A has two identical columns then $\det A = 0$*; and that *if B is obtained from A by interchanging two columns then $\det B = -\det A$*.

If now $\sigma, \vartheta \in P_n$ then for every index i there is a unique index j such that

$$a_{\sigma(i),i} = a_{\sigma[\vartheta(j)],\vartheta(j)}$$

and so, since R is commutative, we have

$$a_{\sigma(1),1} \cdots a_{\sigma(n),n} = a_{\sigma[\vartheta(1)],\vartheta(1)} \cdots a_{\sigma[\vartheta(n)],\vartheta(n)}.$$

Taking in particular $\vartheta = \sigma^{-1}$ and using the fact that $\epsilon_{\sigma^{-1}} = \epsilon_\sigma$, we deduce that

$$\det A = \sum_{\sigma \in P_n} \epsilon_\sigma a_{1,\sigma(1)} \cdots a_{n,\sigma(n)}.$$

Comparing this with the previous expression for $\det A$, we deduce immediately that

$$\det A = \det A^t.$$

It follows that *if A has two identical rows then* $\det A = 0$; and that *if B is obtained from A by interchanging two rows then* $\det B = -\det A$.

Suppose now that A is an $n \times n$ matrix over R. We shall denote by A_{ij} the $(n-1) \times (n-1)$ matrix obtained from A by deleting the i-th row and the j-th column of A. Denoting by $\{e_1, \ldots, e_n\}$ the natural basis of $\text{Mat}_{n \times 1}(R)$, we let

$$f_{ij} : \bigoplus_{k \neq i} Re_k \to \bigoplus_{k \neq j} Re_k$$

be the R-morphism whose matrix, relative to the bases $(e_k)_{k \neq i}, (e_k)_{k \neq j}$ is A_{ij}. Then we have

$(\det A)(e_1 \wedge \ldots \wedge e_n)$
$= f_A(e_1) \wedge \ldots \wedge f_A(e_n)$
$= f_A(e_1) \wedge \ldots \wedge \sum_j a_{ji} e_j \wedge \ldots \wedge f_A(e_n)$
$= \sum_j (f_A(e_1) \wedge \ldots \wedge a_{ji} e_j \wedge \ldots \wedge f_A(e_n))$
$= \sum_j \left(\sum_t a_{t1} e_t \wedge \ldots \wedge a_{ji} e_j \wedge \ldots \wedge \sum_t a_{tn} e_t \right)$
$= \sum_j \left(\sum_{t \neq j} a_{t1} e_t \wedge \ldots \wedge a_{ji} e_j \wedge \ldots \wedge \sum_{t \neq j} a_{tn} e_t \right)$
$= \sum_j ((-1)^{i-1} a_{ji} e_j \wedge f_{ij}(e_1) \wedge \ldots \wedge f_{ij}(e_{i-1}) \wedge f_{ij}(e_{i+1}) \wedge \ldots \wedge f_{ij}(e_n))$
$= \sum_j ((-1)^{i-1} a_{ji} e_j \wedge (\det A_{ji})(e_1 \wedge \ldots \wedge e_{j-1} \wedge e_{j+1} \wedge \ldots \wedge e_n))$
$= \sum_j (-1)^{i+j} a_{ji} \det A_{ji} (e_1 \wedge \ldots \wedge e_n),$

from which we deduce that

$$(i = 1, \ldots, n) \quad \det A = \sum_j (-1)^{i+j} a_{ji} \det A_{ji}.$$

Interchanging i and j, we obtain

$$(j = 1, \ldots, n) \quad \det A = \sum_i (-1)^{i+j} a_{ij} \det A_{ij}.$$

This is called the *Laplace expansion of $\det A$ by the j-th column of A*. Note from the above that this is independent of the row chosen. Since $\det A = \det A^t$ we deduce that also

$$(i = 1, \ldots, n) \quad \det A = \sum_j (-1)^{i+j} a_{ij} \det A_{ij}.$$

This is called the *Laplace expansion of $\det A$ by the i-th row of A*.

Exterior algebras; determinants

We call $(-1)^{i+j} \det A_{ij}$ the *cofactor* of the element a_{ij}. The *adjugate* of A is defined to be the $n \times n$ matrix $\operatorname{Adj} A$ given by

$$[\operatorname{Adj} A]_{ij} = (-1)^{i+j} \det A_{ji}.$$

With this terminology, we have the following result.

Theorem 17.15 *If A is an $n \times n$ matrix over a commutative unitary ring R then*

$$A \cdot \operatorname{Adj} A = (\det A) I_n.$$

Proof The (i,j)-th element of $A \cdot \operatorname{Adj} A$ is given by

$$[A \cdot \operatorname{Adj} A]_{ij} = \sum_{k=1}^{n} a_{ik} (-1)^{k+j} \det A_{jk} = \begin{cases} \det A & \text{if } j = i; \\ 0 & \text{if } j \neq i, \end{cases}$$

for when $j \neq i$ the sum represents the determinant of an $n \times n$ matrix whose i-th row and j-th row are equal, whence it is 0. Thus $A \cdot \operatorname{Adj} A$ is the diagonal matrix every diagonal entry of which is $\det A$. $\Diamond$

Corollary *An $n \times n$ matrix A is invertible if and only if $\det A \neq 0$, in which case*

$$A^{-1} = \frac{1}{\det A} \operatorname{Adj} A.$$

Proof This is immediate from the above and Corollary 1 of Theorem 17.14. $\Diamond$

We end our discussion of determinants with the following useful result.

Theorem 17.16 *Let A be an $n \times n$ matrix over a commutative unitary ring R and suppose that A has the partitioned form*

$$A = \left[\begin{array}{c|c} X & Z \\ \hline 0 & Y \end{array} \right]$$

where X is of size $p \times p$, Y is of size $(n-p) \times (n-p)$, and the matrix in the south-west corner is a zero matrix. Then

$$\det A = \det X \cdot \det Y.$$

Proof If $(e_i)_{1 \leq i \leq n}$ is the natural ordered basis of $\operatorname{Mat}_{n \times 1}(R)$, we let

$$fx : \bigoplus_{i=1}^{p} Re_i \to \bigoplus_{i=1}^{p} Re_i$$

be the R-morphism whose matrix relative to the ordered basis $(e_i)_{1\le i\le p}$ is X and

$$f_Y : \bigoplus_{i=p+1}^{n} Re_i \to \bigoplus_{i=p+1}^{n} Re_i$$

the R-morphism whose matrix relative to the ordered basis $(e_i)_{p+1\le i\le n}$ is Y. Then we have

$$\begin{aligned}
&(\det A)(e_1 \wedge \ldots \wedge e_n) \\
&= f_A(e_1) \wedge \ldots \wedge f_A(e_n) \\
&= \sum_i a_{i1}e_i \wedge \ldots \wedge \sum_i a_{in}e_i \\
&= \sum_{i\le p} x_{i1}e_i \wedge \ldots \wedge \sum_{i\le p} x_{ip}e_i \wedge \sum_i a_{i,p+1}e_i \wedge \ldots \wedge \sum_i a_{in}e_i \\
&= f_X(e_1) \wedge \ldots \wedge f_X(e_p) \wedge \sum_i a_{i,p+1}e_i \wedge \ldots \wedge \sum_i a_{in}e_i \\
&= (\det X)\left(e_1 \wedge \ldots \wedge e_p \wedge \sum_i a_{i,p+1}e_i \wedge \ldots \wedge \sum_i a_{in}e_i\right) \\
&= (\det X)\left(e_1 \wedge \ldots \wedge e_p \wedge \sum_{i>p} a_{i,p+1}e_i \wedge \ldots \wedge \sum_{i>p} a_{in}e_i\right) \\
&= (\det X)(e_1 \wedge \ldots \wedge e_p \wedge f_Y(e_{p+1}) \wedge \ldots \wedge f_Y(e_n)) \\
&= (\det X)(e_1 \wedge \ldots \wedge e_p \wedge (\det Y)(e_{p+1} \wedge \ldots \wedge e_n)) \\
&= (\det X)(\det Y)(e_1 \wedge \ldots \wedge e_n),
\end{aligned}$$

whence the result follows. ◊

We conclude this section with some results on exterior algebras that will be useful to us later.

Theorem 17.17 *Let R be a commutative unitary ring and let $f : M \to N$ be an R-epimorphism. If $f^\wedge : \bigwedge M \to \bigwedge N$ is the unique R-algebra morphism such that the diagram*

$$\begin{array}{ccc} M & \xrightarrow{f} & N \\ \iota_M \downarrow & & \downarrow \iota_N \\ \bigwedge M & \xrightarrow{f^\wedge} & \bigwedge N \end{array}$$

is commutative then $\mathrm{Ker}\, f^\wedge$ is the ideal of $\bigwedge M$ that is generated by $\mathrm{Ker}\, f$.

Proof Given any positive integer n, consider the following diagram in which the notation is as follows :

I denotes the ideal of $\bigwedge M$ generated by $\mathrm{Ker}\, f$;
$I_n = I \cap \bigwedge^n M$;

Exterior algebras; determinants

the unmarked arrows are the natural ones;
$\bigtimes^n f$ is the cartesian product morphism, given by

$$(\bigtimes^n f)(x_1, \ldots, x_n) = (f(x_1), \ldots, f(x_n));$$

ϑ_n is the unique R-morphism that makes the top rectangle commutative (definition of $\bigwedge^n M$);

t is the unique R-algebra morphism that makes the parallelogram commutative (definition of $\bigwedge M$);

α_n is the unique R-morphism that makes the diamond commutative (Theorem 4.3).

Ignore for the moment the three arrows g, g_n, h_n.

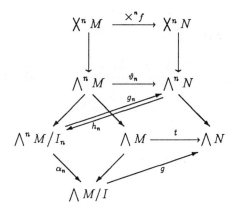

We show first that in fact $t = f^\wedge$. For this purpose, we note that since

$$\bigwedge M = \bigoplus_{n \in \mathbb{N}} \bigwedge^n M$$

it follows from the defining property of direct sums that *the R-morphism t is the same for every n.* In particular, taking $n = 1$ and recalling that $\bigwedge^1 M = M$, we see that $\vartheta_1 = f$ and consequently that $t = f^\wedge$.

We now establish the existence of a unique R-morphism

$$g : \bigwedge M/I \to \bigwedge N$$

such that the diagram

is commutative. Since $(f^\wedge)^\to(\operatorname{Ker} f) = \{0\}$ it follows that $(f^\wedge)^\to(I) = \{0\}$ and so $I \subseteq \operatorname{Ker} f^\wedge$. The existence and uniqueness of g follow by Theorem 3.4. Now, also by Theorem 3.4, we have that $I = \operatorname{Ker} f^\wedge$ if and only if g is injective. The result will follow, therefore, if we can show that g is injective.

For this purpose, some preliminaries are required. Note first that since $I \subseteq \operatorname{Ker} f^\wedge$, and since $t = f^\wedge$ implies that

$$\vartheta_n^\to(\operatorname{Ker} f^\wedge \cap \textstyle\bigwedge^n M) = \{0\},$$

we have $\vartheta_n^\to(I_n) = \{0\}$ and so, by Theorem 3.4, there is a unique R-morphism

$$g_n : \textstyle\bigwedge^n M/I_n \to \textstyle\bigwedge^n N$$

such that the diagram

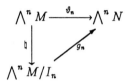

is commutative. We now construct an R-morphism

$$h_n : \textstyle\bigwedge^n N \to \textstyle\bigwedge^n M/I_n$$

such that $g_n \circ h_n$ is the identity map on $\bigwedge^n M/I_n$, whence it will follow immediately that each g_n is injective.

For every $x \in N$ denote by $f^*(x)$ any $y \in M$ such that $f(y) = x$ (recall that f is surjective by hypothesis). Then the reader will readily verify that the assignment

$$(x_1, \ldots, x_n) \mapsto \big(f^*(x_1) \wedge \ldots \wedge f^*(x_n)\big)/I_n$$

yields an n-linear alternating mapping from $\bigtimes^n N$ to $\bigwedge^n M/I_n$. To see that it is well defined, observe that if $x_i = x'_i$ for $i = 1, \ldots, n$ and if $y_i, y'_i \in M$ are such that $f(y_i) = x_i$ and $f(y'_i) = x'_i$ then $f(y_i - y'_i) = x_i - x'_i = 0$ so that $f(y_i) = f(y'_i)$ and consequently

$$f^\wedge(y_1 \wedge \ldots \wedge y_n) = f(y_1) \wedge \ldots \wedge f(y_n) = f(y'_1) \wedge \ldots \wedge f(y'_n) = f^\wedge(y'_1 \wedge \ldots \wedge y'_n),$$

thus showing that

$$(y_1 \wedge \ldots \wedge y_n) - (y'_1 \wedge \ldots \wedge y'_n) \in \textstyle\bigwedge^n M \cap \operatorname{Ker} f^\wedge \subseteq I_n.$$

Exterior algebras; determinants 255

There is therefore a unique R-morphism
$$h_n : \bigwedge^n N \to \bigwedge^n M/I_n$$
such that
$$h_n(x_1 \wedge \ldots \wedge x_n) = \big(f^\star(x_1) \wedge \ldots \wedge f^\star(x_n)\big)/I_n.$$
It is now readily seen that $h_n \circ g_n$ coincides with the identity map on a set of generators of $\bigwedge^n M/I_n$ whence $h_n \circ g_n$ is the identity map and so g_n is injective.

We shall now show that g is injective. For this purpose, we observe that every element of I is a sum of elements of the form $a \wedge x \wedge b$ where $x \in \operatorname{Ker} f$ and $a, b \in \bigwedge M$. Since $\bigwedge M = \bigoplus_{n \in \mathbb{N}} \bigwedge^n M$, it is clear that $I = \bigoplus_{n \in \mathbb{N}} (I \cap \bigwedge^n M)$. The R-morphism
$$\zeta : \bigoplus_{n \in \mathbb{N}} \bigwedge^n M \to \bigoplus_{n \in \mathbb{N}} \bigwedge^n M/I$$
given by the prescription
$$\zeta\big((x_i)_{i \in \mathbb{N}}\big) = (x_i + I)_{i \in \mathbb{N}}$$
is clearly surjective with kernel $\bigoplus_{n \in \mathbb{N}} I_n$. Thus we have
$$\bigwedge M/I = \bigoplus_{n \in \mathbb{N}} \bigwedge^n M \Big/ \bigoplus_{n \in \mathbb{N}} i_n \simeq \bigoplus_{n \in \mathbb{N}} \bigwedge^n M/I_n,$$
and hence $(\bigwedge M/I, (\alpha_n)_{n \in \mathbb{N}})$ is a coproduct of the family $\big(\bigwedge^n M/I_n\big)_{n \in \mathbb{N}}$. Writing $A_n = \bigwedge^n M/I_n$ and $B_n = \bigwedge^n N$, we therefore have a family of commutative diagrams

$$\begin{array}{ccccc} 0 & \longrightarrow & A_n & \xrightarrow{g_n} & B_n \\ & & \downarrow{i_n^A} & & \downarrow{i_n^B} \\ & & \bigoplus_{n \in \mathbb{N}} A_n & \xrightarrow{g} & \bigoplus_{n \in \mathbb{N}} B_n \end{array}$$

in which the top row is split exact (a splitting morphism being h_n), and i_n^A, i_n^B are the natural inclusions. By the definition of $\bigoplus_{n \in \mathbb{N}} B_n$ there is a unique R-morphism
$$h : \bigoplus_{n \in \mathbb{N}} B_n \to \bigoplus_{n \in \mathbb{N}} A_n$$

such that $h \circ i_n^B = i_n^A \circ h_n$ for every $n \in \mathbb{N}$. Consequently,

$(\forall n \in \mathbb{N}) \qquad h \circ g \circ i_n^A = h \circ i_n^B \circ g_n = i_n^A \circ h_n \circ g_n = i_n^A.$

For every $x \in \bigoplus\limits_{n \in \mathbb{N}} A_n$ we then have

$$\begin{aligned}(h \circ g)(x) &= \left(h \circ g \circ \sum_{n \in \mathbb{N}}(i_n^A \circ \mathrm{pr}_n^\oplus)\right)(x) \\ &= \left(\sum_{n \in \mathbb{N}}(h \circ g \circ i_n^A \circ \mathrm{pr}_n^\oplus)\right)(x) \\ &= \sum_{n \in \mathbb{N}}(i_n^A \circ \mathrm{pr}_n^\oplus)(x) \\ &= x,\end{aligned}$$

and so $h \circ g$ is the identity map on $\bigoplus\limits_{n \in \mathbb{N}} A_n$. Thus g is injective. ◇

- Note that since f is surjective in the above, so also are ϑ_n and $f^\wedge$. It follows that g_n and g are also surjective, whence they are R-isomorphisms; moreover, so is h_n with $h_n = g_n^{-1}$.

Theorem 17.18 *Let R be a commutative unitary ring and let $I_1, \ldots, I_n$ be ideals of R. For every $p \in \{1, \ldots, n\}$ let S_p denote the collection of all subsets of $\{1, \ldots, n\}$ consisting of p elements and for every $J \in S_p$ let $I_J = \sum\limits_{k \in J} I_k$. Then*

$$\wedge^p\left(\bigoplus_{k=1}^n R/I_k\right) \simeq \bigoplus_{J \in S_p} R/I_J.$$

Proof Let M be a free R-module of dimension n and let $\{e_i \ ; \ 1 \leq r \leq n\}$ be a basis of M. Let $f : M \to \bigoplus\limits_{k=1}^n R/I_k$ be the R-epimorphism such that $f(e_k) = 1 + I_k$ for $k = 1, \ldots, n,$ and let

$$f^\wedge : \wedge M \to \wedge\left(\bigoplus_{k=1}^n R/I_k\right)$$

be the induced epimorphism. Since f is described by

$$\lambda_1 e_1 + \ldots + \lambda_n e_n \mapsto (\lambda_1 + I_1, \ldots, \lambda_n + I_n)$$

it is clear that $\mathrm{Ker}\, f = \bigoplus\limits_{k=1}^n I_k e_k$. By Theorem 17.17, $\mathrm{Ker}\, f^\wedge$ is the ideal of $\wedge M$ that is generated by $\bigoplus\limits_{k=1}^n I_k e_k$. Every element of $\mathrm{Ker}\, f^\wedge$ is therefore

Exterior algebras; determinants

a sum of elements of the form $a \wedge x \wedge b$ where $a, b \in \bigwedge M$ and $x = \sum_{k=1}^{n} x_k e_k$ with $x_k \in I_k$ for each k. Since $\bigwedge M$ is free and of finite dimension (by Corollary 1 of Theorem 17.11), we can express a, b in terms of the natural basis of $\bigwedge M$. It follows that every element of $\operatorname{Ker} f^{\wedge}$ can be expressed uniquely in the form $\alpha(e_{k_1} \wedge \ldots \wedge e_{k_m})$ where $\alpha \in \sum_{j=1}^{n} I_{k_j}$. We thus see that

$$\operatorname{Ker} f^{\wedge} = \bigoplus_{J \in S_p} I_J e_J$$

where $\{e_J\,;\, J \in S_p\}$ denotes the natural basis of $\bigwedge^p M$.

Consider now the complex diagram in the proof of Theorem 17.17. In this, take

$$N = \bigoplus_{k=1}^{n} R/I_k, \quad \bigwedge^p M = \bigoplus_{J \in S_p} Re_J, \quad I = \operatorname{Ker} f^{\wedge}.$$

Then since h_p is an isomorphism (see the remark following Theorem 17.17), we have

$$\bigwedge^p \left(\bigoplus_{k=1}^{n} R/I_k \right) \simeq \bigoplus_{J \in S_p} Re_J \Big/ \left(\bigoplus_{J \in S_p} Re_J \cap \bigoplus_{J \in S_p} I_J e_J \right).$$

Applying the third isomorphism theorem (the Corollary to Theorem 4.6) to the right-hand side, we see that this becomes

$$\simeq \left(\bigoplus_{J \in S_p} Re_J + \bigoplus_{J \in S_p} I_J e_J \right) \Big/ \bigoplus_{J \in S_p} I_J e_J$$

$$= \bigoplus_{J \in S_p} Re_J \Big/ \bigoplus_{J \in S_p} I_J e_J.$$

Now the mapping

$$\vartheta : \bigoplus_{J \in S_p} Re_J \to \bigoplus_{J \in S_p} (Re_J / I_J e_J)$$

given by

$$\vartheta \left(\sum_{J \in S_p} re_J \right) = \sum_{J \in S_p} (re_J + I_J e_J)$$

is clearly an R-epimorphism with $\operatorname{Ker} \vartheta = \bigoplus_{J \in S_p} I_J e_J$. By the first isomorphism theorem (Theorem 4.4), the above quotient of direct sums is therefore

$$\simeq \bigoplus_{J \in S_p} (Re_J / I_J e_J).$$

Now the mapping $\zeta_J : Re_J \to R/I_J$ given by $\zeta_J(re_J) = r + I_J$ is also an R-epimorphism with $\text{Ker}\,\zeta_J = I_J e_J$. We therefore deduce, again by the first isomorphism theorem, that the above direct sum is

$$\simeq \bigoplus_{J \in S_p} R/I_J.$$

In conclusion, therefore, we see that

$$\wedge^p \left(\bigoplus_{k=1}^n R/I_k \right) \simeq \bigoplus_{J \in S_p} R/I_J. \quad \diamond$$

EXERCISES

In each of the following exercises, R is a commutative unitary ring and M is an R-module.

17.1 If $n \geq 2$ consider the *alternator*

$$\alpha_n = \frac{1}{n!} \sum_{\sigma \in P_n} \epsilon_\sigma \sigma^\star : \otimes^n M \to \otimes^n M,$$

Prove that $\text{Ker}\,\alpha_n \subseteq A(\otimes^n M)$.

[*Hint*. For every $x \in \otimes^n M$ consider $\alpha_n(x) - x$ and use Theorem 17.2.]

Show also that

$$(\forall \vartheta \in P_n) \qquad \alpha_n \circ \vartheta^\star = \epsilon_\vartheta \alpha^n.$$

Deduce that if R is not of characteristic 2 then $\text{Ker}\,\alpha_n = A(\otimes^n M)$.
[*Hint*. Take ϑ to be a transposition.]

Show further that α_n is idempotent and deduce that

$$\otimes^n M = A(\otimes^n M) \oplus \text{Im}\,\alpha^n.$$

[*Hint*. Use Exercise 6.11.]

17.2 If $n \geq 2$ let $S(\otimes^n M)$ be the submodule of $\otimes^n M$ generated by the set of elements of the form $x - \tau^\star(x)$ where $x \in \otimes^n M$ and $\tau \in P_n$ is a transposition. Show that $S(\otimes^n M)$ is $\vartheta^\star$-stable for every transposition $\vartheta \in P_n$.

[*Hint*. Use the equality

$$\vartheta^\star(x - \tau^\star(x)) = (x - \tau^\star(x)) - (x - \vartheta^\star(x)) + (\tau^\star(x) - (\vartheta \circ \tau)^\star(x)).]$$

Deduce that

$$(\forall x \in \bigotimes^n M)(\forall \sigma \in P_n) \qquad x - \sigma^*(x) \in S(\bigotimes^n M).$$

[*Hint.* Argue by induction as in Theorem 17.2.]

Consider now the *symmetriser*

$$\beta_n = \frac{1}{n!} \sum_{\sigma \in P_n} \sigma^* : \bigotimes^n M \to \bigotimes^n M.$$

Show that $\operatorname{Ker} \beta_n \subseteq S(\bigotimes^n M)$.

[*Hint.* For every $x \in \bigotimes^n M$ consider $\beta_n(x) - x$.]

Show also that, for every $x \in \bigotimes^n M$ and every transposition $\tau \in P_n$,

$$\beta_n[x - \tau^*(x)] = 0$$

and deduce that $S(\bigotimes^n M) \subseteq \operatorname{Ker} \beta_n$. Show finally that β_n is idempotent and deduce that

$$\bigotimes^n M = S(\bigotimes^n M) \oplus \operatorname{Im} \beta_n.$$

17.3 Let $I(\bigotimes M)$ be the ideal of $\bigotimes M$ that is generated by $\{x \otimes x \; ; \; x \in M\}$. Show that the quotient algebra $\bigotimes M / I(\bigotimes M)$ is isomorphic to $\bigwedge M$.

17.4 Let N be a submodule of M and let Δ be the submodule of $\bigwedge^n M$ that is generated by

$$\{x_1 \wedge \ldots \wedge x_n \; ; \; (\exists i) \; x_i \in N\}.$$

Show that the assignment

$$(m_1, \ldots, m_n) \mapsto m_1 + N \wedge \ldots \wedge m_n + N$$

defines an n-linear alternating mapping. Using Theorem 3.4, produce an R-morphism

$$\alpha : \bigwedge^n M / \Delta \to \bigwedge^n (M/N).$$

Now show that the assignment

$$(x_1 + N, \ldots, x_n + N) \mapsto (x_1 \wedge \ldots \wedge x_n) + \Delta$$

defines an n-linear alternating mapping. Hence produce an R-morphism

$$\beta : \bigwedge^n (M/N) \to \bigwedge^n M / \Delta.$$

Show that α and β are mutually inverse R-isomorphisms.

17.5 Show that the $\mathbb{Z}$-module $\bigwedge^2 \mathbb{Q}$ is zero.

17.6 Let V be a vector space of dimension n over a field F. Show that $x_1, \ldots, x_p$ are linearly independent in V if and only if $x_1 \wedge \ldots \wedge x_p \neq 0$.

17.7 Let M be a free R-module of dimension n. Identifying $(\bigotimes^p M)^d$ with $\bigotimes^p M^d$ for every $p \in \mathbb{N}$ (see Exercises 16.2 and 16.3), prove that
$$\bigwedge^p M^d \simeq (\bigwedge^p M)^d$$
under the R-isomorphism which to every $x_1^d \wedge \ldots \wedge x_p^d$ in $\bigwedge^p M^d$ assigns the element f of $(\bigwedge^p M)^d$ such that
$$f(x_1 \wedge \ldots \wedge x_p) = \det[\langle x_i, x_j^d \rangle],$$
where the right-hand side is the determinant of the $n \times n$ matrix whose (i,j)-th entry is $\langle x_i, x_j^d \rangle$.

[*Hint.* Using Exercise 16.3, write $\det[\langle x_i, x_j^d \rangle]$ in terms of the alternator α_p on $\bigotimes^p m^d$.]

17.8 Identifying $\bigwedge^p M^d$ with $(\bigwedge^p M)^d$ under the isomorphism of the previous exercise, observe that we have
$$\langle x_1 \wedge \ldots \wedge x_p, x_1^d \wedge \ldots \wedge x_p^d \rangle = \det[\langle x_i, x_j^d \rangle].$$
Use this identity to prove that if M, N are free R-modules of finite dimensions then, for every R-morphism $f : M \to N$,
$$(\bigwedge^p f)^t = \bigwedge^p f^t.$$

17.9 Show that $\bigoplus_{k \in \mathbb{N}} \bigwedge^{2k} M$ is a commutative subalgebra of $\bigwedge M$.

17.10 Let A be an R-algebra. An R-morphism $f : A \to A$ is called a *derivation* if
$$(\forall a, b \in A) \quad f(ab) = f(a)\,b + a\,f(b).$$

[The terminology comes from the standard example of the differentiation map $D : \mathbb{R}[X] \to \mathbb{R}[X]$.]

Prove that for every R-morphism $f : M \to M$ there is a unique derivation $Df : \bigwedge M \to \bigwedge M$ such that the diagram

$$\begin{array}{ccc} M & \xrightarrow{f} & M \\ {\scriptstyle \iota_M}\downarrow & & \downarrow{\scriptstyle \iota_M} \\ \bigwedge M & \xrightarrow{Df} & \bigwedge M \end{array}$$

is commutative.

[*Hint.* Consider the family $(\vartheta_p)_{p\in\mathbb{N}}$ of mappings

$$\vartheta_p : \bigtimes^p M \to \bigwedge^p M$$

given by $\vartheta_0 = 0$, $\vartheta_1 = f$ and, for $p \geq 2$,

$$\vartheta_p(x_1, \ldots, x_p) = \sum_{i=1}^{p}(x_1 \wedge \ldots \wedge f(x_i) \wedge \ldots \wedge x_p).$$

Show that each ϑ_p ($p \geq 2$) is p-linear and alternating. Now construct Df in the obvious way from $(\vartheta_p)_{p\in\mathbb{N}}$.]

17.11 If M is a cyclic R-module prove that $\bigwedge M = R \oplus M$.

17.12 If $\mathbf{x} \in \text{Mat}_{n\times 1}(R)$ prove that

$$\det(I_n + \mathbf{xx}^t) = 1 + \mathbf{x}^t\mathbf{x}.$$

17.13 [*Pivotal condensation*] Let $A = [a_{ij}] \in \text{Mat}_{n\times n}(R)$ and suppose that $a_{pq} \neq 0$. Let $B = [b_{ij}] \in \text{Mat}_{(n-1)\times(n-1)}(R)$ be defined as follows:

$$b_{ij} = \begin{cases} \det \begin{bmatrix} a_{ij} & a_{iq} \\ a_{pj} & a_{pq} \end{bmatrix} & \text{if } 1 \leq i \leq p-1, 1 \leq j \leq q-1; \\[2mm] \det \begin{bmatrix} a_{iq} & a_{ij} \\ a_{pq} & a_{pj} \end{bmatrix} & \text{if } 1 \leq i \leq p-1, q+1 \leq j \leq n; \\[2mm] \det \begin{bmatrix} a_{pj} & a_{pq} \\ a_{ij} & a_{iq} \end{bmatrix} & \text{if } p+1 \leq i \leq n, 1 \leq j \leq q-1; \\[2mm] \det \begin{bmatrix} a_{pq} & a_{pj} \\ a_{iq} & a_{ij} \end{bmatrix} & \text{if } p+1 \leq i \leq n, q+1 \leq j \leq n. \end{cases}$$

Prove that

$$\det A = \frac{1}{a_{pq}^{n-2}} \det B.$$

[B is called the matrix obtained from A by *pivotal condensation using* a_{pq} *as pivot*. This is a useful result for computing determinants of matrices with integer entries, the calculations being easier on choosing a 1 as a pivot whenever possible.]

17.14 If A and B are square matrices of the same order prove that

$$\det \begin{bmatrix} A & B \\ B & A \end{bmatrix} = \det(A+B)\det(A-B).$$

17.15 Given a matrix M of the form

$$M = \begin{bmatrix} P & Q \\ R & S \end{bmatrix}$$

where P, Q, R, S are square matrices of the same order with P invertible, find a matrix N of the form

$$N = \begin{bmatrix} A & 0 \\ B & C \end{bmatrix}$$

such that

$$NM = \begin{bmatrix} I & P^{-1}Q \\ 0 & S - RP^{-1}Q \end{bmatrix}.$$

Hence show that if $PR = RP$ then $\det M = \det(PS - RQ)$, and that if $PQ = QP$ then $\det M = \det(SP - RQ)$.

18 Modules over a principal ideal domain; finitely generated abelian groups

We now turn our attention to modules over a particular type of ring, namely a *principal ideal domain*. Our aim will be to establish a structure theorem for finitely generated modules over such a ring. At the end of this chapter we shall apply this structure theorem to obtain a description of all finitely generated abelian groups (and hence all finite abelian groups). In the chapter that follows, we shall apply the structure theorem to obtain some vector space decomposition theorems that lead to a study of canonical forms for matrices.

Definition An R-module M is said to be *cyclic* if it is generated by a singleton subset; in other words, if there exists $x \in M$ such that $M = Rx$.

Example 18.1 Every simple R-module is cyclic. This follows immediately from Theorem 5.5.

Example 18.2 Let I be an ideal of R. Then the R-module R/I is cyclic, for it is generated by $1 + I$.

Definition By a *principal ideal domain* we shall mean a commutative integral domain every ideal I of which is *principal* in the sense that $I = Ra$ for some $a \in R$; in other words, every ideal is a cyclic R-module.

Theorem 18.1 *A commutative unitary ring R is a principal ideal domain if and only if, whenever M is a cyclic R-module, every submodule of M is cyclic.*

Proof Suppose that R is a principal ideal domain and that $M = Rx$ is a cyclic R-module. If N is a submodule of M then the mapping $\vartheta : R \to M$ given by $\vartheta(r) = rx$ is clearly an R-epimorphism and so, by Theorem 3.1, $\vartheta^{\leftarrow}(N)$ is a submodule of R whence $\vartheta^{\leftarrow}(N) = Ra$ for some $a \in R$. Now since ϑ is surjective we have, by the Corollary to Theorem 3.2,

$$N = \vartheta^{\rightarrow}[\vartheta^{\leftarrow}(N)] = \vartheta^{\rightarrow}(Ra) = R\vartheta(a) = Rax,$$

whence we see that N is cyclic.

The converse is clear from the fact that R itself is a cyclic R-module, being generated by $\{1_R\}$. ◊

Definition For every non-empty subset X of an R-module M we define the *annihilator of X in R* by

$$\text{Ann}_R X = \{r \in R \,;\, (\forall x \in X)\, rx = 0\}.$$

In the case where $X = \{x\}$ we write $\text{Ann}_R X$ as $\text{Ann}_R x$ and, by abuse of language, call this the *annihilator of x in R*. We say that x is a *torsion element* of M when $\text{Ann}_R x \neq \{0\}$; and that x is *torsion free* when $\text{Ann}_R x = \{0\}$. We say that M is a *torsion module* if every element of M is a torsion element; and that M is a *torsion free module* if every non-zero element of M is torsion free.

Example 18.3 If V is a vector space over a field F then V is torsion free. In fact, if $\lambda \in \text{Ann}_R x$ then from $\lambda x = 0$ we deduce, since λ^{-1} exists, that $x = \lambda^{-1} \lambda x = \lambda^{-1} 0 = 0$.

Example 18.4 The R-module R is torsion-free if and only if R is an integral domain. In fact, $\text{Ann}_R x = 0$ for every non-zero $x \in R$ if and only if R has no zero divisors.

Example 18.5 Every finite abelian group G is a torsion $\mathbb{Z}$-module. In fact, every element x of G is of finite order and so there is a positive integer n such that $nx = 0$ whence $\text{Ann}_{\mathbb{Z}} x \neq \{0\}$.

We begin our discussion of modules over a principal ideal domain by considering torsion-free modules. The first result that we establish is the following generalisation of Theorem 15.20.

Theorem 18.2 *If R is a principal ideal domain then an R-module M is flat if and only if it is torsion free.*

Proof Let F be a free R-module and $\pi : F \to M$ an epimorphism. Then, by the left/right analogue of Theorem 15.19, M is flat if and only if

$$(\forall a \in R) \qquad Ra \cdot F \cap \text{Ker}\, \pi \subseteq Ra \cdot \text{Ker}\, \pi;$$

in other words, if and only if

$$(\forall a \in R)(\forall x \in F) \qquad ax \in \text{Ker}\, \pi \Rightarrow ax \in a(\text{Ker}\, \pi).$$

Since R has no zero divisors, we see that this is equivalent to the condition

$$(\forall a \in R)(\forall x \in F) \qquad ax \in \text{Ker}\, \pi \Rightarrow x \in \text{Ker}\, \pi.$$

On passing to quotients modulo $\text{Ker}\, \pi$, we see that this condition is equivalent to the condition

$$(\forall a \in R)(\forall x \in M \simeq F/\text{Ker}\,\pi) \qquad am = 0 \Rightarrow m = 0,$$

Modules over a principal ideal domain

which is precisely the condition that M be torsion free. $\diamond$

Since every free module is projective (Theorem 8.6) and every projective module is flat (Corollary to Theorem 15.16), it follows from the above result that *every free module over a principal ideal domain is torsion free*. Our objective now is to establish the converse of this for *finitely generated* modules. For this purpose, we require the following result.

Theorem 18.3 *If R is a principal ideal domain and M is a free R-module then every submodule N of M is free with* $\dim N \leq \dim M$.

Proof Let $\{x_i \; i \in I\}$ be a basis of M. Then for every subset J of I we can define a submodule N_J of N by

$$N_J = N \cap \bigoplus_{j \in J} Rx_j.$$

Ler Γ denote the set of all pairs (J, B_J) where $J \subseteq I$ and B_J is a basis for N_J with $|B_J| \leq |J|$.

We note first that $\Gamma \neq \emptyset$. In fact, consider a singleton subset $\{i\}$ of I. Here we have

$$N_{\{i\}} = N \cap Rx_i.$$

Now $\{r \in R \; ; \; rx_i \in N\}$ is clearly an ideal of R and so is generated by a singleton subset, $\{r_i\}$ say, of R. If $r_i = 0$ then clearly $N_{\{i\}} = \{0\}$ and $\emptyset$ is a basis for $N_{\{i\}}$ with

$$0 = |\emptyset| < |\{i\}| = 1.$$

On the other hand, if $r_i \neq 0$ then $N_{\{i\}} = Rr_ix_i \neq \{0\}$ and $B_{\{i\}} = \{r_ix_i\}$ is a basis of $N_{\{i\}}$ with

$$|B_i| = 1 = |\{i\}|.$$

Thus we see that $\Gamma \neq \emptyset$.

Let us now order Γ by setting

$$(J, B_J) \sqsubseteq (K, B_K) \iff J \subseteq K, \; B_J \subseteq B_K.$$

We show as follows that Γ is inductively ordered.

Let $\{(J_\alpha, B_{J_\alpha}) \; ; \; \alpha \in A\}$ be a totally ordered subset of Γ and let $J^\star = \bigcup_{\alpha \in A} J_\alpha$ and $B^\star = \bigcup_{\alpha \in A} B_{J_\alpha}$. Then we observe that $(J^\star, B^\star) \in \Gamma$. In fact, since $\{J_\alpha \; ; \; \alpha \in A\}$ is totally ordered so also is $\{\bigoplus_{j \in J_\alpha} Rx_j \; ; \; \alpha \in A\}$ and hence so is $\{N_{J_\alpha} \; ; \; \alpha \in A\}$. Now for a totally ordered set $\{X_\alpha \; ; \; \alpha \in A\}$ of submodules it is readily seen that $\bigcup_{\alpha \in A} X_\alpha$ is an R-module, whence it coincides with $\sum_{\alpha \in A} X_\alpha$. We therefore have

$$\sum_{\alpha \in A} N_{J_\alpha} = \bigcup_{\alpha \in A} N_{J_\alpha} = \bigcup_{\alpha \in A} \left(N \cap \bigoplus_{j \in J_\alpha} Rx_j \right)$$
$$= N \cap \bigcup_{\alpha \in A} \bigoplus_{j \in J_\alpha} Rx_j$$
$$= N \cap \sum_{\alpha \in A} \bigoplus_{j \in J_\alpha} Rx_j$$
$$= N \cap \bigoplus_{j \in J^*} Rx_j$$
$$= N_{J^*}.$$

Now, since B_{J_α} is a basis for N_{J_α} for every $\alpha \in A$, we see that B^* is a linearly independent subset of $\bigcup_{\alpha \in A} N_{J_\alpha} = N_{J^*}$. To see that B^* is in fact a basis of N_{J^*}, we observe that if $x \in N_{J^*}$ then $x \in N_{J_\alpha}$ for some $\alpha \in A$ whence $x \in \mathrm{LC}(B_\alpha) \subseteq \mathrm{LC}(B^*)$. Thus $N_{J^*} \subseteq \mathrm{LC}(B^*)$, whence we have equality. To show that $(J^*, B^*) \in \Gamma$ it remains to show that $|B^*| \le |J^*|$; and this is clear from the fact that, for every $\alpha \in A$, $|B_{J_\alpha}| \le |J_\alpha|$.

We can now apply Zorn's axiom to the inductively ordered set Γ to deduce that Γ contains a maximal element, say (K, B_K). Our objective now is to show that $K = I$.

Suppose, by way of obtaining a contradiction, that $K \subset I$ and let $j \in I \setminus K$. Defining $L = K \cup \{j\}$, we have $N_K \subseteq N_L$. If $N_K = N_L$ then clearly $(L, B_K) \in \Gamma$, contradicting the maximality of (K, B_K). Thus we have $N_K \subset N_L$ and so, for every $y \in N_L \setminus N_K$, there exists a non-zero $a \in R$ such that $y - ax_j \in N_K$. It follows that

$$I(N_L) = \{a \in R \ ; \ (\exists y \in N_L) \ y - ax_j \in N_K\}$$

is a non-zero ideal of R, whence it is of the form Ra_j for some non-zero $a_j \in R$. Since $a_j \in I(N_L)$ there exists $y_1 \in N_L$ such that $y_1 - a_j x_j \in N_K$. We shall show that $B_k \cup \{y_1\}$ is a basis for N_L.

Given $y \in N_L$, there exists $r \in R$ such that $y - rx_j \in N_K$. Since $r \in I(N_L)$ we have $r = sa_j$ for some $s \in R$. Then $y - sy_1 \in N_K$ and so

$$y \in \mathrm{LC}(N_K \cup \{y_1\}) = \mathrm{LC}(B_K \cup \{y_1\}),$$

and consequently $B_K \cup \{y_1\}$ generates N_L. This set is also linearly independent; for $y_1 - a_j x_j \in \mathrm{LC}(B_K)$ with $a_j \ne 0$, so that no non-zero multiple of y_1 can belong to $\mathrm{LC}(B_K)$. Thus we see that $B_K \cup \{y_1\}$ is a basis of N_L.

Now since $|B_K| \le |K|$ it is clear that $|B_K \cup \{y_1\}| \le |L|$ so that $(L, B_K \cup \{y_1\}) \in \Gamma$, contradicting the maximality of (K, B_K).

Modules over a principal ideal domain 267

The sum-total of these observations is that $K = I$.
It now follows that
$$N_K = N_I = N \cap \bigoplus_{i \in I} Rx_i = N \cap M = N$$
and so B_K is a basis of N. In conclusion, therefore, N is free with
$$\dim N = |B_K| \leq |I| = \dim M. \quad \diamond$$

Corollary *The flat $\mathbb{Z}$-module $\mathbb{Q}$ is not projective.*

Proof If $\mathbb{Q}$ were projective it would be a direct summand of a free $\mathbb{Z}$-module and so, by the above result, $\mathbb{Q}$ would be free; and this is not the case (see the remark that follows Theorem 7.10). $\diamond$

Theorem 18.4 *Let R be a principal ideal domain. Then every finitely generated torsion free R-module M is free and of finite dimension.*

Proof Let $G = \{x_1, \ldots, x_m\}$ be a set of generators of M and let $H = \{y_1, \ldots, y_n\}$ be a maximal linearly independent subset of G. Given $x_j \in G \setminus H$, it follows from the fact that $\{x_j\} \cup H$ is not linearly independent that there exist $\alpha_j, \beta_1, \ldots, \beta_n \in R$, not all of which are zero, such that
$$\alpha_j x_j + \beta_1 y_1 + \ldots + \beta_n y_n = 0.$$
Moreover, $\alpha_j \neq 0$ since otherwise H is not linearly independent. We thus have
$$\alpha_j x_j \in \bigoplus_{i=1}^{n} Ry_i.$$
It follows that there exist non-zero $\alpha_1, \ldots, \alpha_m \in R$ such that
$$(j = 1, \ldots, m) \quad \alpha_j x_j \in \bigoplus_{i=1}^{n} Ry_i.$$
Now let $r = \prod_{j=1}^{m} \alpha_j$. Then clearly $r \neq 0$ and $rM \subseteq \bigoplus_{i=1}^{n} Ry_i$. The mapping
$$\vartheta : M \to \bigoplus_{i=1}^{n} Ry_i$$
given by $\vartheta(m) = rm$ is then an R-monomorphism with $\text{Im}\,\vartheta$ a submodule of the free R-module $\bigoplus_{i=1}^{n} Ry_i$. By Theorem 18.3, $\text{Im}\,\vartheta$ is free and of dimension less than or equal to n, whence so also is M. $\diamond$

Corollary *For a finitely generated module over a principal ideal domain, the conditions of being free, projective, flat, torsion free are equivalent.* $\diamond$

- Note that the restriction that M be finitely generated is essential in Theorem 18.4. For example, $\mathbb{Q}$ is a non-finitely generated torsion free $\mathbb{Z}$-module that is not free.

Theorem 18.5 *Let R be a principal ideal domain. Then an R-module is finitely generated if and only if it is noetherian.*

Proof $\Leftarrow$: By Theorem 5.1, every noetherian module is finitely generated.

$\Rightarrow$: Suppose that M is a finitely generated R-module, R being a principal ideal domain. In order to prove that M is noetherian, we note first that R itself is noetherian; this follows by Theorem 5.1 and the fact that every ideal of R is generated by a singleton.

We establish the result by induction on the number of generators of M. If M is cyclic, say $M = Rx$, then $f : R \to M$ given by $f(r) = rx$ is an R-epimorphism and so

$$M \simeq R/\operatorname{Ker} f = R/\operatorname{Ann}_R(x).$$

It follows from this that M is noetherian; for, by Theorem 5.3, every quotient module of a noetherian module is noetherian. Suppose, by way of induction, that the result holds for all R-modules with less than or equal to k generators. Let M have $k+1$ generators $x_1, \ldots, x_{k+1}$ and consider the submodule N that is generated by $\{x_1, \ldots, x_k\}$. By the induction hypothesis, N is noetherian; and so also is M/N since it is generated by $\{x_{k+1} + N\}$. It now follows by Theorem 5.4 that M is noetherian. $\diamond$

Corollary *If R is a principal ideal domain then every submodule of a finitely generated R-module is finitely generated.* $\diamond$

For every non-zero module M over a principal ideal domain R we shall denote by $T(M)$ the subset of M consisting of the torsion elements of M. It is readily seen that $T(M)$ is a submodule of M. In fact, if $x, y \in T(M)$ then $rx = 0 = sy$ for some non-zero $r, s \in R$ whence, since R is commutative,

$$rs(x - y) = rsx - rsy = rsx - sry = 0 - 0 = 0$$

with $rs \neq 0$, and so $x - y \in T(M)$; and for every $p \in R$ we have

$$rpx = prx = p0 = 0$$

so that $px \in T(M)$. We call $T(M)$ the *torsion submodule* of M.

We shall now show that, in the case where M is finitely generated, $T(M)$ is a direct summand of M.

Modules over a principal ideal domain

Theorem 18.6 *Let M be a non-zero finitely generated module over a principal ideal domain R. Then $M/T(M)$ is free, and there is a free submodule F of M such that*

$$M = T(M) \oplus F.$$

Moreover, the dimension of such a submodule F is uniquely determined.

Proof We note first that $M/T(M)$ is torsion free. In fact, if $a \in R$ is such that $a \neq 0$ and $a(x + T(M)) = 0 + T(M)$ then $ax \in T(M)$ and so $bax = 0$ for some non-zero $b \in R$ whence, since $ba \neq 0$, we have $x \in T(M)$ and so $x + T(M) = 0 + T(M)$.

Since M is finitely generated, it is clear that so also is every quotient module of M; and in particular so is $M/T(M)$.

It now follows by Theorem 18.4 that $M/T(M)$ is free and of finite dimension.

Since in particular $M/T(M)$ is projective, the natural short exact sequence

$$0 \longrightarrow T(M) \stackrel{\iota}{\longrightarrow} M \stackrel{\natural}{\longrightarrow} M/T(M) \longrightarrow 0$$

splits and so we have that

$$M = T(M) \oplus F$$

where F is a submodule of M that is isomorphic to $M/T(M)$, whence it is free; in fact, $F = \operatorname{Im} \vartheta$ where ϑ is a splitting morphism associated with $\natural$.

The final statement is now clear from the fact that

$$\dim F = \dim M/T(M). \quad \diamond$$

- The dimension of the free R-module F in the above decomposition is often called the *rank* of M.

Since we know, by the Corollary to Theorem 7.6, that every free R-module is isomorphic to a direct sum of copies of R, we shall now concentrate on torsion R-modules. Once we know their structure, we can use Theorem 18.6 to determine that of all finitely generated modules over a principal ideal domain. Theorem 18.6 represents the first blow struck in this direction; we shall require two others to complete our task.

In order to carry out this investigation, we mention some basic facts concerning principal ideal domains; the reader who is unfamiliar with these should consult a standard algebra text.

- Every principal ideal domain R is a *unique factorisation domain*. Every $a \in R$ that is neither zero nor a unit can be expressed in the form
$$a = u \prod_{i=1}^{t} p_i^{\alpha_i}$$
where u is a unit and $p_1, \ldots, p_t$ are non-associated primes in R with each α_i a positive integer. Such a decomposition is unique to within association.

- If $a_1, \ldots, a_n \in R$ then d is a greatest common divisor of $a_1, \ldots, a_n$ if and only if
$$Rd = \sum_{i=1}^{n} Ra_i.$$
In particular, $a_1, \ldots, a_n$ are relatively prime, in the sense that 1_R is a greatest common divisor of $a_1, \ldots, a_n$ if and only if there exist $x_1, \ldots, x_n \in R$ such that
$$a_1 x_1 + \ldots + a_n x_n = 1_R.$$

Suppose now that M is an R-module. Given $x \in M$ and $r \in R$, we shall say that x is *annihilated* by r if $r \in \text{Ann}_R x$. We shall denote by $M(r)$ the set of elements x in M that are annihilated by r. It is clear that $M(r)$ forms a submodule of M for every $r \in R$.

Suppose now that $r, s \in R$ are such that $r|s$ (i.e. $s = tr$ for some $t \in R$). Then clearly we have $M(r) \subseteq M(s)$. In particular, for every $r \in R$ we have the ascending chain
$$M(r) \subseteq M(r^2) \subseteq \ldots \subseteq M(r^n) \subseteq M(r^{n+1}) \subseteq \ldots$$
of submodules of M. It is clear that
$$M_r = \bigcup_{n \geq 1} M(r^n)$$
is also a submodule of M, and consists of those elements of M that are annihilated by some power of r. Moreover, for every submodule N of M we have
$$N_r = N \cap M_r.$$

Definition Let R be a principal ideal domain and let $p \in R$ be prime. An R-module M is said to be a *p-module* (or to be *p-primary*) if $M = M_p$.

- Note that since M is a p-module if and only if, for every $x \in M$, there is an integer $n \geq 1$ such that $p^n x = 0$, every p-module is a torsion module.

Modules over a principal ideal domain

Example 18.6 If $p \in R$ is prime and $n \geq 1$ then R/Rp^n is a cyclic p-module.

Theorem 18.7 *If R is a principal ideal domain and $a_1, \ldots a_n \in R$ are pairwise prime then, for every torsion R-module M,*

$$M\left(\prod_{i=1}^n a_i\right) = \bigoplus_{i=1}^n M(a_i).$$

Proof There being nothing to prove when $n = 1$, consider the case $n = 2$. There exist $x_1, x_2 \in R$ such that $a_1 x_1 + a_2 x_2 = 1$. For every $y \in M(a_1 a_2)$ we therefore have

$$y = a_1 x_1 y + a_2 x_2 y$$

with $a_1 x_1 y \in M(a_2)$ since

$$a_2 a_1 x_1 y = x_1 a_1 a_2 y = x_1 0 = 0,$$

and likewise $a_2 x_2 y \in M(a_1)$. Thus we see that

$$M(a_1 a_2) = M(a_1) + M(a_2).$$

Suppose now that $z \in M(a_1) \cap M(a_2)$; then

$$z = a_1 x_1 z + a_2 x_2 z = x_1 a_1 z + x_2 a_2 z = x_1 0 + x_2 0 = 0.$$

Consequently we have $M(a_1 a_2) = M(a_1) \oplus M(a_2)$.

The result is now immediate by induction; for a_n and $\prod_{i=1}^{n-1} a_i$ are relatively prime, so that

$$M\left(\prod_{i=1}^n a_i\right) = M\left(\prod_{i=1}^{n-1} a_i\right) \oplus M(a_n). \quad \diamond$$

Corollary *Let $a \in R$ have the prime factorisation*

$$a = u \prod_{i=1}^n p_i^{\alpha_i}$$

where u is a unit. Then, for $i = 1, \ldots, n$ we have

$$[M(a)]_{p_i} = M(p_i^{\alpha_i}).$$

Proof For each i, $p_i^{\alpha_i}$ is relatively prime to $\prod_{j \neq i} p_j^{\alpha_j}$ and so, by the above, we have

$$M(a) = M(p_i^{\alpha_i}) \oplus M\left(\prod_{j \neq i} p_j^{\alpha_j}\right).$$

Using the fact that $M(p_i^{\alpha_i}) \subseteq M_{p_i}$ we then have, using the modular law (Theorem 2.4),

$$\begin{aligned}[M(a)]_{p_i} &= M(a) \cap M_{p_i} \\ &= \left[M(p_i^{\alpha_i}) \oplus \left(\prod_{j \neq i} p_j^{\alpha_j}\right)\right] \cap M_{p_i} \\ &= \left[M_{p_i} \cap M\left(\prod_{j \neq i} p_j^{\alpha_j}\right)\right] + M(p_i^{\alpha_i}).\end{aligned}$$

We shall now show that

$$M_{p_i} \cap M\left(\prod_{j \neq i} p_j^{\alpha_j}\right) = \{0\},$$

whence the result will follow. Now it is readily seen that, for every positive integer β, p_i^β is relatively prime to $\prod_{j \neq i} p_j^{\alpha_j}$ and so there exist $r, t \in R$ such that

$$rp_i^\beta + t\prod_{j \neq i} p_j^{\alpha_j} = 1.$$

For every $y \in M_{p_i} \cap M\left(\prod_{j \neq i} p_j^{\alpha_j}\right)$ we then have, using a sufficiently large β,

$$y = 1_R y = rp_i^\beta y + t\prod_{j \neq i} p_j^{\alpha_j} y = r0 + t0 = 0. \quad \diamond$$

We are now ready to strike the second blow.

Theorem 18.8 *Let R be a principal ideal domain and let M be a non-zero finitely generated torsion R-module. Then there are primes $p_1, \ldots, p_n$ that are unique to within association, such that $M_{p_i} \neq \{0\}$ for each i, and*

$$M = \bigoplus_{i=1}^n M_{p_i}.$$

Proof Consider $\mathrm{Ann}_R M$. This is an ideal of R. It is not the whole of R since M is non-zero; and it is not zero since if $\{x_1, \ldots, x_n\}$ is a set of generators of M then there are non-zero elements $r_1, \ldots, r_n$ of R such that $r_i x_i = 0$ for each i and consequently $r = \prod_{i=1}^n r_i$ is a non-zero element of $\mathrm{Ann}_R M$. We thus have $\mathrm{Ann}_R M = Rg$ where g is neither zero nor a unit. Let

$$g = u\prod_{i=1}^m p_i^{\alpha_i}$$

Modules over a principal ideal domain

be a factorisation of g. Since $M = M(g)$ it follows from Theorem 18.7 that

$$M = M(g) = \bigoplus_{i=1}^{m} M(p_i^{\alpha_i}).$$

By the Corollary to Theorem 18.7 we deduce that, for every i,

$$M_{p_i} = [m(g)]_{p_i} = M(p_i^{\alpha_i}).$$

We thus have $M = \bigoplus_{i=1}^{m} M_{p_i}$.

We now observe that each M_{p_i} is non-zero. In fact, suppose that we had $M_{p_j} = \{0\}$ for some j. Then we would have

$$M = \bigoplus_{i \neq j} M_{p_i} = \bigoplus_{i \neq j} M(p_i^{\alpha_i}) = M\left(\prod_{i \neq j} p_i^{\alpha_i}\right)$$

from which it would follow that

$$\prod_{i \neq j} p_i^{\alpha_i} \in \text{Ann}_R M = Rg$$

whence g, and in particular p_j, would divide $\prod_{i \neq j} p_i^{\alpha_i}$, a contradiction.

That each M_{p_i} is finitely generated follows immediately from the Corollary to Theorem 18.5.

As for uniqueness, suppose that $M_q \neq \{0\}$ where q is a prime. Let x be a non-zero element of M_q so that $\text{Ann}_R x = Rd$ for some $d \in R$ with d not a unit (for otherwise $\text{Ann}_R x = R$ and we have the contradiction $x = 0$). Now this ideal Rd clearly contains q^β for some $\beta > 0$ and also contains g. Thus $d | q^\beta$ and so $d = q^\gamma$ where $1 \leq \gamma \leq \beta$. Since $d | g$ we deduce that $q | g$. Thus q is an associate of a unique prime divisor p_i of g; in other words, we have $M_q = M_{p_i}$. ◊

The above result reduces our problem to a study of non-zero finitely generated p-modules over a principal ideal domain. In order to tackle this, we require some preliminary results.

Theorem 18.9 *Let R be a principal ideal domain, let M be a free R-module, and let N be a non-zero submodule of M. Let $f \in M^d$ be such that $f^{\rightarrow}(N)$ is maximal in the subset $\{\vartheta^{\rightarrow}(N) \,;\, \vartheta \in M^d\}$ of ideals of R. Then there exists $d \in R$ and $x \in M$ such that $d \neq 0$, $f(x) = 1$ and*

(1) $M = Rx \oplus \text{Ker } f$;
(2) $N = Rd \oplus (N \cap \text{Ker } f)$.

Proof We note first that R is noetherian and therefore satisfies the maximum condition on ideals (Theorem 5.1); such a choice of $f \in M^d$ is therefore possible.

Since $f^{\rightarrow}(N)$ is a principal ideal of R, there exists $d \in R$ such that $f^{\rightarrow}(N) = Rd$. We note first that $d \neq 0$; for otherwise we would have $f^{\rightarrow}(N) = \{0\}$ and the maximality of $f^{\rightarrow}(N)$ would then imply that $\vartheta^{\rightarrow}(N) = \{0\}$ for every $\vartheta \in M^d$, which is nonsense. [For example, if $\{e_i \; ; \; i \in I\}$ is a basis of M and x is a non-zero element of N then for some coordinate form e_j^d on M we have $e_j^d(x) \neq 0$; see the Corollary of Theorem 9.1.]

Now let $y \in N$ be such that $f(y) = d$. Then we note that there exists $x \in M$ such that $y = dx$. [In fact, for every $g \in M^d$ we have $g(y) \in Rd$. For, the ideal $Rd + Rg(y)$ is principal, say $Rd + Rg(y) = Ra$, so that $rd + sg(y) = a$ for some $r, s \in R$, whence

$$(rf + sg)(y) = rf(y) + sg(y) = rd + sg(y) = a,$$

giving $a \in (rf + sg)^{\rightarrow}(N)$ and therefore

$$Rd \subseteq Ra \subseteq (rf + sg)^{\rightarrow}(N).$$

The maximality of $Rd = f^{\rightarrow}(N)$ now yields $Rd = Ra$ whence we obtain

$$g(y) \in Rg(y) \subseteq Rd.$$

It follows from this observation that the coordinates of y relative to any basis of M all belong to Rd (see Corollary 1 to Theorem 9.1) whence $y = dx$ for some $x \in M$.] Furthermore, since $f = f(y) = f(dx) = df(x)$ and $d \neq 0$, we have $f(x) = 1$.

We now establish (1). Since $f(x) = 1$ we have

$$Rx \cap \mathrm{Ker}\, f = \{0\};$$

for $f(rx) = 0$ implies that $r = r1 = rf(x) = f(rx) = 0$. Also, given $m \in M$, we have

$$f[m - f(m)x] = f(m) - f(m)f(x) = f(m) - f(m) = 0$$

and so $m - f(m)x \in \mathrm{Ker}\, f$, showing that

$$M = Rx + \mathrm{Ker}\, f.$$

As for (2), we note that

$$Rdx \cap N \cap \mathrm{Ker}\, f \subseteq Rx \cap \mathrm{Ker}\, f = \{0\}.$$

Moreover, if $n \in N$ then $f(n) = rd$ for some $r \in R$, so that

$$f(n - rdx) = f(n) - rdf(x) = rd - rd = 0$$

whence
$$n = rdx + (n - rdx) \in Rdx + (N \cap \operatorname{Ker} f).$$
Thus we have $N \subseteq Rdx + (N \cap \operatorname{Ker} f)$; and since the reverse inclusion is obvious, (2) follows. ◊

Corollary *For all $g \in M^d$ we have*
$$g^{\rightarrow}(N) \subseteq f^{\rightarrow}(N);$$
in other words, $f^{\rightarrow}(N)$ is maximum in the subset $\{\vartheta^{\rightarrow}(N) \ ; \ \vartheta \in M^d\}$ of ideals of R.

Proof We note first that, for every $g \in M^d$,
$$g^{\rightarrow}(N \cap \operatorname{Ker} f) \subseteq f^{\rightarrow}(N).$$
In fact, suppose that that $g^{\rightarrow}(N \cap \operatorname{Ker} f) \not\subseteq f^{\rightarrow}(N)$. Since $M = Rx \oplus \operatorname{Ker} f$ we can define $h \in M^d$ by the following prescription : given $z = z_1 + z_2$ with $z_1 \in Rx$ and $z_2 \in \operatorname{Ker} f$, let $h(z) = f(z_1) + g(z_2)$. Then from
$$N = Rdx \oplus (N \cap \operatorname{Ker} f)$$
we have
$$\begin{aligned}h^{\rightarrow}(N) &= h^{\rightarrow}(Rdx) + h^{\rightarrow}(N \cap \operatorname{Ker} f) \\ &= f^{\rightarrow}(Rdx) + g^{\rightarrow}(N \cap \operatorname{Ker} f) \\ &= f^{\rightarrow}(N) + g^{\rightarrow}(N \cap \operatorname{Ker} f) \\ &\supset f^{\rightarrow}(N),\end{aligned}$$
which contradicts the maximality of $f^{\rightarrow}(N)$.

It follows immediately from this that
$$\begin{aligned}g^{\rightarrow}(N) &= g^{\rightarrow}(Rdx) + g^{\rightarrow}(N \cap \operatorname{Ker} f) \\ &\subseteq Rg(dx) + f^{\rightarrow}(N) \\ &= Rg(y) + f^{\rightarrow}(N) \\ &\subseteq Rd + f^{\rightarrow}(N) \\ &= f^{\rightarrow}(N).\end{aligned}\quad \diamond$$

Theorem 18.10 *Let R be a principal ideal domain. If M is a free R-module and N is a submodule of M of finite dimension n then there is a basis B of M, a subset $\{x_1, \ldots, x_n\}$ of B, and non-zero elements $d_1, \ldots, d_n$ of R such that*

(1) $\{d_1 x_1, \ldots, d_n x_n\}$ *is a basis of N*;
(2) $(i = 1, \ldots, n)$ $d_{i+1} | d_i$.

Moreover, the principal ideals $Rd_1, \ldots, Rd_n$ are uniquely determined by these conditions.

Proof We proceed by induction. If $n = 0$ then $N = \{0\}$ and $\emptyset$ is a basis for N. There is therefore nothing to prove in this case. Suppose then that the result holds for submodules of dimension $n - 1$ and let N be of dimension n. By Theorem 18.9 and its Corollary, there exist $d_n \in R$ and $x_n \in M$ such that

$$d_n \neq 0, \quad f(x_n) = 1, \quad M = Rx \oplus \mathrm{Ker}\, f, \quad N = Rd_n x_n \oplus (N \cap \mathrm{Ker}\, f),$$

where $f \in M^d$ is such that $f^\rightarrow(N) = Rd_n$ is the greatest element in the subset $\{\vartheta^\rightarrow(N)\,;\,\vartheta \in M^d\}$ of ideals of R.

Since $f(x_n) = 1$ and $d_n \neq 0$, we see that $\{d_n x_n\}$ is linearly independent and so, by Theorem 7.8 applied to the direct sum

$$Rd_n x_n \oplus (N \cap \mathrm{Ker}\, f),$$

we see that the submodule $N \cap \mathrm{Ker}\, f$ of $\mathrm{Ker}\, f$ has dimension $n - 1$. By the induction hypothesis, there is therefore a basis B_1 of $\mathrm{Ker}\, f$, a subset $\{x_1, \ldots, x_{n-1}\}$ of B_1, and non-zero elements $d_1, \ldots, d_{n-1}$ of R such that $\{d_1 x_1, \ldots, d_{n-1} x_{n-1}\}$ is a basis of $N \cap \mathrm{Ker}\, f$ with $d_{i+1} | d_i$ for $1 \leq i \leq n-2$. The direct sum decomposition $M = Rx_n \oplus \mathrm{Ker}\, f$ now shows that $B = \{x_n\} \cup B_1$ is a basis of M; and $N = Rd_n x_n \oplus (N \cap \mathrm{Ker}\, f)$ shows that

$$\{d_1 x_1, \ldots, d_{n-1} x_{n-1}, d_n x_n\}$$

is a basis of N. We have to show that $d_n | d_{n-1}$. For this purpose, define $\vartheta \in M^d$ by

$$\vartheta(z) = \begin{cases} 1 & \text{if } z = x_{n-1}; \\ 0 & \text{if } z \in B \setminus \{x_{n-1}\}. \end{cases}$$

Then we have

$$Rd_{n-1} = \vartheta^\rightarrow(N) \subseteq f^\rightarrow(N) = Rd_n$$

and so $d_n | d_{n-1}$.

As for the last statement, we note from (2) that the ideals Rd_i form the ascending chain

$$\{0\} \subset Rd_1 \subseteq Rd_2 \subseteq \ldots \subseteq Rd_n \subseteq R.$$

If now any d_i is a unit then so is every d_j with $j > i$. Suppose then that $d_1, \ldots, d_k$ are the non-units in the list of d_i, so that we have the chain

$$\{0\} \subset Rd_1 \subseteq Rd_2 \subseteq \ldots \subseteq Rd_k \subset R.$$

Modules over a principal ideal domain

Clearly, we have

$$\bigoplus_{i=1}^{n} R/Rd_i = \bigoplus_{i=1}^{k} R/Rd_i. \tag{1}$$

Moreover, since the assignment

$$(r_1 x_1, \ldots, r_n x_n) \mapsto (r_1 + Rd_1, \ldots, r_n + Rd_n)$$

defines an R-epimorphism

$$\zeta : \bigoplus_{i=1}^{n} Rx_i \to \bigoplus_{i=1}^{n} R/Rd_i$$

with kernel $\bigoplus_{i=1}^{n} Rd_i x_i$, we have

$$\bigoplus_{i=1}^{n} Rx_i / N = \bigoplus_{i=1}^{n} Rx_i \Big/ \bigoplus_{i=1}^{n} Rd_i x_i \simeq \bigoplus_{i=1}^{n} R/Rd_i. \tag{2}$$

To establish uniqueness, we shall invoke Theorem 17.18. By that result, for $q = 1, \ldots, k$ the q-th exterior power of $\bigoplus_{i=1}^{k} R/Rd_i$ is such that

$$\bigwedge^q \left(\bigoplus_{i=1}^{k} R/Rd_i \right) \simeq \bigoplus_{J \in S_q} R/I_J \tag{3}$$

where S_q denotes the collection of all subsets of $\{1, \ldots, k\}$ that consist of q elements and, for $J \in S_q$, $I_J = \sum_{i \in J} Rd_i$. Now in the case under consideration, the ideals in question form an ascending chain. For every $J \in S_q$ we therefore have $I_J = Rd_{q(J)}$ where $q(J)$ denotes the greatest integer in J. Since $q(J) \geq q$ with $q(J) = q$ only when $J = \{1, \ldots, q\}$, it follows that

$$(q = 1, \ldots, k) \quad Rd_q = \bigcap_{J \in S_q} Rd_{q(J)} = \bigcap_{J \in S_q} I_J.$$

We therefore conclude from (1), (2), (3) that, for each q,

$$Rd_q = \bigcap_{J \in S_q} I_J = \operatorname{Ann}_R \left(\bigoplus_{J \in S_q} R/I_J \right)$$

$$= \operatorname{Ann}_R \bigwedge^q \left(\bigoplus_{i=1}^{k} R/Rd_i \right)$$

$$= \operatorname{Ann}_R \bigwedge^q \left(\bigoplus_{i=1}^{n} Rx_i / N \right),$$

which establishes the required uniqueness. ◊

We are now ready to strike the third blow.

Theorem 18.11 *Let M_p be a non-zero finitely generated p-module over a principal ideal domain R. Then M_p is a copoduct of a finite number of uniquely determined non-zero cyclic p-modules.*

More precisely, there are uniquely determined ideals $Rp^{\alpha_1}, \ldots, Rp^{\alpha_k}$ of R such that

$$\{0\} \subset Rp^{\alpha_1} \subseteq Rp^{\alpha_2} \subseteq \ldots \subseteq Rp^{\alpha_k} \subset R$$

and an isomorphism

$$M_p \simeq \bigoplus_{j=1}^{k} R/Rp^{\alpha_j}.$$

Proof Let F be a free R-module of dimension n such that there is an epimorphism $\pi : F \to M_p$. By Theorem 18.3, Ker π is free and of dimension m where $m \leq n$. By Theorem 18.10, there is a basis $B = \{x_1, \ldots, x_n\}$ of F, a subset $\{x_1, \ldots, x_m\}$ of B, and $d_1, \ldots, d_m \in R$ such that

(1) $\{d_1 x_1, \ldots, d_m x_m\}$ is a basis of Ker π;
(2) $(i = 1, \ldots, n-1)$ $d_{i+1} | d_i$.

We note from (1) that, for each i,

$$Rd_i = \operatorname{Ann}_R \pi(x_i).$$

In fact, if $r \in \operatorname{Ann}_R \pi(x_i)$ then $\pi(rx_i) = r\pi(x_i) = 0$ gives $rx_i \in \operatorname{Ker} \pi$ whence, by (1), there exists $\lambda_i \in R$ such that $rx_i = \lambda_i d_i x_i$. It follows that $r = \lambda_i d_i \in Rd_i$ and so $\operatorname{Ann}_R \pi(x_i) \subseteq Rd_i$. On the other hand, $d_i \pi(x_i) = \pi(d_i x_i) = 0$ so $d_i \in \operatorname{Ann}_R \pi(x_i)$ and we have the reverse inclusion.

We also note that some of the d_i may be units; this will happen when $x_i \in \operatorname{Ker} \pi$, for then $Rd_i = \operatorname{Ann}_R \{0\} = R$.

We note further that if d_j is not a unit then $Rd_j = Rp^{\alpha_j}$ for some $\alpha_j \geq 1$. For, since M_p is a p-module, $\operatorname{Ann}_R \pi(x_i)$ contains a power of p, say p^{β_j}, whence $Rp^{\beta_j} \subseteq \operatorname{Ann}_R \pi(x_j)$ and consequently $\operatorname{Ann}_R \pi(x_j) = Rp^{\alpha_j}$ where $\alpha_j \leq \beta_j$; and $\alpha_j \geq 1$ since otherwise $\alpha_j = 0$ and $Rd_j = \operatorname{Ann}_R \pi(x_j) = R$ which contradicts the assumption that d_j is not a unit.

We shall now show that in fact $m = n$. Suppose, by way of obtaining a contradiction, that $m < n$ and let $\bigoplus^{n-m} R$ denote a direct sum of $n-m$ copies of R. Consider the R-morphism

$$\vartheta : \bigoplus_{i=1}^{N} Rx_i \to \bigoplus_{i=1}^{m} R/Rd_i \oplus \bigoplus^{n-m} R$$

given by the prescription

$$\vartheta(r_1 x_1, \ldots, r_n x_n) = (r_1 + Rd_1, \ldots, r_n + Rd_n, r_{m+1}, \ldots, r_n).$$

Modules over a principal ideal domain   279

This is clearly an R-epimorphism with kernel $\bigoplus_{i=1}^{m} Rd_i x_i = \operatorname{Ker} \pi$. Consequently,
$$\bigoplus_{i=1}^{M} R/Rd_i \oplus \bigoplus^{n-m} R \simeq F/\operatorname{Ker} \pi \simeq M_p.$$

But M_p is a p-module and so is a torsion module; and in the above isomorphisms the module on the left has torsion submodule $\bigoplus_{i=1}^{m} R/Rd_i$. This contradiction shows that we must have $m = n$.

Since $m = n$, we deduce from the above that
$$M_p \simeq \bigoplus_{i=1}^{n} R/Rd_i.$$

On deleting any units in the list $d_1, \ldots, d_n$ we obtain a chain of ideals
$$\{0\} \subset Rd_1 \subseteq Rd_2 \subseteq \ldots \subseteq Rd_k \subset R$$
with $R/Rd_i \neq \{0\}$ for $i = 1, \ldots, k$ and
$$M_p \simeq \bigoplus_{i=1}^{n} R/Rd_i = \bigoplus_{i=1}^{k} R/Rd_i = \bigoplus_{i=1}^{k} R/Rp^{\alpha_i}.$$

The uniqueness statement is immediate from Theorem 18.10. ◇

Combining Theorems 18.6, 18.8 and 18.11, we can now state the fundamental structure theorem for finitely generated modules over a principal ideal domain.

Theorem 18.12 *Every non-zero finitely generated module M over a principal ideal domain R is a coproduct of a finite number of cyclic modules*:
$$M \simeq \bigoplus_{i=1}^{n} \bigoplus_{j=1}^{k_i} R/Rp_i^{\alpha_{ij}} \oplus \bigoplus^{m} R.$$

Moreover, such a direct sum representation is unique in the sense that the number m (the rank of M) of torsion-free cyclic modules (copies of R) is the same for all such representations, and the cyclic p_i-modules $R/Rp_i^{\alpha_{ij}}$ are determined to within isomorphism.

Proof Let T be the torsion submodule of M. Then by Theorem 18.6 we have
$$M \simeq T \oplus \bigoplus^{m} R$$

where m is the torsion-free rank of M. Now let g be a generator of $\text{Ann}_R T$. If
$$g = u \prod_{i=1}^{n} p_i^{\alpha_i}$$
is a prime factorisation of g then, by Theorem 18.8, $T = \bigoplus_{i=1}^{n} T_{p_i}$ where each T_{p_i} is a p_i-module. By Theorem 18.11, we have
$$T_{p_i} \simeq \bigoplus_{j=1}^{k_i} R/Rp_i^{\alpha_{ij}}$$
whence the result, together with the uniqueness properties, follows. ◊

Our objective now is to show that the above result allows no further refinement of the direct summands involved. For this purpose, consider the following notion.

Definition An R-module M is said to be *indecomposable* if it cannot be expressed as a direct sum of two non-zero submodules.

For a principal ideal domain R the finitely generated R-modules that are indecomposable are completely determined by the following result.

Theorem 18.13 *A non-zero finitely generated module M over a principal ideal domain R is indecomposable if and only if M is either a torsion-free cyclic module (hence isomorphic to R) or a cyclic p-module (hence isomorphic to R/Rp^n for some prime p and some $n > 0$).*

Proof The necessity follows immediately from Theorem 18.12.

As for sufficiency, suppose that M is cyclic and that $M = N \oplus Q$. If m generates M, let $n \in N$ and $q \in Q$ be such that $m = n + q$. Since $M = Rm$ there exist $\alpha, \beta \in R$ such that $n = \alpha m$ and $q = \beta m$. Consequently we have $\alpha \beta m \in N \cap Q = \{0\}$ and so $\alpha \beta m = 0$.

Suppose first that M is torsion free. In this case $\alpha \beta = 0$ so either $\alpha = 0$ or $\beta = 0$. If $\alpha = 0$ then $n = 0$ and $m = q \in Q$ whence $M = Q$; and if $\beta = 0$ then $q = 0$ and $m = n \in N$ whence $M = N$. Thus M is indecomposable.

Suppose now that M is a p-module. In this case $\text{Ann}_R m = Rp^r$ for some $r \geq 1$. Since $\alpha \beta m = 0$ we therefore have $p^r | \alpha \beta$. Thus there exist $s, t \in \mathbb{N}$ such that $s + t = r$, $p^s | \alpha$, and $p^t | \beta$. Let α', β' be such that $\alpha = p^s \alpha'$ and $\beta = p^t \beta'$, and let $a = r - \min\{s, t\}$. Then we have
$$p^a m = p^a(n + q) = p^a(\alpha m + \beta n) = p^a(p^s \alpha' m + p^t \beta' m) = 0;$$
for $a + s \geq r$ and $a + t \geq r$ so that
$$p^{a+s}, p^{a+t} \in Rp^r = \text{Ann}_R m.$$

Modules over a principal ideal domain

We thus have $p^a \in \mathrm{Ann}_R m = Rp^r$. It follows from the definition of a that $\min\{s,t\} = 0$. Thus either $s = 0$ in which case $t = r$, $\beta m = 0$, $q = 0$ and $m = n \in B$; or $t = 0$ in which case $s = r$, $\alpha m = 0$, $n = 0$ and $m = q \in Q$. Thus in this case M is again indecomposable. ◇

It is clear from Theorem 18.13 that no further refinement of the direct sum decomposition of Theorem 18.12 is possible. An important consequence of this is the following cancellation property.

Theorem 18.14 *Suppose that M, N, P are finitely generated modules over a principal ideal domain R. Then*

$$M \oplus N \simeq M \oplus P \implies N \simeq P.$$

Proof It clearly suffices to consider only the case where M is indecomposable. Expressing $M \oplus N$ and $M \oplus P$ as coproducts of indecomposables as in Theorem 18.12, we obtain the result from the uniqueness up to isomorphism of the summands. ◇

- It should be noted that Theorem 18.14 does not hold when M is not finitely generated. For example, consider the mapping

$$\vartheta : \mathbb{Z}^{\mathbb{N}} \oplus \mathbb{Z} \to \mathbb{Z}^{\mathbb{N}}$$

described by $(f, m) \mapsto \vartheta_{f,m}$ where

$$\vartheta_{f,m}(n) = \begin{cases} m & \text{if } n = 0; \\ f(n-1) & \text{if } n \neq 0. \end{cases}$$

Put less formally, $\vartheta_{f,m}$ is the sequence whose first term is m and whose other terms are those of f. We leave to the reader the task of showing that ϑ is a $\mathbb{Z}$-isomorphism. We therefore have

$$\mathbb{Z}^{\mathbb{N}} \oplus \mathbb{Z} \simeq \mathbb{Z}^{\mathbb{N}} \simeq \mathbb{Z}^{\mathbb{N}} \oplus \{0\},$$

and so the cancellation property of Theorem 18.14 fails for $M = \mathbb{Z}^{\mathbb{N}}$.

Suppose now that M is a finitely generated torsion module over a principal ideal domain R. By the structure theorem we have

$$M \simeq \bigoplus_{i=1}^{n} \bigoplus_{j=1}^{k_i} R/Rp_i^{\alpha_{ij}}.$$

The unique ideals $Rp_i^{\alpha_{ij}}$ appearing in this decomposition are called the *elementary divisor ideals* associated with M. Our aim now is to rearrange the above direct sum in a way that is essentially the same as inverting the order of summation. In so doing, we shall bring to light other important ideals of R.

For this purpose, we require the following properties of a principal ideal domain R.

- If $a, b \in R$ let d be a greatest common divisor of a, b and let m be a least common multiple of a, b. Then md and ab are associates. In particular, if a and b are relatively prime then ab is a least common multiple. A simple inductive proof shows that if $a_1, \ldots, a_n$ are pairwise prime then $\prod_{i=1}^{n} a_i$ is a least common multiple of $a_1, \ldots, a_n$. Since m is a least common multiple of a, b if and only if $Rm = Ra \cap Rb$, it follows that if $a_1, \ldots, a_n$ are pairwise prime then

$$R\left(\prod_{i=1}^{n} a_i\right) = \bigcap_{i=1}^{n} Ra_i.$$

- We also have, for a and b relatively prime,

$$R/(Ra \cap Rb) \simeq R/Ra \times R/Rb.$$

In fact, consider the R-morphism $f : R \to R/Ra \times R/Rb$ given by

$$f(r) = (r + Ra, r + Rb).$$

It is clear that $\operatorname{Ker} f = Ra \cap Rb$. We show as follows that f is surjective. Since a, b are relatively prime, there exist $x, y \in R$ such that $xa + yb = 1$. Consequently,

$$\begin{aligned}
f(txa + syb) &= (txa + syb + Ra, txa + syb + Rb) \\
&= (syb + Ra, txa + Rb) \\
&= (s - sxa + Ra, t - tyb + Rb) \\
&= (s + Ra, t + Rb),
\end{aligned}$$

and so f is surjective. It follows that

$$R/(Ra \cap Rb) = R/\operatorname{Ker} f \simeq R/Ra \times R/Rb.$$

A simple inductive proof shows that if $a_1, \ldots, a_n$ are pairwise prime then

$$R/\bigcap_{i=1}^{n} Ra_i \simeq \bigoplus_{i=1}^{n} R/Ra_i.$$

Modules over a principal ideal domain 283

Let us now return to our consideration of the elementary divisor ideals of a given torsion R-module M. By Theorem 18.11 we have the chains

$$\{0\} \subset Rp_1^{\alpha_{11}} \subseteq Rp_1^{\alpha_{12}} \subseteq \ldots \subseteq Rp_1^{\alpha_{1k_1}} \subset R;$$
$$\{0\} \subset Rp_2^{\alpha_{21}} \subseteq Rp_2^{\alpha_{22}} \subseteq \ldots \subseteq Rp_2^{\alpha_{2k_2}} \subset R;$$
$$\vdots$$
$$\{0\} \subset Rp_n^{\alpha_{n1}} \subseteq Rp_n^{\alpha_{n2}} \subseteq \ldots \subseteq Rp_n^{\alpha_{nk_n}} \subset R.$$

Let $t = \max\{k_i \ ; \ 1 \leq i \leq n\}$ and for each i define $\alpha_{ij} = 0$ for $j = k_i+1, \ldots, t$. [This is simply a convenient device whereby the above array may be regarded as having n rows and t columns of ideals $Rp_i^{\alpha_{ij}}$ some of which are equal to R.] Now, for $j = 1, \ldots, t$ define

$$q_j = \prod_{i=1}^n p_i^{\alpha_{ij}};$$

in other words, q_j is formed by taking the product of the entries in the j-th column of the array. Since for each j the elements $p_1^{\alpha_{1j}}, \ldots, p_n^{\alpha_{nj}}$ are pairwise prime, their product q_j is a least common multiple and so

$$Rq_j = R\left(\prod_{i=1}^n p_i^{\alpha_{ij}}\right) = \bigcap_{i=1}^n Rp_i^{\alpha_{ij}}.$$

The various inclusions in the above array now show that we have the chain

$$\{0\} \subset Rq_1 \subseteq Rq_2 \subseteq \ldots \subseteq Rq_t \subset R.$$

Now since $p_1^{\alpha_{1j}}, \ldots, p_n^{\alpha_{nj}}$ are pairwise prime, it follows from the second of the above observations concerning principal ideal domains that

$$(j = 1, \ldots, t) \qquad R/Rq_j = R/\bigcap_{i=1}^n Rp_i^{\alpha_{ij}} \simeq \bigoplus_{i=1}^n R/Rp_i^{\alpha_{ij}}.$$

We therefore see, by Theorem 18.12 and the associativity and commutativity of coproducts, that

$$M \simeq \bigoplus_{j=1}^t R/Rq_j.$$

These observations lead to the following result which can be regarded as a generalisation of Theorem 18.11.

Theorem 18.15 *Let M be a finitely generated torsion module over a principal ideal domain R. Then there are uniquely determined ideals $Rq_1, \ldots, Rq_t$ of R such that*

$$\{0\} \subset \mathrm{Ann}_R M = Rq_1 \subseteq Rq_2 \subseteq \ldots \subseteq Rq_t \subset R$$

and an isomorphism
$$M \simeq \bigoplus_{i=1}^{t} R/Rq_i.$$

Proof In view of the preceding observations and the fact that

$$\mathrm{Ann}_R M = \mathrm{Ann}_R \bigoplus_{i=1}^{t} R/Rq_i = \bigcap_{i=1}^{t} Rq_i = Rq_1,$$

it suffices to establish uniqueness.

Suppose then that $Ra_1, \ldots, Ra_m$ are ideals of R such that

$$\{0\} \subset Ra_1 \subseteq Ra_2 \subseteq \ldots \subseteq Ra_m \subset R$$

and $M \simeq \bigoplus_{i=1}^{m} R/Ra_i$. Then $\mathrm{Ann}_R M = \bigcap_{i=1}^{m} Ra_i = Ra_1$ and so $Ra_1 = Rq_1$. We thus have

$$R/Rq_1 \oplus \bigoplus_{i=2}^{t} R/Rq_i \simeq M \simeq R/Rq_1 \oplus \bigoplus_{i=2}^{m} R/Ra_i.$$

It follows by Theorem 18.14 that

$$\bigoplus_{i=2}^{t} R/Rq_i \simeq \bigoplus_{i=2}^{m} R/Ra_i.$$

Now the annihilator of the module on the left is $\bigcap_{i=2}^{t} Rq_i = Rq_2$, whereas that of the module on the right is $\bigcap_{i=2}^{m} Ra_i = Ra_2$. Consequently $Ra_2 = Rq_2$ and we can repeat the above argument. The outcome is that $m = t$ and $Ra_i = Rq_i$ for $i = 1, \ldots, t$ and this establishes the uniqueness. ◊

Definition The uniquely determined ideals $Rq_1, \ldots, Rq_t$ of R are called the *invariant factor ideals* associated with M.

We shall now rephrase some of the above results in terms of submodules. If R is a principal ideal domain and M is an R-module then a finite sequence $(M_i)_{1 \le i \le n}$ of submodules of M is called a *normal sequence* for M if each M_i is a non-zero cyclic submodule such that $M = \bigoplus_{i=1}^{n} M_i$ and

$$\mathrm{Ann}_R M_1 \subseteq \mathrm{Ann}_R M_2 \subseteq \ldots \subseteq \mathrm{Ann}_R M_n.$$

If M is finitely generated then it follows from the above results that *there is a normal sequence for M that is unique in the sense that the elemnts* of any such sequence are determined to within R-isomorphism. In fact, let $M = T(M) \oplus F$ where $T(M)$ is the torsion submodule of M and F is free (see Theorem 18.6). If $\{a_1, \ldots, a_n\}$ is a basis of F and if

$$\vartheta : T(M) \to \bigoplus_{i=1}^{t} R/Rq_i$$

is an R-isomorphism (Theorem 18.15), then for $i = 1, \ldots, t$ we see that

$$\vartheta^{\leftarrow}(R/Rq_i) = M_i$$

is a non-zero cyclic submodule of $T(M)$ with $\text{Ann}_R M_i = Rq_i$, from which it follows that $(m_i)_{1 \leq i \leq t}$ is a normal sequence for $T(M)$ and hence that

$$Ra_1, \ldots, Ra_n, M_1, \ldots, M_n$$

is a normal sequence for M. The uniqueness up to isomorphism of these submodules is assured by the previous results.

We end this section by considering the particular case where the principal ideal domain in question is $\mathbb{Z}$.

Now it is clear that an abelian group is a torsion $\mathbb{Z}$-module when every element is of finite order (Example 18.5), and is a torsion-free $\mathbb{Z}$-module when every non-zero element is of infinite order. The general structure theorem therefore yields the following particular case.

Theorem 18.16 *Every finitely generated abelian group G is a coproduct of a finite number of cyclic groups :*

$$G \simeq \bigoplus_{i=1}^{n} \bigoplus_{j=1}^{k_i} \mathbb{Z}/\mathbb{Z}p_i^{\alpha_{ij}} \oplus \bigoplus^m \mathbb{Z}.$$

Moreover, such a direct sum decomposition is unique in the sense that the number m (the rank of G) of torsion-free cyclic groups (copies of $\mathbb{Z}$) is the same for all such decompositions, and the cyclic p_i-groups $\mathbb{Z}/\mathbb{Z}p_i^{\alpha_{ij}}$ are unique to within isomorphism. ◊

Corollary *Every finite abelian group G is a coproduct of a finite number of uniquely determined cyclic groups :*

$$G \simeq \bigoplus_{i=1}^{n} \bigoplus_{j=1}^{k_i} \mathbb{Z}/\mathbb{Z}p_i^{\alpha_{ij}}.$$ ◊

Of course, every finitely generated abelian group also has a normal sequence $(G_i)_{1 \leq i \leq n}$ of subgroups of which it is the direct sum. Note that in this case, if $\text{Ann}_{\mathbb{Z}} G_k = \mathbb{Z}q_k$ then q_k is precisely the order of G_k.

Recalling how the elementary divisor ideals were obtained, we can also establish the following result.

Theorem 18.17 *Let n be a positive integer with prime factorisation*

$$n = \prod_{i=1}^{m} p_i^{\alpha_i}.$$

For each exponent α_i let $S(\alpha_i)$ denote the set of all decreasing chains of integers

$$\beta_{i1} \geq \beta_{i2} \geq \ldots \geq \beta_{in_i} > 0$$

such that $\sum_{j=1}^{n_i} \beta_{ij} = \alpha_i$. Then the number of pairwise non-isomorphic abelian groups of order n is

$$\prod_{i=1}^{m} |S(\alpha_i)|.$$

Proof Every element of $\underset{i=1}{\overset{m}{\times}} S(\alpha_i)$ gives rise to an abelian group of order n, namely

$$\bigoplus_{i=1}^{m} \bigoplus_{j=1}^{n_i} \mathbb{Z}/\mathbb{Z}p_i^{\beta_{ij}}.$$

Moreover, by the preceding results and the various uniqueness properties, every abelian group of order n is isomorphic to such a coproduct; and distinct elements of $\underset{i=1}{\overset{m}{\times}} S(\alpha_i)$ yield non-isomorphic groups. ◊

EXERCISES

18.1 If R is a commutative unitary ring and N is a cyclic R-module prove that a mapping $f : N \to N$ is an R-morphism if and only if f is a homothety.

18.2 Let M be a finitely generated module over a principal ideal domain R and let the finite sequence $(Ra_i)_{1 \leq i \leq n}$ of cyclic modules be a normal sequence for M. Given $i, j \in \{1, \ldots, t\}$ with $i \leq j$, prove that there is a unique R-morphism $f_{ij} : Ra_i \to Ra_j$ such that $f_{ij}(a_i) = a_j$.

[Hint. Let $\alpha_i : R \to Ra_i$ be the epimorphism $r \mapsto ra_i$. Let $\zeta_i : R/\text{Ann}_R a_i \to Ra_i$ be the isomorphism induced by α_i and let $f :$

$R/\operatorname{Ann}_R a_i \to R/\operatorname{Ann}_R a_j$ be the morphism induced by id_R. Consider $f_{ij} = \zeta_j \circ f \circ \zeta_i^{-1}$.]

Deduce that there is a unique R-morphism $\alpha_{ij} : M \to M$ such that

$$\alpha_{ij}(a_k) = \begin{cases} a_j & \text{if } k = i; \\ 0 & \text{if } k \neq i. \end{cases}$$

[*Hint.* Try $\alpha_{ij} = \operatorname{in}_j \circ f_{ij} \circ \operatorname{pr}_i$.]

Hence show that the ring $(\operatorname{Mor}_R(M, M), +, \circ)$ is commutative if and only if M is cyclic.

[*Hint.* $\Rightarrow$: Show that $t \geq 2$ is impossible by considering $(\alpha_{12} \circ \alpha_{21})(a)$.

$\Leftarrow$: Use Exerise 18.1.]

18.3 Let M be a finitely generated p-module over a principal ideal domain R so that, by Theorem 18.11,

$$M \simeq \bigoplus_{j=1}^k R/Rp^{\alpha_j}.$$

Prove that

$$M(p) \simeq \bigoplus_{j=1}^k Rp^{\alpha_j - 1}/Rp^{\alpha_j} \simeq \bigoplus^k R/Rp$$

and deduce that $M(p)$ is a vector space of dimension k over the field R/Rp. Observing that every submodule of M is also a finitely generated p-module, deduce that if H is a submodule of M then

$$H \simeq \bigoplus_{j=1}^h R/Rp^{\beta_j}$$

where $h \leq k$ and $\beta_j \leq \alpha_j$ for $j = 1, \ldots, h$.

[*Hint.* Observe that $H(p)$ is also a vector space over R/Rp, of dimension h. Suppose, by way of obtaining a contradiction, that there is a smallest j such that $\alpha_j < \beta_j$. Consider $M'_{j-1} = \{p^{\alpha_j} x \ ; \ x \in M\}$ and $H' = \{p^{\alpha_j} x \ ; \ x \in H\}$; show that

$$M' \simeq \bigoplus_{i=1}^{j-1} Rp^{\alpha_j}/Rp^{\alpha_i},$$

a direct sum of $j - 1$ non-zero cyclic modules, and likewise for H'.]

18.4 If M is a non-zero torsion module over a principal ideal domain R, prove that M is cyclic if and only if there are pairwise non-associated primes $p_1,\ldots,p_k \in R$ and positive integers $\alpha_1,\ldots,\alpha_k$ such that
$$M \simeq \bigoplus_{i=1}^{k} R/Rp_i^{\alpha_i}.$$

18.5 Let G be an abelian group of order p^n where p is a prime, and let its chain of elementary divisor ideals be
$$\mathbb{Z}p^{\alpha_1} \subseteq \mathbb{Z}p^{\alpha_2} \subseteq \ldots \subseteq \mathbb{Z}p^{\alpha_k}.$$
Show that G is generated by $\{g_1,\ldots,g_k\}$ where the order of g_i is p^{α_i}.

[*Hint*. Use Theorem 18.11.]

For $r = 0,\ldots,n$ define $G_r = \{x \in G\ ;\ p^r x = 0\}$. Show that G_r is a subgroup of G and that it is generated by
$$\{p^{\alpha_1-r}g_1,\ldots,p^{\alpha_i-r}g_i, g_{i+1},\ldots,g_k\}$$
where i is the integer such that $\alpha_{i+1} < r \le \alpha_i$, with the convention that $\alpha_0 = n$ and $\alpha_{k+1} = 0$.

[*Hint*. Show that $\sum_{j=1}^{k} m_j g_j \in G_r$ if and only if $p^{\alpha_j-r}|m_j$ for $j = 1,\ldots,i$.]

Deduce that the order of G_r is $p^{ri+\alpha_{i+1}+\ldots+\alpha_k}$. Hence show that the number of elements of order p^r in G is
$$p^{ri+\alpha_{i+1}+\ldots+\alpha_k} - p^{(r-1)j+\alpha_{j+1}+\ldots+\alpha_k}$$
where i,j are given by $\alpha_{i+1} < r \le \alpha_i$ and $\alpha_{j+1} < r-1 \le \alpha_j$.

[*Hint*. $x \in G$ has order p^r when $x \in G_r$ and $x \notin G_{r-1}$.]

Conclude that the number of elements of order p in G is $p^k - 1$.

18.6 Determine the number of pairwise non-isomorphic abelian groups of order (*a*) 1 000; (*b*) 1 001; (*c*) 1 000 000.

19 Vector space decomposition theorems; canonical forms under similarity

In the previous chapter we saw how the structure theorem for finitely generated modules over a principal ideal domain could be applied to obtain the structure of all finitely generated abelian groups. We shall now show how this structure theorem can be applied to study of the decomposition of a vector space modulo a given linear transformation, and to the problem of determining canonical forms for matrices.

We begin by describing a generalisation of Theorem 10.6.

Theorem 19.1 *Let M, N be free R-modules of dimensions m, n respectively over a principal ideal domain R. If $f : M \to N$ is an R-morphism then there are bases $\{a_1, \ldots, a_m\}$ of M and $\{b_1, \ldots, b_n\}$ of N, and non-zero elements $d_1, \ldots, d_r$ of R, unique to within association, such that $d_{i+1}|d_i$ and*

$$(i = 1, \ldots, m) \qquad f(a_i) = \begin{cases} d_i b_i & \text{if } 1 \leq i \leq r; \\ 0 & \text{if } r+1 \leq i \leq m. \end{cases}$$

Proof By Theorem 18.3, $\operatorname{Im} f$ is free of dimension $r \leq n$. By Theorem 18.10, there is a basis B of N, a subset $\{b_1, \ldots, b_r\}$ of B, and non-zero elements $d_1, \ldots, d_r$ of R, unique to within association, such that

(1) $\{d_1 b_1, \ldots, d_r b_r\}$ is a basis of $\operatorname{Im} f$;
(2) $d_{i+1}|d_i$.

Let $a_1, \ldots, a_r \in M$ be such that $f(a_i) = d_i b_i$ for each i. Then $\{a_i \; ; \; 1 \leq i \leq r\}$ is linearly independent; for if $\sum_{i=1}^{r} \lambda_i a_i = 0$ then

$$0 = f\left(\sum_{i=1}^{r} \lambda_i a_i\right) = \sum_{i=1}^{r} \lambda_i f(a_i) = \sum_{i=1}^{r} \lambda_i d_i b_i$$

whence $\lambda_i d_i = 0$ for each i, and consequently every $\lambda_i = 0$ since $d_i \neq 0$. Let $M' = \operatorname{LC}\{a_i \; ; \; 1 \leq i \leq r\}$ and note that, since the restriction of f to M' carries a basis of M' to a basis of $\operatorname{Im} f$, we have

$$M' \simeq \operatorname{Im} f \simeq M/\operatorname{Ker} f.$$

Since $\operatorname{Im} f$ is free, hence projective, the canonical exact sequence

$$0 \longrightarrow \operatorname{Ker} f \longrightarrow M \longrightarrow M/\operatorname{Ker} f \longrightarrow 0$$

splits, and consequently we have

$$M = \operatorname{Ker} f \oplus M'.$$

Now let $\{a_i \; ; \; r+1 \le i \le m\}$ be a basis of $\operatorname{Ker} f$; then clearly a basis of M is $\{a_i \; ; \; 1 \le i \le m\}$ and the result follows. $\diamond$

Corollary 1 *If A is a non-zero $n \times m$ matrix over a principal ideal domain R then A is equivalent to an $n \times m$ matrix of the form*

$$\left[\begin{array}{c|c} X & 0 \\ \hline 0 & 0 \end{array}\right]$$

where X is the diagonal matrix

$$\begin{bmatrix} d_1 & & & \\ & d_2 & & \\ & & \ddots & \\ & & & d_r \end{bmatrix}$$

in which each $d_i \ne 0$ and $d_{i+1} | d_i$. Moreover, $d_1, \ldots, d_r$ are unique to within association. $\diamond$

- The non-zero ideals $Rd_1, \ldots, Rd_r$ of R are called the *invariant factor ideals* of the matrix A or of the R-morphism f. Using this terminology, we have the following criterion for matrices to be equivalent.

Corollary 2 *If R is a principal ideal domain and if A, B are $n \times m$ matrices over R then A and B are equivalent if and only if they have the same invariant factor ideals.* $\diamond$

We now turn our attention to what is by far the most surprising application of the structure theorem of the previous section, namely to a study of what some authors call the theory of a single linear transformation. More precisely, we shall be concerned with the decomposition of a finite-dimensional vector space as a direct sum of particular subspaces related to a given linear transformation. The results we shall obtain may be interpreted in terms of matrices and lead naturally to a study of canonical forms.

Our immediate aim, therefore, is to express a vector space over a field F in some way as a module over a principal ideal domain. We shall in fact be concerned with the ring $F[X]$ of polynomials with coefficients in F. It is well known that $F[X]$ is a euclidean domain and so is a

Vector space decomposition theorems

principal ideal domain. As we shall see, the key to the entire theory will be the simple, yet very profound, observation that *relative to a given non-zero linear transformation $f : V \to V$, a vector space V can be given the structure of a finitely generated torsion $F[X]$-module*. Before proceeding to establish this, we require some additional notation.

If A is a unitary associative algebra over a commutative unitary ring R then for every $a \in A$ we shall denote by $R[a]$ the subalgebra of A that is generated by $\{1_R, a\}$. It is clear that the elements of $R[a]$ are those of the form
$$r_0 + r_1 a + \ldots + r_n a^n$$
where each $r_i \in R$. Given
$$p = r_0 + r_1 X + \ldots + r_n X^n \in R[X]$$
and $a \in A$, we define
$$p(a) = r_0 + r_1 a + \ldots + r_n a^n$$
and say that $p(a)$ is obtained from p by *substituting a for X in p*.

Theorem 19.2 *Let A be a unitary associative R-algebra over a commutative unitary ring R. Given $a \in A$, the mapping $\zeta_a : R[X] \to R[a]$ described by $\zeta_a(p) = p(a)$ is an R-algebra epimorphism.*

Proof Let $p = \sum_{i=0}^{n} \alpha_i X^i$ and $q = \sum_{i=1}^{m} \beta_i X^i$. Then we have
$$pq = \sum_{k=0}^{n+m} \left(\sum_{j=0}^{k} \alpha_j \beta_{k-j} \right) X^k$$
and so
$$\zeta_a(pq) = \sum_{k=0}^{n+m} \left(\sum_{j=0}^{k} \alpha_j \beta_{k-j} \right) a^k = p(a) q(a).$$
Likewise we can show that
$$\zeta_a(p+q) = p(a) + q(a), \quad \zeta_a(\lambda p) = \lambda p(a).$$
Since $\zeta_a(X) = a$ and $\zeta_a(X^0) = 1_A$, it follows that ζ_a is an R-algebra morphism. Moreover, $\text{Im}\,\zeta_a$ is a subalgebra of $R[a]$ containing both a and 1_A and so $\text{Im}\,\zeta_a = R[a]$ whence ζ_a is an epimorphism. $\Diamond$

- The R-algebra morphism ζ_a is called the *substitution morphism* associated with $a \in A$.

In what follows we shall be concerned with the case where A is the unitary associative algebra $\text{Mor}_F(V, V)$ where V is a finite-dimensional vector space over a field F.

Suppose then that V is a non-zero vector space over a field F and let $f : V \to V$ be a non-zero linear transformation. Let V_f denote the algebraic structure $(V, +, \cdot_f)$ where $+$ denotes the usual addition on V and $\cdot_f : F[X] \times V \to V$ is the action given by

$$(p, x) \mapsto p \cdot_f x = [p(f)](x).$$

It is readily verified that V_f is an $F[X]$-module. Now the module obtained from V_f by restricting the action $\cdot_f$ to the subset $F \times V$ of $F[X] \times V$ is simply the F-vector space V; for, the elements of F regarded as elements of $F[X]$ are the constant polynomials. It follows that if B is a basis of V then B is a set of generators of V_f; for, every linear combination with coefficients in F is a linear combination with coefficients in $F[X]$. In particular, *if V is a vector space of finite dimension over F then V_f is a finitely generated $F[X]$-module.*

We now consider some elementary properties of the $F[X]$-module V_f.

Definition If M, N are R-modules then we say that $f \in \text{Mor}_R(M, M)$ and $g \in \text{Mor}_R(N, N)$ are *similar* if there is an R-isomorphism $\vartheta : M \to N$ such that the diagram

$$\begin{array}{ccc} M & \xrightarrow{f} & M \\ \vartheta \downarrow & & \downarrow \vartheta \\ N & \xrightarrow{g} & N \end{array}$$

is commutative.

Theorem 19.3 *Let V, W be non-zero finite-dimensional vector spaces over a field F and let $f : V \to V, g : W \to W$ be linear transformations. Then f and g are similar if and only if the $F[X]$-modules V_f, W_g are isomorphic.*

Proof $\Rightarrow$: Suppose that f, g are similar and let $\vartheta : V \to W$ be an isomorphism such that $\vartheta \circ f = g \circ \vartheta$. Then if, for some $n \geq 1$, we have $\vartheta \circ f^n = g^n \circ \vartheta$ it follows that

$$\vartheta \circ f^{n+1} = \vartheta \circ f^n \circ f = g^n \circ \vartheta \circ f = g^n \circ g \circ \vartheta = g^{n+1} \circ \vartheta.$$

Thus we see by induction that

$$(\forall n \geq 1) \qquad \vartheta \circ f^n = g^n \circ \vartheta.$$

Vector space decomposition theorems 293

To show that ϑ is an isomorphism from V_f onto W_g, it suffices to show that ϑ is $F[X]$-linear. This we can establish using the above observation and the fact that, for any $\sum_{i=0}^{m} \alpha_i X^i \in F[X]$,

$$\vartheta\left(\sum_{i=0}^{m} \alpha_i X^i \cdot_f x\right) = \vartheta\left(\sum_{i=0}^{m} \alpha_i f^i(x)\right)$$
$$= \sum_{i=0}^{m} \alpha_i (\vartheta \circ f^i)(x)$$
$$= \sum_{i=0}^{m} \alpha_i (g^i \circ \vartheta)(x)$$
$$= \sum_{i=0}^{m} \alpha_i X^i \cdot_f \vartheta(x).$$

$\Leftarrow$: Conversely, suppose that $\vartheta : V_f \to W_g$ is an $F[X]$-isomorphism. Then

$$(\forall x \in V)(\forall \lambda \in F) \qquad \vartheta(\lambda x) = \vartheta(\lambda X^0 \cdot_f x) = \lambda X^0 \cdot_g \vartheta(x) = \lambda \vartheta(x),$$

so that ϑ is an F-isomorphism from V onto W. Moreover,

$$(\forall x \in V) \qquad \vartheta[f(x)] = \vartheta(X \cdot_f x) = X \cdot_g \vartheta(x) = g[\vartheta(x)]$$

and so $\vartheta \circ f = g \circ \vartheta$. Hence f and g are similar. $\Diamond$

Definition Let M be an R-module and $f : M \to M$ be an R-morphism. Then we shall say that a submodule N of M is *f-stable*, or *invariant under f*, whenever $f^{\to}(N) \subseteq N$.

Theorem 19.4 *Let V be a non-zero vector space over a field F and let $f : V \to V$ be a linear transformation. Then a subset M of V is a submodule of the $F[X]$-module V_f if and only if M is an f-stable subspace of V.*

Proof $\Rightarrow$: Suppose that M is a submodule of V_f. Since $F \subseteq F[X]$ it is clear that M is a subspace of V. Now for every $x \in M$ we have

$$f(x) = X \cdot_f x \in M$$

and so M is f-stable.

$\Leftarrow$: Suppose conversely that M is an f-stable subspace of V. Then clearly, for every $x \in M$ and every positive integer k, a simple inductive argument yields $f^k(x) \in M$. It follows that, for every $p = \sum_{k=0}^{n} \alpha_k X^k \in F[X]$ and every $x \in M$,

$$p \cdot_f x = [p(f)](x) = \left(\sum_{k=0}^{n} \alpha_k f^k\right)(x) = \sum_{k=0}^{n} f^k(x) \in M,$$

whence M is a submodule of V_f. ◊

Theorem 19.5 *Let V be a non-zero finite-dimensional vector space over a field F and let $f : V \to V$ be a linear transformation. Then the finitely generated $F[X]$-module V_f is a torsion module.*

Proof Suppose, by way of obtaining a contradiction, that V_f is not a torsion $F[X]$-module. Then by the fundamental structure theorem (18.12) we see that V_f contains a submodule W that is $F[X]$-isomorphic to $F[X]$. Since W is then a subspace of V that is F-isomorphic to the F-vector space $F[X]$, we obtain a contradiction to the finite dimensionality of V on noting that $F[X]$ is an infinite-dimensional F-space. ◊

Since, as we have just seen, V_f is a torsion $F[X]$-module, we can consider the annihilator of V_f in $F[X]$, namely the principal ideal

$$\{p \in F[X] \; ; \; (\forall x \in V) \, [p(f)](x) = 0\}$$

of $F[X]$. The unique monic generator g of this ideal is called the *minimum polynomial* of f.

- The reason for the terminology *minimum* polynomial is that g is the monic polynomial of least degree such that $g(f)$ is the zero morphism. Roughly speaking, g is the monic polynomial of least degree that is satisfied by f, in the sense that $g(f) = 0$.

- That f is thus a root of a polynomial equation is really quite remarkable. Equally remarkable, of course, is the fact that if V is of dimension n then every $n \times n$ matrix that represents f is also a root of a polynomial equation.

We shall now translate some of the results of the previous section into the language of the present section. With the above dictionary of terms, the reader will have no trouble in verifying that Theorem 18.8 applied to V_f yields the following fundamental result.

Theorem 19.6 [Primary Decompositon Theorem] *Let V be a non-zero finite-dimensional vector space over a field F and let $f : V \to V$ be a linear transformation with minimum polynomial $g = \prod_{i=1}^{n} (p_i)^{\alpha_i}$ where $p_1, \ldots, p_n$ are distinct irreducible polynomials and $\alpha_1, \ldots, \alpha_n$ are positive integers. For $i = 1, \ldots, n$ let*

$$M_{p_i} = \{x \in V \; ; \; (\exists k \geq 1) \, [p_i(f)]^k(x) = 0\}.$$

Then every M_{p_i} is an F-stable subspace of V, $p_i^{\alpha_i}$ is the minimum polynomial of the F-morphism induced on M_{p_i} by f, and

$$V = \bigoplus_{i=1}^{n} M_{p_i}. \quad \diamond$$

- Note that since $p_i^{\alpha_i}$ is the minimum polynomial of the F-morphism induced by f on M_{p_i}, we have $x \in \operatorname{Ker} p_i^{\alpha_i}$ for every $x \in M_{p_i}$. We thus see that

$$M_{p_i} = \operatorname{Ker} p_i^{\alpha_i}.$$

We call M_{p_i} the *i-th primary component* of M.

Definition We shall say that a linear transformation $f : V \to V$ is *cyclic* if the $F[X]$-module V_f is cyclic.

With this terminology, the reader can deduce from Theorem 18.15 the following fundamental result.

Theorem 19.7 [Rational Decomposition Theorem] *Let V be a nonzero finite-dimensional vector space over a field F and let $f : V \to V$ be a linear transformation. Then there is a unique sequence $(f_i)_{1 \leq i \leq n}$ of monic polynomials over F such that*

$$(i = 1, \ldots, n-1) \quad f_{i+1} | f_i,$$

and a sequence $(W_k)_{1 \leq k \leq n}$ of f-stable subspaces of V such that

$$V = \bigoplus_{k=1}^{n} W_k$$

and, for every k, the F-morphism induced by f on W_k is cyclic with minimum polynomial f_k. Moreover, the minimum polynomial of f is f_1. $\diamond$

Definition In the case where V_f is a cyclic $F[X]$-module, generated by $\{c\}$ say, we shall call such an element c of V a *cyclic vector* for f.

- If c is a cyclic vector for f then we have

$$V_f = \{p \cdot_f c \,;\, p \in F[X]\};$$
$$V = \{[p(f)](c) \,;\, p \in F[f]\}.$$

Our objective now is to use the above two decomposition theorems to obtain canonical forms under similarity for square matrices over a field. Their description hinges on the following basic result.

Theorem 19.8 *Let V be a non-zero finite-dimensional vector space over a field F and let $f : V \to V$ be a cyclic linear transformation with cyclic vector c. Then the mapping*

$$\vartheta_c : F[f] \to V$$

given by $\vartheta_c(p) = [p(f)](c)$ is an F-isomorphism. Moreover, if

$$g = \alpha_0 + \alpha_1 X + \ldots + \alpha_{n-1} X^{n-1} + X^n$$

is the minimum polynomial of f then $\deg g = \dim V$,

$$\{c, f(c), f^2(c), \ldots, f^{n-1}(c)\}$$

is a basis of V, and the matrix of f relative to the ordered basis $(a_i)_n$ where $a_i = f^{i-1}(c)$ is

$$\begin{bmatrix} 0 & 0 & 0 & \ldots & 0 & -\alpha_0 \\ 1 & 0 & 0 & \ldots & 0 & -\alpha_1 \\ 0 & 1 & 0 & \ldots & 0 & -\alpha_2 \\ 0 & 0 & 1 & \ldots & 0 & -\alpha_3 \\ \vdots & \vdots & \vdots & & \vdots & \vdots \\ 0 & 0 & 0 & \ldots & 1 & -\alpha_{n-1} \end{bmatrix}.$$

Proof Since c is a cyclic vector for f, it is clear that ϑ_c is an epimorphism. Suppose now that $p \in \operatorname{Ker} f_c$. Then

$$0 = [p(f)](c) = p \cdot_f c$$

and so, since

$$\operatorname{Ann}_{F[X]} V_f = \operatorname{Ann}_{F[X]}(c) = F[X]g,$$

we see that $p = qg$ for some $q \in F[X]$. Applying the substitution morphism ζ_f, we deduce that, in $F[f]$,

$$p(f) = q(f)g(f) = q(f)0 = 0.$$

Thus $\operatorname{Ker} \vartheta_c = \{0\}$ and so ϑ_c is an F-isomorphism.

Since the $F[X]$-module V_f is generated by $\{c\}$, every $x \in V_f$ can be written in the form $x = p \cdot_f c$ for some $p = \sum_{i=0}^{m} \beta_i X^i \in F[X]$. Then every $x \in V$ can be written as

$$x = \left(\sum_{i=0}^{m} \beta_i f^i \right)(c) = \sum_{i=0}^{m} \beta_i f^i(c)$$

Vector space decomposition theorems

and so $\{f^i(c)\ ;\ i\in\mathbb{N}\}$ generates V. But $[g(f)](c) = 0$ and so we deduce that V is generated by

$$\{c, f(c), f^2(c), \ldots, f^{n-1}(c)\}$$

where $n = \deg g$.

Suppose now that $\sum_{i=0}^{n-1} \lambda_i f^i(c) = 0$. Then clearly we have $[h(f)](c) = 0$ where $h = \sum_{i=0}^{n-1} \lambda_i X^i$. Suppose that $\lambda_{n-1} \neq 0$ and let

$$h_1 = \sum_{i=0}^{n-1} \lambda_i \lambda_{n-1}^{-1} X^i.$$

Then clearly $[h_1(f)](c) = 0$, which contradicts the fact that g is the monic polynomial of least degree such that $[g(f)](c) = 0$. We thus have $\lambda_{n-1} = 0$, and clearly a repetition of this argument shows that every $\lambda_i = 0$. This then shows that $\{c, f(c), \ldots, f^{n-1}(c)\}$ is linearly independent, whence it is a basis for V, the dimension of which is then $n = \deg g$. The final statement is clear. $\Diamond$

Definition The matrix displayed in Theorem 19.8 is called the *companion matrix* of the monic polynomial

$$\alpha_0 + \alpha_1 X + \ldots + \alpha_{n-1} X^{n-1} + X^n.$$

- Note that when V is of dimension 1 the minimum polynomial of f has degree 1, say $g = \alpha_0 + X$. In this case $g(f) = \alpha + 0\,\mathrm{id}_V + f$ and so

$$0 = [g(f)](c) = \alpha_0 c + f(c).$$

Since $\{c\}$ is a basis of V, it follows that the companion matrix of f in this case is the 1×1 matrix $[-\alpha_0]$.

The converse of Theorem 19.8 is the following.

Theorem 19.9 *Let V be a non-zero finite-dimensional vector space over a field F and let $f : V \to V$ be a linear transformation. Suppose that there is an ordered basis $(a_i)_n$ of V relative to which the matrix of f is of the form*

$$\begin{bmatrix} 0 & 0 & 0 & \ldots & 0 & -\alpha_0 \\ 1 & 0 & 0 & \ldots & 0 & -\alpha_1 \\ 0 & 1 & 0 & \ldots & 0 & -\alpha_2 \\ 0 & 0 & 1 & \ldots & 0 & -\alpha_3 \\ \vdots & \vdots & \vdots & & \vdots & \vdots \\ 0 & 0 & 0 & \ldots & 1 & -\alpha_{n-1} \end{bmatrix}.$$

Then a_1 is a cyclic vector for f, and the minimum polynomial of f is

$$g = \alpha_0 + \alpha_1 X + \ldots + \alpha_{n-1} X^{n-1} + X^n.$$

Proof It is clear that $f^{k-1}(a_1) = a_k$ for $k = 1, \ldots, n$. Thus the f-stable subspace generated by $\{a_1\}$ contains the basis $\{a_1, \ldots, a_n\}$ and so coincides with V. Consequently, a_1 is a cyclic vector for f. Now

$$f^n(a_1) = f[f^{n-1}(a_1)] = f(a_n) = \sum_{i=0}^{n-1} -\alpha_i a_{i+1} = \sum_{i=0}^{n-1} -\alpha_i f^i(a_1),$$

and so we see that $[g(f)](a_1) = 0$. Since $\{a_1\}$ generates V_f, it follows that $g(f) = 0$. But a_1 is a cyclic vector for f and so, by Theorem 19.8, the degree of the minimum polynomial of f is the dimension of V. The minimum polynomial of f therefore has the same degree as the monic polynomial g and, since it divides g, must coincide with g. ◇

Using the above results, we can translate directly in terms of matrices the rational decomposition theorem and derive some useful consequences. For this purpose, we make the simple observation that, if $V = \bigoplus_{i=1}^n M_i$ where each M_i is an f-stable subspace of V and if B_i is an ordered basis of M_i for each i, then the matrix of f relative to the basis $\bigcup_{i=1}^n B_i$ (ordered in the obvious way) is of the form

$$\begin{bmatrix} \boxed{A_1} & & & \\ & \boxed{A_2} & & \\ & & \ddots & \\ & & & \boxed{A_n} \end{bmatrix}$$

in which A_i is the matrix relative to B_i of the F-morphism induced on M_i by f. By the rational decomposition theorem, we therefore have:

Theorem 19.10 *Let V be a non-zero finite-dimensional vector space over a field F and let $f : V \to V$ be a linear transformation. Then there is a unique sequence $(f_i)_{1 \leq i \leq n}$ of monic polynomials over F such that*

$$(i = 1, \ldots, n) \qquad f_{i+1} | f_i,$$

and an ordered basis of V with respect to which the matrix of f is of the form

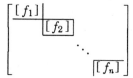

in which $[f_i]$ denotes the companion matrix of f_i. ◊

Definition The monic polynomials f_i of Theorems 19.7 and 19.10 are called the *invariant factors* of f. A matrix of the form exhibited in Theorem 19.10 (in which $f_{i+1}|f_i$ for each i) will be called a *rational (invariant factor) canonical matrix*.

- A rational (invariant factor) canonical matrix is often described as a direct sum of the companion matrices associated with the invariant factors.

Corollary 1 *If V is a non-zero finite-dimensional vector space over a field F and if $f : V \to V$ is a linear transformation then there is a unique rational (invariant factor) canonical matrix of f.*

Proof Suppose that

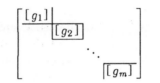

is also a rational (invariant factor) canonical matrix of f. For $i = 1, \ldots, m$ let V_i be the f-stable subspace of V associated with the companion matrix $[g_i]$. Since $g_{i+1}|g_i$ with

$$\mathrm{Ann}_{F[X]}(V_i)_f = F[X]g_i,$$

it is clear that $(V_i)_{1 \le i \le m}$ is a normal sequence for V_f. By Theorem 19.9, g_i is the minimum polynomial of the F-morphism induced on V_i by f. The result now follows by the uniqueness up to F-isomorphism of the subspaces in a normal sequence. ◊

Corollary 2 *The linear transformations $f, g : V \to V$ are similar if and only if they have the same invariant factors.*

Proof By Theorem 19.3, f and g are similar if and only if the $F[X]$-modules V_f and V_g are isomorphic. Now by the uniqueness of normal sequences it is clear that V_f and V_g are isomorphic if and only if, in these normal sequences, corresponding terms are isomorphic; and this is the case if and only if f and g have the same invariant factors. ◊

- Because of the above result, we can define the invariant factors of a square matrix A over F to be the invariant factors of any F-morphism $f : V \to V$ that is represented by A relative to some ordered basis. Then Corollary 2 may be rephrased as follows.

Corollary 3 *Two $n \times n$ matrices over a field F are similar if and only if they have the same invariant factors.* ◊

Corollary 4 *If $f_1, \ldots, f_n \in F[X]$ are the invariant factors of a square matrix A over a field F then these are also the invariant factors of A when A is considered as a matrix over any field K with $F \subseteq K$.*

Proof To say that $f_1, \ldots, f_n$ are the invariant factors of A over F is equivalent to saying that there is an invertible matrix P over F such that PAP^{-1} is a direct sum of companion matrices associated with the f_i. The result therefore follows from the fact that the elements of P and the coefficients of the f_i belong to every extension field K. ◊

Corollary 5 *Let F and K be fields with $F \subseteq K$, and let A, B be $n \times n$ matrices over F. Then if A and B are similar over K they are similar over F.*

Proof This is immediate from Corollaries 3 and 4. ◊

From the discussion in Chapter 18, the buiding blocks in the general structure theorem are the indecomposable modules. In the case under consideration, these are the cyclic p_i-modules where $p_i \in F[X]$ is prime.

Definition We shall say that an F-morphism $f : V \to V$ is *p-linear* for some prime polynomial p if the corresponding $F[X]$-module V_f is a p-module.

For such mappings, we have the following particular case of the rational decomposition theorem.

Theorem 19.11 *Let V be a non-zero finite-dimensional vector space over a field F, let $p \in F[X]$ be prime, and let $f : V \to V$ be p-linear. Then there is a unique sequence $(m_i)_{1 \le i \le n}$ of integers such that*

$$0 < m_1 \le m_2 \le \ldots \le m_n,$$

and a sequence $(W_k)_{1 \le k \le n}$ of f-stable subspaces of V such that

$$V = \bigoplus_{k=1}^{n} W_k$$

and, for each k, the F-morphism induced on W_k is cyclic with minimum polynomial p^{m_k}. Moreover, the minimum polynomial of f is p^{m_1}. ◊

- Recall that the ideals $F[X]p^{m_i}$ are simply the elementary divisor ideals associated with the p-module V_f. In what follows, we shall refer to the polynomials p^{m_i} as the *elementary divisors* of the p-linear mapping f. We shall denote the companion matrix of p^{m_k} by $[p]^{m_k}$. By Theorem 19.8, the induced F-morphism on W_k may be represented by $[p]^{m_k}$ and so there is an ordered basis of V with respect to which the matrix of f is

$$\begin{bmatrix} [p]^{m_1} & & & \\ & [p]^{m_2} & & \\ & & \ddots & \\ & & & [p]^{m_k} \end{bmatrix}.$$

Definition Let F be a field and let $p \in F[X]$ be prime. By a *rational p-matrix* over F we shall mean a matrix A such that, for some sequence $(m_i)_{1 \le i \le n}$ of integers with $0 < m_1 \le m_2 \le \ldots \le m_n$,

$$A = \begin{bmatrix} [p]^{m_1} & & & \\ & [p]^{m_2} & & \\ & & \ddots & \\ & & & [p]^{m_k} \end{bmatrix}.$$

It is immediate from Theorems 19.8, 19.9, and 19.11 that *if $f : V \to V$ is p-linear where $p \in F[X]$ is prime then there is a unique rational p-matrix of f.*

Definition By a *rational (elementary divisor) canonical matrix* over a field F we shall mean a square matrix of the form

$$\begin{bmatrix} A_1 & & & \\ & A_2 & & \\ & & \ddots & \\ & & & A_n \end{bmatrix}$$

in which, for distinct prime polynomials $p_1, \ldots, p_n$ over F, A_i is a rational p_i-matrix.

Now since, in the primary decomposition theorem (Theorem 19.6), each M_{p_i} is a p_i-submodule of V_f, we observe that the morphism induced on M_{p_i} by f is p_i-linear. We therefore deduce the following result.

Theorem 19.12 *Let V be a non-zero finite-dimensional vector space over a field F and let $f : V \to V$ be a linear transformation. Then there is a* rational (elementary divisor) canonical matrix of f. Moreover, if

$$A = \begin{bmatrix} \boxed{A_1} & & & \\ & \boxed{A_2} & & \\ & & \ddots & \\ & & & \boxed{A_n} \end{bmatrix}, \quad B = \begin{bmatrix} \boxed{B_1} & & & \\ & \boxed{B_2} & & \\ & & \ddots & \\ & & & \boxed{B_n} \end{bmatrix}$$

are rational (elementary divisor) canonical matrices of f then $n = m$ and there is a permutation σ on $\{1, \ldots, n\}$ such that $A_i = B_{\sigma(i)}$ for each i.

Proof The existence is clear from what has gone before. The uniqueness, up to a rearrangement of the p_i-matrices down the diagonal, follows from the uniqueness of the p_i-modules in the representation as a direct sum. ◊

- A rational (elementary divisor) canonical matrix of f is often described as a direct sum of the companion matrices associated with the elementary divisors of f.

In the above discussion we had occasion to deal with cyclic morphisms whose minimal polynomials are of the form p^{m_k} for some prime $p \in F[X]$. We shall now show that we can also represent such morphisms by a matrix that is constructed from the companion matrix of p rather than the companion matrix of p^{m_k}. This leads to yet another canonical form. The fundamental result in this direction is the following.

Theorem 19.13 *Let V be a non-zero finite-dimensional vector space over a field F and let $f : V \to V$ be a cyclic F-morphism with cyclic vector c. Suppose that the minimum polynomial of f is g^r where*

$$g = \alpha_0 + \alpha_1 X + \ldots + \alpha_{n-1} X^{n-1} + X^n.$$

Then the rn elements

c	$f(c)$	$f^2(c)$	$\ldots$	$f^{n-1}(c)$
$[g(f)](c)$	$[g(f)][f(c)]$	$[g(f)][f^2(c)]$	$\ldots$	$[g(f)][f^{n-1}(c)]$
$\vdots$	$\vdots$	$\vdots$		$\vdots$
$[g(f)]^{r-1}(c)$	$[g(f)]^{r-1}[f(c)]$	$[g(f)]^{r-1}[f^2(c)]$	$\ldots$	$[g(f)]^{r-1}[f^{n-1}(c)]$

Vector space decomposition theorems

constitute an ordered basis of V with respect to which the matrix of f is the $rn \times rn$ matrix

$$\begin{bmatrix} [g] & & & & \\ A & [g] & & & \\ & A & [g] & & \\ & & & \ddots & \ddots & \\ & & & & A & [g] \end{bmatrix}$$

in which $[g]$ is the $n \times n$ companion matrix of g and A is the $n \times n$ matrix

$$\begin{bmatrix} & & 1 \\ & & \\ & & \end{bmatrix}.$$

Proof By Theorem 19.8, the dimension of V is the degree of the minimum polynomial of f, namely rn. To show that the given set of elements is a basis, it therefore suffices to show that that this set is linearly independent. Suppose, by way of obtaining a contradiction, that this is not so. Then some linear combination of these elements is zero with not all the coefficients zero. This implies that there is a polynomial h with $[h(f)](c) = 0$ and $\deg h < rn = \deg g^r$, contradicting the fact that g^r is the minimum polynomial of f. Thus the given set constitutes a basis of V. We order this basis by taking the elements in the order in which we normally read them, namely the elements in the first row, then those in the second row, and so on.

Now we observe that f maps each basis vector onto the next one in the same row in the above array, except for those at the end of a row; for, f commutes with every power of g. As for the elements at the end of a row, we observe that

$$f[f^{n-1}(c)] = f^n(c) = -\alpha_0 c - \alpha_1 f(c) - \ldots - \alpha_{n-1} f^{n-1}(c) + [g(f)](c)$$

and similarly

$$f[g(f)]^{r-m}[f^{n-1}(c)]$$
$$= [g(f)]^{r-m}[f^n(c)]$$
$$= [g(f)]^{r-m}[-\alpha_0 c - \ldots - \alpha_{n-1} f^{n-1}(c) + [g(f)](c)]$$
$$= -\alpha_0 [g(f)]^{r-m}(c) - \ldots - \alpha_{n-1}[g(f)]^{r-m}[f^{n-1}(c)] + [g(f)]^{r-m+1}(c).$$

It follows immediately from this that the matrix of f relative to the above ordered basis is of the form stated. ◊

- Note that in the matrix of Theorem 19.13 every entry immediately below a diagonal entry is 1.

The converse of Theorem 19.13 is the following.

Theorem 19.14 *Let V be a non-zero finite-dimensional vector space over a field F and let $f : V \to V$ be a linear transformation. Suppose that there is a monic polynomial g of degree n and an ordered basis $(a_i)_{1 \leq i \leq rn}$ relative to which the matrix of f has the form given in Theorem 19.13. Then a_1 is a cyclic vector for f and the minimum polynomial of f is g^r.*

Proof Write the basis elements in the array

$$
\begin{array}{cccc}
a_1 & a_2 & \cdots & a_n \\
a_{n+1} & a_{n+2} & \cdots & a_{2n} \\
\vdots & \vdots & & \vdots \\
a_{(r-1)n+1} & a_{(r-1)n+2} & \cdots & a_{rn}
\end{array}
$$

and observe that, in any given row, each element except the first is the image under f of the previous element. Observe also that

$$
\begin{aligned}
f^n(a_1) &= f[f^{n-1}(a_1)] \\
&= f(a_n) \\
&= -\alpha_0 a_1 - \alpha_1 a_2 - \cdots - \alpha_n a_n + a_{n+1} \\
&= -\alpha_0 a_1 - \alpha_1 f(a_1) - \cdots - = \alpha_{n-1} f^{n-1}(a_1) + a_{n+1}
\end{aligned}
$$

whence we obtain

$$a_{n+1} = [g(f)](a_1);$$

and that, for $1 \leq k \leq r-2$,

$$
\begin{aligned}
f^n(a_{kn+1}) &= f(a_{(k+1)n}) \\
&= -\alpha_0 a_{kn+1} - \alpha_1 a_{kn+2} - \cdots - \alpha_{n-1} a_{(k+1)n} + a_{(k+1)n+1} \\
&= -\alpha_0 a_{kn+1} - \alpha_1 f(a_{kn+1}) - \cdots - \alpha_{n-1} f^{n-1}(a_{kn+1}) + a_{(k+1)n+1}
\end{aligned}
$$

whence we obtain

$$a_{(k+1)n+1} = [g(f)](a_{kn+1}).$$

It now follows easily by induction that

$$(k = 1, \ldots, r-1) \qquad a_{kn+1} = [g(f)]^k(a_1).$$

We thus see that the f-stable subspace generated by $\{a_1\}$ contains the basis elements $a_1, \ldots, a_{rn}$ whence it coincides with V. Thus a_1 is a cyclic vector for f.

We next observe that $[g(f)]^r(a_1) = 0$. In fact,

$$\begin{aligned}
[g(f)]^r(a_1) &= [g(f)]([g(f)]^{r-1}(a_1)) \\
&= [g(f)](a_{(r-1)n+1}) \\
&= \sum_{k=0}^{n-1} \alpha_k f^k(a_{(r-1)n+1}) + f^n(a_{(r-1)n+1}) \\
&= \sum_{k=0}^{n-1} \alpha_k a_{(r-1)n+k+1} + f(a_{rn}) \\
&= 0.
\end{aligned}$$

Since $\{a_1\}$ generates V_f, it follows that g^r belongs to $\text{Ann}_{F[X]} V_f$ whence it is divisible by the minimum polynomial of f. Now by Theorem 19.8 the degree of the minimum polynomial of f is the dimension of V, which is the degree of g^r. Since each of these polynomials is monic, it follows that g^r is the minimum polynomial of f. $\Diamond$

In what follows we shall denote a matrix of the form exhibited in Theorem 19.13 by $[g]_r$. With this notation, we note that in Theorem 19.11 the F-morphism induced on W_k by f may be represented by $[p]_{m_k}$ and so there is a basis of V with respect to which the matrix of f is

$$\begin{bmatrix} [p]_{m_1} & & & \\ & [p]_{m_2} & & \\ & & \ddots & \\ & & & [p]_{m_n} \end{bmatrix}.$$

Definition Let F be a field and let $p \in F[X]$ be a prime. By a *classical p-matrix* we shall mean a matrix A of the form

$$A = \begin{bmatrix} [p]_{m_1} & & & \\ & [p]_{m_2} & & \\ & & \ddots & \\ & & & [p]_{m_n} \end{bmatrix}$$

in which $0 < m_1 \leq m_2 \leq \ldots \leq m_n$.

It is immediate from Theorems 19.11, 19.13, and 19.14 that *if $f : V \to V$ is p-linear where p is prime then there is a unique classical p-matrix of f.*

Definition By a *classical canonical matrix* over a field F we shall mean a square matrix of the form

$$\begin{bmatrix} \boxed{A_1} & & & \\ & \boxed{A_2} & & \\ & & \ddots & \\ & & & \boxed{A_n} \end{bmatrix}$$

in which, for distinct prime polynomials $p_1, \ldots, p_n$ over F, A_i is a classical p_i-matrix.

The following analogue of Theorem 19.12 is now immediate.

Theorem 19.15 *Let V be a non-zero finite-dimensional vector space over a field F and let $f : V \to V$ be a linear transformation. Then there is a classical canonical matrix of f. Moreover, if*

$$A = \begin{bmatrix} \boxed{A_1} & & & \\ & \boxed{A_2} & & \\ & & \ddots & \\ & & & \boxed{A_n} \end{bmatrix}, \quad B = \begin{bmatrix} \boxed{B_1} & & & \\ & \boxed{B_2} & & \\ & & \ddots & \\ & & & \boxed{B_n} \end{bmatrix}$$

are classical canonical matrices of f then $n = m$ and there is a permutation σ on $\{1, \ldots, n\}$ such that $A_i = B_{\sigma(i)}$ for each i. ◊

A particularly important special case of classical canonical matrices arises when the corresponding prime polynomials p_i are linear, say $p_i = X - \lambda$ where $\lambda \in F$. In this case, the matrix exhibited in Theorem 19.13 is the $r \times r$ matrix

$$\begin{bmatrix} \lambda & & & & \\ 1 & \lambda & & & \\ & 1 & \lambda & & \\ & & \ddots & \ddots & \\ & & & 1 & \lambda \end{bmatrix}.$$

We shall denote this matrix by $[\,\lambda\,]_r$ and call it the *elementary $r \times r$ Jordan matrix* determined by λ.

Definition By a *Jordan $(X - \lambda)$-matrix* we shall mean a matrix of the form

$$\begin{bmatrix} \boxed{[\lambda]_{m_1}} & & & \\ & \boxed{[\lambda]_{m_2}} & & \\ & & \ddots & \\ & & & \boxed{[\lambda]_{m_n}} \end{bmatrix}$$

where $0 < m_1 \leq m_2 \leq \ldots \leq m_n$.

Vector space decomposition theorems

- A Jordan $(X-\lambda)$-matrix is thus a direct sum of elementary Jordan matrices associated with λ.

By a *Jordan canonical matrix* we shall mean a square matrix of the form

$$\begin{bmatrix} \boxed{A_1} & & & \\ & \boxed{A_2} & & \\ & & \ddots & \\ & & & \boxed{A_n} \end{bmatrix}$$

in which, for distinct scalars $\lambda_1, \ldots \lambda_n \in F$, A_i is a Jordan $(X - \lambda_i)$-matrix.

Definition We shall say that an F-morphism $f : V \to V$ is a *Jordan morphism* if its minimum polynomial factorises into a product of linear polynomials.

In the case where the ground field F is algebraically closed (i.e. every non-zero polynomial over F factorises into a product of linear polynomials; for example, when F is the field $\mathbb{C}$ of complex numbers), it is clear that every linear transformation $f : V \to V$ is a Jordan morphism and consequently that every square matrix over F is similar to a Jordan canonical matrix (which, by Theorem 19.15, is unique up to the arrangement of the $(X - \lambda_i)$-matrices down the diagonal).

- It can be shown that if F is a field then there is a smallest algebraically closed field F^* that contains F as a subfield. This algebraically closed field F^* is called the *algebraic closure* of F. For example, the algebraic closure of $\mathbb{R}$ is $\mathbb{C}$. In a very loose sense, therefore, every F-morphism can be regarded as a Jordan morphism; more precisely, as a Jordan F^*-morphism.

At this juncture it is both appropriate and instructive to consider some examples to illustrate the above results.

Example 19.1 Let V be a real vector space of dimension 10 and suppose that $f : V \to V$ is a linear transformation whose sequence of invariant factors is

$$(X^3 + 1)^2, \quad X^3 + 1, \quad X + 1.$$

Recalling that the rational (invariant factor) canonical matrix of f is the direct sum of the companion matrices associated with the invariant factors, we see that this matrix is

$$\begin{bmatrix} 0 & 0 & 0 & 0 & 0 & -1 & & & & \\ 1 & 0 & 0 & 0 & 0 & 0 & & & & \\ 0 & 1 & 0 & 0 & 0 & 0 & & & & \\ 0 & 0 & 1 & 0 & 0 & -2 & & & & \\ 0 & 0 & 0 & 1 & 0 & 0 & & & & \\ 0 & 0 & 0 & 0 & 1 & 0 & & & & \\ & & & & & & 0 & 0 & -1 & \\ & & & & & & 1 & 0 & 0 & \\ & & & & & & 0 & 1 & 0 & \\ & & & & & & & & & -1 \end{bmatrix}$$

In order to determine a rational (elementary divisor) canonical matrix of f we must first determine the elementary divisors. We recall that these are of the form $p_i^{m_i}$ where p_i is a prime polynomial and m_i is a positive integer. These may be arranged in rows, a typical row being

$$p_i^{m_{i_1}} \quad p_i^{m_{i_2}} \quad \ldots \quad p_i^{m_{i_n}}$$

where $m_{i_1} \geq m_{i_2} \geq \ldots \geq m_{i_n} > 0$. Allowing some exponents to be zero, we can form a rectangular array in which the invariant factors are the column products (see Chapter 18). Since the invariant factors and the elementary divisors are uniquely determined, it is readily seen that in the example under consideration such an array is

$$(X^2 - X + 1)^2 \quad X^2 - X + 1 \quad 1$$
$$(X + 1)^2 \quad\quad\quad X + 1 \quad\quad\quad X + 1.$$

Now the rational $(X^2 - X + 1)$-matrix associated with the top row is the direct sum of the associated companion matrices, and likewise so is the rational $(X + 1)$-matrix associated with the bottom row. Hence a rational (elementary divisor) canonical matrix of f is

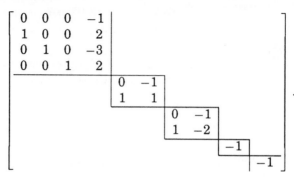

Vector space decomposition theorems

Let us now determine a classical canonical matrix of f. This is easily derived from the above rational (elementary divisor) canonical matrix by replacing the rational p-matrices involved by the corresponding classical p-matrices. It is therefore clear that a classical canonical matrix of f is

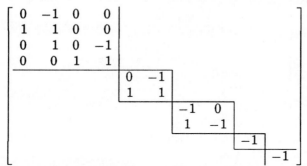

Since the minimum polynomial of f is the first invariant factor, namely $(X^3 + 1)^2$, we see that f is not a Jordan morphism.

Example 19.2 Consider the same situation as in Example 19.1 but now let V be a vector space over $\mathbb{C}$. Then the rational (invariant factor) canonical matrix of f is the same as before. Different rational (elementary divisor) canonical matrices arise, however, since over the field $\mathbb{C}$ the polynomial $X^2 - X + 1$ is not prime; in fact we have

$$X^2 - X + 1 = (X - \alpha)(X - \beta)$$

where $\alpha = \frac{1}{2}(1 + i\sqrt{3})$ and $\beta = \frac{1}{2}(1 - i\sqrt{3})$. In this case the array of elementary divisors is readily seen to be

$$(X - \alpha)^2 \quad X - \alpha \quad 1$$
$$(X - \beta)^2 \quad X - \beta \quad 1$$
$$(X + 1)^2 \quad X + 1 \quad X + 1$$

A rational (elementary divisor) canonical matrix of f is therefore

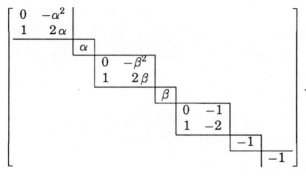

A classical canonical matrix of f is then

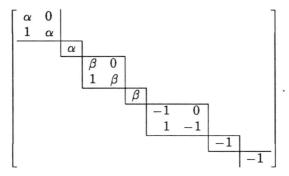

Since $\mathbb{C}$ is algebraically closed, the above classical canonical matrix is a Jordan canonical matrix.

Example 19.3 Let V be a real vector space of dimension 6 and let $f : V \to V$ be a linear transformation with elementary divisors $X + 2$ of multiplicity 3, $X - 2$ of multiplicity 2, and $X + 3$. Arranging these in the usual way, we obtain the array

$$\begin{array}{ccc} X+2 & X+2 & X+2 \\ X-2 & X-2 & 1 \\ X+3 & 1 & 1 \end{array}$$

whence the sequence of invariant factors of f is

$$(X+2)(X-2)(X+3), \quad (X+2)(X-2), \quad X+2.$$

The rational (invariant factor) canonical matrix of f is then

$$\left[\begin{array}{ccc|cc|c} 0 & 0 & 12 & & & \\ 1 & 0 & 4 & & & \\ 0 & 1 & -3 & & & \\ \hline & & & 0 & 4 & \\ & & & 1 & 0 & \\ \hline & & & & & -2 \end{array}\right]$$

A rational (elementary divisor) canonical matrix of f is then

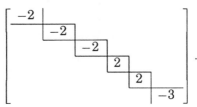

Vector space decomposition theorems

This is clearly also a classical canonical matrix and a Jordan canonical matrix.

Our aim now is to bring to light another important polynomial associated with a linear transformation $f : V \to V$. For this purpose, we note the following result.

Theorem 19.16 *If g is a monic polynomial of degree n over a field F and if $[g]$ denotes the companion matrix of g then*

$$g = \det(X I_n - [g]).$$

Proof We note first that if $n = 1$, so that $g = \alpha_0 + X$, then $[g]$ is the 1×1 matrix $[-\alpha_0]$ and the result is trivial. Suppose, by way of induction, that the result holds for monic polynomials of degree $n - 1$ and let

$$g = \alpha_0 + \alpha_1 X + \ldots + \alpha_{n-1} X^{n-1} + X^n.$$

Then we have

$$\det(X I_n - [g]) = \det \begin{bmatrix} X & & & & \alpha_0 \\ -1 & X & & & \alpha_1 \\ & -1 & X & & \alpha_2 \\ & & \ddots & \ddots & \vdots \\ & & & -1 & X + \alpha_{n-1} \end{bmatrix}.$$

Now let h be the monic polynomial of degree $n - 1$ given by

$$h = \alpha_1 + \alpha_2 X + \ldots + \alpha_{n-1} X^{n-2} + X^{n-1}.$$

Considering the Laplace expansion of the above determinant by the first column, we see that $\det(X I_n - [g])$ can be written as

$$X \det \begin{bmatrix} X & & & \alpha_1 \\ -1 & X & & \alpha_2 \\ & -1 & X & \alpha_3 \\ & & \ddots & \ddots & \vdots \\ & & & -1 & X + \alpha_{n-1} \end{bmatrix} + \det \begin{bmatrix} 0 & 0 & & \alpha_1 \\ -1 & X & & \alpha_2 \\ & -1 & X & \alpha_3 \\ & & \ddots & \ddots & \vdots \\ & & & -1 & X + \alpha_{n-1} \end{bmatrix}.$$

Now by the induction hypothesis the first of these determinants is h; and considering the Laplace expansion of the second via the first row, and using the fact that the determinant of a triangular matrix is simply the product of its diagonal entries, we see that the second determinant is

$$(-1)^n \alpha_0 (-1)^{n-2} = \alpha_0.$$

Thus we see that

$$\det(XI_n - [g]) = Xh + \alpha + 0 = g. \quad \diamond$$

Definition Let V be a non-zero finite-dimensional vector space over a field F. Let $f : V \to V$ be a linear transformation with elementary divisors $q_1, \ldots, q_n$. Then by the *characteristic polynomial* of f we mean the polynomial

$$\chi_f = \prod_{i=1}^{n} q_i.$$

We define also the characteristic polynomial χ_A of a square matrix A over a field F to be that of any linear transformation which is represented by A relative to some ordered basis.

Theorem 19.17 *If A is an $n \times n$ matrix over a field F then*

$$\chi_A = \det(XI_n - A).$$

Proof Since similar matrices have the same invariant factors they have the same characteristic polynomials. It suffices, therefore, by Corollary 1 of Theorem 19.10, to consider the case where A is a rational (invariant factor) canonical matrix. The result then follows from Theorems 19.16 and 17.16, the latter implying via an inductive argument that the determinant of a direct sum of square matrices is the product of their determinants. $\diamond$

Corollary 1 *Let g be a monic polynomial over a field F. Then the characteristic polynomial of the companion matrix of g is g.*

Proof This is simply a restatement of Theorem 19.16. $\diamond$

Corollary 2 *The constant term in the characteristic polynomial of an $n \times n$ matrix A is $(-1)^n \det A$.*

Proof $\chi_A(0) = \det(-A) = (-1)^n \det A$. $\diamond$

Corollary 3 *A square matrix is invertible if and only if the constant term in its characteristic polynomial is non-zero.* $\diamond$

The basic connection between the minimum polynomial and the characteristic polynomial is the following.

Theorem 19.18 *Let V be a non-zero finite-dimensional vector space over a field F and let $f : V \to V$ be a linear transformation. Let m be the*

Vector space decomposition theorems 313

number of invariant factors of f, let χ_f be the characteristic polynomial of f, and let m_f be the minimum polynomial of f. Then

$$m_f \mid \chi_f \mid m_f^m.$$

Proof Since m_f is the first invariant factor of f and χ_f is the product of the invariant factors of f, it is clear that $m_f \mid \chi_f$. Moreover, since each invariant factor of f divides the first invariant factor m_f, their product χ_f divides m_f^m. ◊

Corollary 1 $\chi_f(f) = 0$.

Proof We know that $m_f(f) = 0$; and by the above $\chi_f = pm_f$ for some $p \in F[X]$. ◊

- The above Corollary is often referred to as the *Cayley-Hamilton theorem*. Although the proof that we have given is very simple, the result itself is very deep. In terms of matrices, it says that every $n \times n$ matrix over a field is a zero of its characteristic polynomial (which, by Theorem 19.17, is of degree n).

Corollary 2 m_f and χ_f have the same zeros in F.

Proof This is immediate from Theorem 19.18. ◊

Example 19.4 It is readily seen that the characteristic and minimum polynomials of the real matrix

$$A = \begin{bmatrix} 1 & 2 & 2 \\ 2 & 1 & 2 \\ 2 & 2 & 1 \end{bmatrix}$$

are $\chi_A = (X + 1)^2(X - 5)$ and $m_A = (X + 1)(X - 5)$. Since χ_A is the product of the invariant factors, the first of which is m_A, we see that the invariant factors of A are

$$(X + 1)(X - 5), \quad X + 1.$$

A rational (invariant factor) canonical matrix of A is then

$$\left[\begin{array}{cc|c} 0 & 5 & 0 \\ 1 & 4 & 0 \\ \hline 0 & 0 & -1 \end{array} \right].$$

Since the array of elementary divisors is

$$\begin{array}{cc} X+1 & X+1 \\ X-5 & 1 \end{array}$$

it follows that a rational (elementary divisor) canonical matrix of A is

$$\left[\begin{array}{cc|c} -1 & 0 & 0 \\ 0 & -1 & 0 \\ \hline 0 & 0 & 5 \end{array}\right],$$

this also being a classical canonical matrix and a Jordan canonical matrix.

Example 19.5 Let $f : \mathbb{R}^7 \to \mathbb{R}^7$ be a linear transformation whose characteristic and minimum polynomials are

$$\chi_f = (X-1)^3(X-2)^4, \qquad m_f = (X-1)^2(X-2)^3.$$

The sequence of invariant factors of f is

$$(X-1)^2(X-2)^3 \quad (X-1)(X-2).$$

A rational (invariant factor) canonical matrix for f is then

$$\left[\begin{array}{ccccc|cc} & & & & 8 & & \\ 1 & & & & -28 & & \\ & 1 & & & 38 & & \\ & & 1 & & -25 & & \\ & & & 1 & 8 & & \\ \hline & & & & & & 1 \\ & & & & & & 2 \end{array}\right].$$

The array of elementary divisors is

$$\begin{array}{cc} (X-1)^2 & X-1 \\ (X-2)^3 & X-2 \end{array}$$

and so a rational (elementary divisor) canonical matrix is

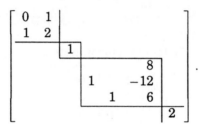

Finally, a classical canonical matrix, which is also a Jordan canonical matrix, is

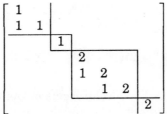

EXERCISES

19.1 For each of the following matrices over $\mathbb{R}$ determine

(a) the characteristic polynomial;

(b) the minimum polynomial;

(c) the invariant factors;

(d) the elementary divisors;

(e) the rational (invariant factor) canonical matrix;

(f) a rational (elementary divisor) canonical matrix;

(g) a classical canonical matrix.

$$\begin{bmatrix} 1 & 1 & 3 \\ 5 & 2 & 6 \\ -2 & -1 & -3 \end{bmatrix} ; \quad \begin{bmatrix} 2 & 2 & 1 \\ 2 & 2 & 1 \\ 2 & 2 & 1 \end{bmatrix} ; \quad \begin{bmatrix} 0 & 0 & 1 & 1 \\ 0 & 0 & 1 & 1 \\ 1 & 1 & 0 & 0 \\ 1 & 1 & 0 & 0 \end{bmatrix}.$$

19.2 Prove that the $F[X]$-module V_f is cyclic if and only if the minimum polynomial of f coincides with the characteristic polynomial of f.

19.3 Let $\mathbb{C}_3[X]$ be the complex vector space of polynomials of degree less than or equal to 3. Let $D : \mathbb{C}_3[X] \to \mathbb{C}_3[X]$ be the differentiation map. Determine a Jordan canonical matrix of D.

19.4 Let A be a 7×7 matrix over $\mathbb{R}$ with minimum polynomial

$$m_A = (X^2 + 2)(X + 3)^3.$$

Find all possible rational (elementary divisor) canonical forms of A.

[*Hint.* Argue that the sequence of invariant factors is one of

(a) $(X^2 + 2)(X + 3)^3$, $\quad X^2 + 2$;
(b) $(X^2 + 2)(X + 3)^3$, $\quad (X + 3)^2$;
(c) $(X^2 + 2)(X + 3)^3$, $\quad X + 3$, $\quad X + 3$.]

19.5 Determine all possible Jordan forms of the matrices whose characteristic and minimum polynomials are

(a) $\chi_A = (X - 7)^5$, $m_A = (X - 7)^2$;
(b) $\chi_A = (X - 3)^4(X - 5)^4$, $m_A = (X - 3)^2(X - 5)^2$.

19.6 If V is a non-zero finite-dimensional vector space over a field F and if $f : V \to V$ is a cyclic transformation, prove that the set of f-stable subspaces of V is equipotent to the set of monic divisors of the minimum polynomial of f.

[*Hint.* Use the correspondence theorem.]

19.7 Prove that every square matrix is similar to its transpose.

[*Hint.* Recall Corollary 5 of Theorem 19.10. By using the algebraic closure of F, reduce the problem to showing that it holds for Jordan (and hence elementary Jordan) matrices.]

19.8 If V is a non-zero finite-dimensional vector space over a field F and if $f, g : V \to V$ are linear transformations such that, for some prime $p \in F[X]$,

$$m_f = p = m_g,$$

prove that f and g are similar.

[*Hint.* Use the rational decomposition theorem; what are the rational (invariant factor) canonical matrices of f, g?]

19.9 Let A be a square matrix over $\mathbb{C}$ and let P, Q be rectangular matrices over $\mathbb{C}$ such that $A = PQ$ and $B = QP$ exist. Given $h \in \mathbb{C}[X]$, prove that $Ah(A) = Ph(B)Q$.

[*Hint.* Argue by induction on $\deg h$.]

Deduce that one of the following holds :

$$m_A = m_B; \quad m_A = Xm_B; \quad m_B = Xm_A.$$

Express the matrix

$$\begin{bmatrix} 1 & 1 & \ldots & 1 \\ 2 & 2 & \ldots & 2 \\ \vdots & \vdots & & \vdots \\ r & r & \ldots & r \end{bmatrix}$$

as a product of a column matrix and a row matrix. Hence determine its minimum polynomial.

20 Diagonalisation; normal transformations

We shall now concentrate on the common zeros of the characteristic and minimum polynomials. These are called the *eigenvalues*. The set of eigenvalues of a linear transformation f is called the *spectrum* of f. Associated with the eigenvalues is the following important notion.

Definition Let V be a vector space of dimension n over a field F. Then a linear transformation $f : V \to V$ is said to be *diagonalisable* if there is an ordered basis $(e_i)_n$ of V with respect to which the matrix of f is

$$\begin{bmatrix} \lambda_1 & & & \\ & \lambda_2 & & \\ & & \ddots & \\ & & & \lambda_n \end{bmatrix}$$

for some $\lambda_1, \ldots, \lambda_n \in F$.

We can characterise diagonalisable transformations as follows.

Theorem 20.1 *Let V be a non-zero finite-dimensional vector space over a field F and let $f : V \to V$ be a linear transformation. Then f is diagonalisable if and only if the minimum polynomial of f is of the form*

$$m_f = (X - \lambda_1)(X - \lambda_2) \cdots (X - \lambda_m)$$

where $\lambda_1, \ldots, \lambda_m$ are distinct elements of F.

Proof $\Rightarrow$: If f is diagonalisable then there is an ordered basis relative to which the matrix of f is the diagonal matrix

$$M = \mathrm{diag}\{\lambda_1, \ldots, \lambda_n\}.$$

Now M has the same invariant factors as f and hence the same minimum polynomial as f. Let the distinct λ_i in the above list be $\lambda_1, \ldots, \lambda_m$. Then it is clear that the minimum polynomial of M, and hence that of f, is

$$(X - \lambda_1)(X - \lambda_2) \cdots (X - \lambda_m).$$

$\Leftarrow$: Suppose that the minimum polynomial m_f of f is of the stated form. Since m_f is the first invariant factor of f, and since all the invariant factors of f divide m_f, it is clear that every elementary divisor of f

Diagonalisation; normal transformations

is of the form $X - \lambda_i$. Consequently, every rational (elementary divisor) canonical matrix of f is diagonal and so f is diagonalisable. ◊

Applying the primary decomposition theorem (Theorem 19.6) to a diagonalisable transformation, we obtain the following result.

Theorem 20.2 *Let V be a non-zero finite-dimensional vector space over a field F and let $f : V \to V$ be a diagonalisable transformation. Let $\lambda_1, \ldots, \lambda_m$ be the distinct eigenvalues of f and for $i = 1, \ldots, m$ let*

$$V_{\lambda_i} = \{x \in V \; ; \; f(x) = \lambda_i x\}.$$

Then each V_{λ_i} is an f-stable subspace of V and

$$V = \bigoplus_{i=1}^{m} V_{\lambda_i}.$$

Proof It suffices to observe that the prime polynomials p_i of Theorem 19.6 are the polynomials $X - \lambda_i$. ◊

With the notation of Theorem 20.2, the *non-zero* elements of V_{λ_i} are called the *eigenvectors* associated with the eigenvalue λ_i. The f-stable subspace v_{λ_i} is called the corresponding *eigenspace*.

- Note that $V_{\lambda_i} = \text{Ker}(f - \lambda_i \text{id}_V)$.

The following result is now immediate; indeed it is often taken as a definition of a diagonalisable transformation.

Theorem 20.3 *Let V be a non-zero finite-dimensional vector space over a field F and let $f : V \to V$ be a linear transformation. Then f is diagonalisable if and only if V has a basis consisting of eigenvectotrs of f.* ◊

A useful criterion for diagonalisability is the following.

Theorem 20.4 *Let V be a non-zero finite-dimensional vector space over a field F and let $f : V \to V$ be a linear transformation. Then f is diagonalisable if and only if there are non-zero projections $p_1, \ldots, p_k : V \to V$ and distinct scalars $\lambda_1, \ldots, \lambda_k \in F$ such that*

(1) $f = \sum_{i=1}^{k} \lambda_i p_i;$

(2) $\sum_{i=1}^{k} p_i = \text{id}_V;$

(3) $(i \neq j)\ p_i \circ p_j = 0$.

Proof $\Rightarrow$: If f is diagonalisable then by Theorem 20.2 we have

$$V = \bigoplus_{i=1}^{k} V_{\lambda_i}$$

where $\lambda_1, \ldots, \lambda_k$ are the distinct eigenvalues of f, and

$$V_{\lambda_i} = \mathrm{Ker}(f - \lambda_i \mathrm{id}_V)$$

is the eigenspace associated with λ_i. If $p_i : V \to V$ denotes the associated projection then (2) and (3) follow immediately from Theorem 6.14. Since $\mathrm{Im}\, p_i = V_{\lambda_i}$, we also have, for every $x \in V$,

$$f(x) = f\left(\sum_{i=1}^{k} p_i(x)\right) = \sum_{i=1}^{k} f[p_i(x)]$$
$$= \sum_{i=1}^{k} \lambda_i p_i(x)$$
$$= \left(\sum_{i=1}^{k} \lambda_i p_i\right)(x),$$

whence (1) follows.

$\Leftarrow$: Conversely, suppose that the conditions hold. Then by Theorem 6.14 we have

$$V = \bigoplus_{i=1}^{k} \mathrm{Im}\, p_i.$$

Now the λ_i appearing in (1) are precisely the distinct eigenvalues of f. For, on the one hand, for every j,

$$f \circ p_j = \left(\sum_{i=1}^{k} \lambda_i p_i\right) \circ p_j = \sum_{i=1}^{k} \lambda_i (p_i \circ p_j)$$
$$= \lambda_j (p_j \circ p_j)$$
$$= \lambda_j p_j$$

so that $(f - \lambda_j \mathrm{id}_V) \circ p_j = 0$ whence

$$\{0\} \neq \mathrm{Im}\, p_j \subseteq \mathrm{Ker}(f - \lambda_j \mathrm{id}_V),$$

showing that each λ_j is indeed an eigenvalue of f. On the other hand, for every $\lambda \in F$ we have

$$f - \lambda \mathrm{id}_V = \sum_{i=1}^{k} \lambda_i p_i - \sum_{i=1}^{k} \lambda p_i = \sum_{i=1}^{k} (\lambda_i - \lambda) p_i$$

so that if x is an eigenvector of f we have $\sum_{i=1}^{k}(\lambda_i - \lambda)p_i(x) = 0$ whence, since $V = \bigoplus_{i=1}^{k} \operatorname{Im} p_i$, we deduce that $(\lambda_i - \lambda)p_i(x) = 0$ for $i = 1, \ldots, k$. If now $\lambda \neq \lambda_i$ for any i then we have $p_i(x) = 0$ for every i and hence the contradiction

$$x = \sum_{i=1}^{k} p_i(x) = 0.$$

Thus $\lambda = \lambda_i$ for some i and consequently $\lambda_1, \ldots, \lambda_k$ are indeed the distinct eigenvalues of f.

We now observe that in fact

$$\operatorname{Im} p_j = \operatorname{Ker}(f - \lambda_j \operatorname{id}_V).$$

For, suppose that $f(x) = \lambda_j x$; then

$$0 = \sum_{i=1}^{k}(\lambda_i - \lambda_j)p_i(x)$$

whence $(\lambda_i - \lambda_j)p_i(x) = 0$ for all i and so $p_i(x) = 0$ for $i \neq j$. Thus

$$x = \sum_{i=1}^{k} p_i(x) = p_j(x) \in \operatorname{Im} p_j.$$

Since $V = \bigoplus_{i=1}^{k} \operatorname{Ker}(f - \lambda_i \operatorname{id}_V)$ it follows by Theorem 7.8 that V has a basis consisting of eigenvalues of f. We conclude by Theorem 20.3 that f is diagonalisable. ◊

Definition For a diagonalisable transformation f the equality

$$f = \sum_{i=1}^{k} \lambda_i p_i$$

of Theorem 20.4 is called the *spectral resolution* of f.

We shall now consider the problem of deciding when two diagonalisable transformations $f, g : V \to V$ are simultaneously diagonalisable, in the sense that there is a basis of V that consists of eigenvectors of both f and g.

Theorem 20.5 *Let V be a non-zero finite-dimensional vector space over a field F and let $f, g : V \to V$ be diagonalisable transformations. Then f and g are simultaneously diagonalisable if and only if $f \circ g = g \circ f$.*

Proof ⇒ : Suppose that there is a basis $\{e_1,\ldots,e_n\}$ of V that consists of eigenvectors of both f and g. If, for each i, we have $f(e_i) = \lambda_i e_i$ and $g(e_i) = \mu_i e_i$ then

$$(f \circ g)(e_i) = f(\mu_i e_i) = \mu_i f(e_i) = \mu_i \lambda_i e_i;$$

$$(g \circ f)(e_i) = g(\lambda_i e_i) = \lambda_i g(e_i) = \lambda_i \mu_i e_i,$$

whence we see that $f \circ g$ and $g \circ f$ agree on a basis of V. It follows that $f \circ g = g \circ f$.

⇐ : Conversely, suppose that $f \circ g = g \circ f$. Since f is diagonalisable, its minimum polynomial is of the form

$$m_f = (X - \lambda_1)(X - \lambda_2) \cdots (X - \lambda_m).$$

Now, by Theorem 20.2,

$$V = \bigoplus_{i=1}^{m} \operatorname{Ker}(f - \lambda_i \operatorname{id}_V).$$

Since f and g commute, we have, for $v_i \in V_i = \operatorname{Ker}(f - \lambda_i \operatorname{id}_V)$,

$$f[g(v_i)] = g[f(v_i)] = g(\lambda_i v_i) = \lambda_i g(v_i)$$

from which it follows that

$$g(v_i) \in \operatorname{Ker}(f - \lambda_i \operatorname{id}_V) = V_i,$$

so that each V_i is g-stable. Let $g_i : V_i \to V_i$ be the F-morphism induced by g. Since g is diagonalisable, so also is each g_i; for, the minimum polynomial of g_i divides that of g. Thus, by Theorem 20.3, we can find a basis B_i of V_i consisting of eigenvectors of g_i. Since every eigenvector of g_i is trivially an eigenvector of g and since every element of V_i is an eigenvector of f, it follows that $\bigcup_{i=1}^{m} B_i$ is a basis of V that consists of eigenvectors of both f and g. ◊

As we shall see, the above result yields an important property of Jordan morphisms that is useful in applications.

We note that if $f : V \to V$ is a Jordan morphism then clearly every Jordan canonical matrix of f can be written as the sum of a diagonal matrix and a matrix of the form

$$N = \begin{bmatrix} 0 & & & & \\ 1 & 0 & & & \\ & 1 & 0 & & \\ & & \ddots & \ddots & \\ & & & 1 & 0 \end{bmatrix}.$$

Diagonalisation; normal transformations

If N is of size $n \times n$ then it is readily seen that $N^n = 0$. This gives rise to the following notion.

Definition A linear transformation $f : V \to V$ (or a square matrix A) is said to be *nilpotent* if, for some positive integer n, $f^n = 0$ (or $A^n = 0$).

Theorem 20.6 [Jordan decomposition theorem] *Let V be a non-zero finite-dimensional vector space over a field F and let $f : V \to V$ be a Jordan transformation. Then there is a diagonalisable transformation $\delta : V \to V$ and a nilpotent transformation $\eta : V \to V$ such that $f = \delta + \eta$ and $\delta \circ \eta = \eta \circ \delta$. Moreover, there exist $p, q \in F[X]$ such that $\delta = p(f)$ and $\eta = q(f)$. Furthermore, δ and η are uniquely determined, in the sense that if $\delta', \eta' : V \to V$ are respectively diagonalisable and nilpotent transformations such that $f = \delta' + \eta'$ and $\delta' \circ \eta' = \eta' \circ \delta'$ then $\delta' = \delta$ and $\eta' = \eta$.*

Proof Since f is a Jordan transformation its minimum polynomial is of the form

$$m_f = (X - \lambda_1)^{m_1}(X - \lambda_2)^{m_2} \cdots (X - \lambda_n)^{m_n}$$

and, by Theorem 20.2, $V = \bigoplus_{i=1}^{n} V_i$ where $V_i = \mathrm{Ker}\,(f - \lambda_i \mathrm{id}_V)^{m_i}$. Let $\delta : V \to V$ be the linear transformation given by

$$\delta = \sum_{i=1}^{n} \lambda_i p_i$$

where $p_i : V \to V$ is the projection on V_i parallel to $\sum_{j \neq i} V_j$. Since, for $v_i \in V_i$,

$$\delta(v_i) = \left(\sum_{j=1}^{n} \lambda_j p_j\right)(v_i) = \lambda_i v_i,$$

it follows by Theorem 7.8 that V has a basis consisting of eigenvectors of δ and so, by Theorem 20.3, δ is diagonalisable.

Now let $\eta = f - \delta$. Then for $v_i \in V_i$ we have

$$\eta(v_i) = f(v_i) - \delta(v_i) = (f - \lambda_i \mathrm{id}_V)(v_i)$$

and consequently

$$\eta^{m_i}(v_i) = (f - \lambda_i \mathrm{id}_V)^{m_i}(v_i) = 0.$$

It follows that, for some k, $\mathrm{Ker}\,\eta^k$ contains a basis of V whence we have that $\eta^k = 0$ and so η is nilpotent.

Since $V = \bigoplus_{i=1}^{n} V_i$, every $v \in V$ can be written uniquely in the form

$$v = v_1 + \ldots + v_n$$

with $v_i \in V_i$ for each i. Since each V_i is f-stable, we deduce that

$$(p_i \circ f)(v) = p_i[f(v_1) + \ldots + f(v_n)] = f(v_i) = (f \circ p_i)(v).$$

Consequently $p_i \circ f = f \circ p_i$ for $i = 1, \ldots, n$ and so

$$\delta \circ f = \left(\sum_{i=1}^{n} \lambda_i p_i\right) \circ f = \sum_{i=1}^{n} \lambda_i (p_I \circ f)$$
$$= \sum_{i=1}^{n} \lambda_i (f \circ p_i)$$
$$= f \circ \left(\sum_{i=1}^{n} \lambda_i p_i\right)$$
$$= f \circ \delta.$$

It follows from this that

$$\delta \circ \eta = \delta \circ (f - \delta) = (\delta \circ f) - \delta^2 = (f \circ \delta) - \delta^2 = (f - \delta) \circ \delta = \eta \circ \delta.$$

We now show that there are polynomials p, q such that $\delta = p(f)$ and $\eta = q(f)$. For this purpose, define

$$(i = 1, \ldots, n) \qquad t_i = \frac{m_f}{(X - \lambda_i)^{m_i}}.$$

Since $t_1, \ldots, t_n$ are relatively prime, there exist polynomials $a_1, \ldots, a_n$ such that

$$t_1 a_1 + \ldots + t_n a_n = 1. \tag{1}$$

Now let $b_i = t_i a_i$ for $i = 1, \ldots, n$ and let $v = v_1 + \ldots + v_n$ be an arbitrary element of V, with of course $v_i \in V_i$ for each i. Observing that if $j \neq i$ then b_j is a multiple of $(X - \lambda_i)^{m_i}$, we deduce that

$$(j \neq i) \qquad [b_j(f)](v_i) = 0.$$

It now follows from (1) that

$$(i = 1, \ldots, n) \qquad [b_i(f)](v_i) = v_i.$$

Consequently, we have

$$(i = 1, \ldots, n) \qquad [b_i(f)](v) = v_i = p_i(v)$$

Diagonalisation; normal transformations 325

and so $b_i(f) = p_i$. It now follows from the definition of δ that $\delta = p(f)$ where $p = \sum_{i=1}^{n}\lambda_i b_i$. Since $\eta = f - \delta$ there is then a polynomial q with $\eta = q(f)$.

As for uniqueness, suppose that $\delta', \eta' : V \to V$ are respectively diagonalisable and nilpotent transformations such that

$$f = \delta' + \eta', \quad \delta' \circ \eta' = \eta' \circ \delta'.$$

We note first that these conditions give $f \circ \delta' = \delta' \circ f$ and $f \circ \eta' = \eta' \circ f$. Now, as we have just seen, there are polynomials p, q such that $\delta = p(f)$ and $\eta = q(f)$. It follows, therefore, that $\delta \circ \delta' = \delta' \circ \delta$ and $\eta \circ \eta' = \eta' \circ \eta$. Consider now the equality

$$\delta - \delta' = \eta' - \eta.$$

Since η, η' commute we can use the binomial theorem to deduce from the fact that η, η' are nilpotent that $\eta' - \eta$ is nilpotent. On the other hand, since δ, δ' commute, it follows by Theorem 20.3 that there is a basis of V consisting of eigenvectors of both δ and δ'. Each of these eigenvectors is then an eigenvector of $\delta - \delta'$; for if $\delta(v) = \lambda v$ and $\delta'(v) = \lambda' v$ then

$$(\delta - \delta')(v) = (\lambda - \lambda')v.$$

Consequently, the matrix of $\delta - \delta'$ with repect to this (suitably ordered) basis is diagonal, say

$$D = \text{diag}\{d_1, \ldots, d_n\}.$$

Since $\delta - \delta' = \eta' - \eta$, this diagonal matrix is also nilpotent. It follows that some power of each d_i is zero whence every d_i is zero and hence

$$\delta - \delta' = \eta' - \eta = 0,$$

from which the uniqueness follows. ◊

- In the above decomposition $f = \delta + \eta$ we call δ the *diagonalisable part* and η the *nilpotent part* of the Jordan transformation f.

- Note from the above proof that for each projection p_i we have $p_i = b_i(f)$. We shall use this observation later.

At this stage we return to our discussion of inner product spaces. Let V be a finite-dimensional inner product space and let $f : V \to V$ be a linear transformation. We shall now consider the question : under what conditions is f *ortho-diagonalisable*, in the sense that there is an *orthonormal* basis consisting of eigenvectors of f? Put another way, under what conditions does there exist an ordered *orthonormal* basis of V relative to which the matrix of f is diagonal? Expressed purely in terms of matrices, this question is equivalent to asking precisely when is a given square matrix (over $\mathbb{R}$ or $\mathbb{C}$) unitarily similar to a diagonal matrix?

In order to tackle this problem, we shall concentrate on direct sum decompositions and the associated projections (see Chapter 6). The first preparatory result that we shall require is the following.

Theorem 20.7 *If W, X are subspaces of a finite-dimensional inner product space V such that $V = W \oplus X$, then $V = W^\perp \oplus X^\perp$. Moreover, if p is the projection on W parallel to X then the adjoint p^* of p is the projection on $X^\perp$ parallel to $W^\perp$.*

Proof We note first that if A, B are subspaces of V then

$$A \subseteq B \Rightarrow B^\perp \subseteq A^\perp.$$

To see this, it suffices to observe that every element that is orthogonal to B is clearly orthogonal to A. Next we note that

$$(A+B)^\perp = A^\perp \cap B^\perp, \qquad (A \cap B)^\perp = A^\perp + B^\perp.$$

In fact, since $A, B \subseteq A + B$ we have $(A+B)^\perp \subseteq A^\perp \cap B^\perp$; and since $A \cap B \subseteq A, B$ we have $A^\perp + B^\perp \subseteq (A \cap B)^\perp$. Since then

$$A \cap B = (A \cap B)^{\perp\perp} \subseteq (A^\perp + B^\perp)^\perp \subseteq A^{\perp\perp} \cap B^{\perp\perp} = A \cap B,$$

we deduce that $A \cap B = (A^\perp + B^\perp)^\perp$, whence $(A \cap B)^\perp = A^\perp + B^\perp$. The other equality can be derived from this by replacing A, B by $A^\perp, B^\perp$.

Using the above observations, we see that if $V = W \oplus X$ then

$$V^\perp = (W + X)^\perp = W^\perp \cap X^\perp;$$
$$V = \{0\}^\perp = (W \cap X)^\perp = W^\perp + X^\perp,$$

whence $V = W^\perp \oplus X^\perp$.

Suppose now that p is the projection on W parallel to X. Since, by Theorem 11.10(3),

$$p^* \circ p^* = (p \circ p)^* = p^*,$$

we see that p^* is a projection. Now since $\text{Im}\, p = W$ and since, from the definition of adjoint, $\langle p(x) \,|\, y \rangle = \langle x \,|\, p^*(y) \rangle$ we see that

$$y \in \text{Ker}\, p^* \iff y \in W^\perp,$$

so that $\operatorname{Ker} p^* = W^\perp$. Now for all $x, y \in V$ we have

$$\begin{aligned} \langle x \mid y - p^*(y) \rangle &= \langle x \mid y \rangle - \langle x \mid p^*(y) \rangle \\ &= \langle x \mid y \rangle - \langle p(x) \mid y \rangle \\ &= \langle x - p(x) \mid y \rangle. \end{aligned}$$

Since, by Theorem 6.12, $\operatorname{Im} p = \{x \in V \; ; \; x = p(x)\}$ and $\operatorname{Ker} p = \{x - p(x) \; ; \; x \in V\}$, it follows that

$$y \in \operatorname{Im} p^* \iff y \in (\operatorname{Ker} p)^\perp,$$

so that $\operatorname{Im} p^* = (\operatorname{Ker} p)^\perp = X^\perp$. Thus we see that p^* is the projection on $X^\perp$ parallel to $W^\perp$. ◊

Definition By an *ortho-projection* on an inner product space V we shall mean a projection $p : V \to V$ such that $\operatorname{Im} p = (\operatorname{Ker} p)^\perp$.

Theorem 20.8 *Let V be a non-zero finite-dimensional inner product space. If p is a projection on V then p is an ortho-projection if and only if p is self-adjoint.*

Proof By Theorem 6.13, p is the projection on $\operatorname{Im} p$ parallel to $\operatorname{Ker} p$; and, by Theorem 20.7, p^* is the projection on $\operatorname{Im} p^* = (\operatorname{Ker} p)^\perp$ parallel to $\operatorname{Ker} p^* = (\operatorname{Im} p)^\perp$. If then p is self-adjoint we have $p = p^*$ and so $\operatorname{Im} p = \operatorname{Im} p^* = (\operatorname{Ker} p)^\perp$; and conversely, if $\operatorname{Im} p = (\operatorname{Ker} p)^\perp$ then $\operatorname{Im} p = \operatorname{Im} p^*$ and

$$\operatorname{Ker} p = (\operatorname{Ker} p)^{\perp\perp} = (\operatorname{Im} p)^\perp = \operatorname{Ker} p^*,$$

from which it follows by Theorem 6.12 that $p = p^*$. ◊

Definition Let $V_1, \ldots, V_n$ be subspaces of the inner product space V. Then we shall say that V is the *ortho-direct sum* of $V_1, \ldots, V_n$ if

(1) $V = \bigoplus_{i=1}^{n} V_i$;

(2) $(i = 1, \ldots, n)$ $V_i^\perp = \sum_{j \neq i} V_j$.

If $V = \bigoplus_{i=1}^{n} V_i$ and if p_i denotes the projection on V_i parallel to $\sum_{j \neq i} V_j$, then it is clear that V is the ortho-direct sum of $V_1, \ldots, V_n$ if and only if each p_i is an ortho-projection. It is also clear that if $V = \bigoplus_{i=1}^{n} V_i$ then V is the ortho-direct sum of $V_1, \ldots, V_n$ if and only if every element of V_i is orthogonal to every element of V_j for $j \neq i$; for then $\sum\limits_{j \neq i} V_j \subseteq V_i^\perp$ whence we have equality since

$$\dim \sum_{j \neq i} V_j = \dim V - \dim V_i = \dim V_i^\perp.$$

The following result is now immediate from the above definitions and Theorem 20.4.

Theorem 20.9 *Let V be a non-zero finite-dimensional inner product space. Then a linear transformation $f : V \to V$ is ortho-diagonalisable if and only if there are ortho-projections $p_1, \ldots, p_k : V \to V$ and distinct scalars $\lambda_1, \ldots, \lambda_k$ such that*

(1) $f = \sum\limits_{i=1}^{k} \lambda_i p_i$;

(2) $\sum\limits_{i=1}^{k} p_I = \mathrm{id}_V$;

(3) $(i \neq j) \quad p_i \circ p_j = 0.$ ◊

Suppose now that $f : V \to V$ is ortho-diagonalisable. Then, applying Theorem 11.10 to the conditions (1), (2), (3) of Theorem 20.9, we obtain, using Theorem 20.8,

$$(1^\star) \quad f^\star = \sum_{i=1}^{k} \overline{\lambda_i} p_i^\star = \sum_{i=1}^{k} \overline{\lambda_i} p_i; \quad (2^\star) = (2), \quad (3^\star) = (3).$$

We deduce from this that $f^\star$ is ortho-diagonalisable and that $(1^\star)$ gives its spectral resolution. Thus $\overline{\lambda_1}, \ldots, \overline{\lambda_k}$ are the distinct eigenvalues of $f^\star$. A simple computation now shows that

$$f \circ f^\star = \sum_{i=1}^{k} |\lambda_i|^2 p_i = f^\star \circ f,$$

so that *ortho-diagonalisable transformations commute with their adjoints*.

This leads to the following important notion.

Definition If V is a finite-dimensional inner product space and if $f : V \to V$ is a linear transformation then we shall say that f is *normal* if it commutes with its adjoint. Similarly, if A is a square matrix over the ground field of V then we say that A is *normal* if $AA^\star = A^\star A$.

We have just observed that a necessary condition for a linear transformation f to be ortho-diagonalisable is that it be normal. It is remarkable that, when the ground field is $\mathbb{C}$, this condition is also sufficient. In

Diagonalisation; normal transformations

order to establish this, we require the following properties of normal transformations.

Theorem 20.10 *Let V be a non-zero finite-dimensional inner product space and let $f : V \to V$ be a normal transformation. Then*

(1) $(\forall x \in V) \quad \|f(x)\| = \|f^*(x)\|$;

(2) *if p is a polynomial with coefficients in the ground field of V then the transformation $p(f) : V \to V$ is also normal;*

(3) $\operatorname{Im} f \cap \operatorname{Ker} f = \{0\}$.

Proof (1) Since f commutes with f^* we have, for all $x \in V$,

$$\langle f(x) \mid f(x) \rangle = \langle x \mid f^*[f(x)] \rangle = \langle x \mid f[f^*(x)] \rangle = \langle f^* \mid f^*(x) \rangle,$$

from which (1) follows.

(2) If $p = a_0 X^0 + a_1 X^1 + \ldots + a_n X^n$ then

$$p(f) = a_0 \operatorname{id}_V + a_1 f + \ldots + a_n f$$

and so, by Theorem 11.10,

$$[p(f)]^* = \overline{a_0} \operatorname{id}_V + \overline{a_1} f^* + \ldots + \overline{a_n} (f^*)^n.$$

Since f and f^* commute, it is clear that so do $p(f)$ and $[p(f)]^*$. Hence $p(f)$ is normal.

(3) If $x \in \operatorname{Im} f \cap \operatorname{Ker} f$ then there exists $y \in V$ such that $x = f(y)$ and $f(x) = 0$. By (1) we have $f^*(x) = 0$ and so

$$0 = \langle f^*(x) \mid y \rangle = \langle x \mid f(y) \rangle = \langle x \mid x \rangle$$

whence we see that $x = 0$. $\Diamond$

Theorem 20.11 *Let V be a non-zero finite-dimensional inner product space. If p is a projection on V then p is normal if and only if it is self-adjoint.*

Proof Clearly, if p is self-adjoint then p is normal. Suppose, conversely, that p is normal. By Theorem 20.10(1), we then have $\|p(x)\| = \|p^*(x)\|$ and so $p(x) = 0$ if and only if $p^*(x) = 0$. Given $x \in V$ let $y = x - p(x)$. Then

$$p(y) = p(x) - p[p(x)] = p(x) - p(x) = 0$$

and so $0 = p^*(y) == p^*(x) - p^*[p(x)]$. Consequently we see that $p^* = p^* \circ p$. It now follows that

$$p = p^{**} = (p^* \circ p)^* = p^* \circ p^{**} = p^* \circ p = p^*. \quad \Diamond$$

Theorem 20.12 *Let V be a non-zero finite-dimensional complex inner product space. If $f : V \to V$ is a linear transformation then the following are equivalent :*
  (1) *f is ortho-diagonalisable;*
  (2) *f is normal.*

Proof We have already seen that $(1) \Rightarrow (2)$. As for $(2) \Rightarrow (1)$, suppose that f is normal. To show that f is diagonalisable it suffices, by Theorem 20.1, to show that the minimum polynomial m_f of f is a product of distinct linear polynomials. Since $\mathbb{C}$ is algebraically closed, m_f is certainly a product of linear polynomials. Suppose, by way of obtaining a contradiction, that $c \in \mathbb{C}$ is a multiple root of m_f, so that

$$m_f = (X - c)^2 g$$

for some polynomial g. Then

$$(\forall x \in V) \qquad 0 = m_f(x) = [(f - c\,\mathrm{id}_V)^2 \circ g(f)](x)$$

and consequently $[(f - c\,\mathrm{id}_V) \circ g(f)](x)$ belongs to both the image and the kernel of $f - c\,\mathrm{id}_V$. Since, by Theorem 20.10(2), $f - c\,\mathrm{id}_V$ is normal, we deduce by Theorem 20.10(3) that

$$(\forall x \in V) \qquad [(f - c\,\mathrm{id}_V) \circ g(f)](x) = 0.$$

Consequently $(f - c\,\mathrm{id}_V) \circ g(f)$ is the zero transformation on V. This contradicts the fact that $(X - c)^2 g$ is the minimum polynomial of f. Thus we see that f is diagonalisable.

To show that f is ortho-diagonalisable, it is enough to show that the projections p_i corresponding to Theorem 20.4 are ortho-projections; and by Theorem 20.8 it is enough to show that these projections are self-adjoint.

Now since f is diagonalisable it is a Jordan transformation and thus coincides with its diagonal part as described in Theorem 20.6. In the proof of that result, we observed that there existed a polynomial b_i such that $b_i(f) = p_i$. We thus see by Theorem 20.10(2) that the projections p_i are normal. It now follows by Theorem 20.11 that each p_i is self-adjoint. ◊

Corollary *If A is a square matrix over $\mathbb{C}$ then A is unitarily similar to a diagonal matrix if and only if A is normal.* ◊

- It should be noted that in the proof of Theorem 20.12 we used the fact that $\mathbb{C}$ is algebraically closed. This is not so for $\mathbb{R}$ and we might

Diagonalisation; normal transformations

expect that the corresponding result in the case where the ground field is $\mathbb{R}$ is false in general. This is indeed the case : there exist normal transformations on a *real* inner product space that are *not* diagonalisable. One way in which this can happen, of course, is when the transformation in question has all its eigenvalues *complex*. For example, the reader will verify that the matrix

$$\begin{bmatrix} -\tfrac{1}{2} & -\tfrac{\sqrt{3}}{2} \\ \tfrac{\sqrt{3}}{2} & -\tfrac{1}{2} \end{bmatrix} = \begin{bmatrix} \cos\tfrac{2\pi}{3} & -\sin\tfrac{2\pi}{3} \\ \sin\tfrac{2\pi}{3} & \cos\tfrac{2\pi}{3} \end{bmatrix}$$

has minimum polynomial $X^2 + X + 1$, which is irreducible over $\mathbb{R}$. In order to obtain an analogue for Theorem 20.12 in the case where the ground field is $\mathbb{R}$, we asre therefore led to consider normal transformations whose eigenvalues are all real.

Theorem 20.13 *Let V be a non-zero finite-dimensional complex inner product space. If $f : V \to V$ is a linear transformation then the following are equivalent :*

(1) *f is normal and every eigenvalue of f is real;*
(2) *f is self-adjoint.*

Proof (1) $\Rightarrow$ (2) : If f is normal then by Theorem 20.12 it is diagonalisable. Let $f = \sum_{i=1}^{k} \lambda_i p_i$ be its spectral resolution. We know that $f^\star$ is also normal with spectral resolution $f^\star = \sum_{i=1}^{k} \overline{\lambda_i} p_i$. Since each λ_i is real by hypothesis, we deduce that $f = f^\star$.

(2) $\Rightarrow$ (1) : If $f = f^\star$ then it is clear that f is normal. If the spectral resolutions are $f = \sum_{i=1}^{k} \lambda_i p_i$ and $f^\star = \sum_{i=1}^{k} \overline{\lambda_i} p_i$ then $f = f^\star$ gives $\sum_{i=1}^{k}(\lambda_i - \overline{\lambda_i})p_i = 0$ and so

$$(\forall x \in V) \quad \sum_{i=1}^{k}(\lambda_i - \overline{\lambda_i})p_i(x) = 0,$$

whence $(\lambda_i - \overline{\lambda_i})p_i(x) = 0$ for each i since $V = \bigoplus_{i=1}^{k} \operatorname{Im} p_i$. Since no p_i is zero, we deduce that $\lambda_i = \overline{\lambda_i}$ for every i. Consequently every eigenvalue of f is real. $\diamond$

Corollary *All eigenvalues of a hermitian matrix are real.* $\diamond$

We can now describe the ortho-diagonalisable transformations on a real inner product space.

Theorem 20.14 *If V is a non-zero finite-dimensional real inner product space and $f : V \to V$ is a linear transformation then f is ortho-diagonalisable if and only if f is self-adjoint.*

Proof $\Rightarrow$: If f is ortho-diagonalisable then, as in Theorem 20.9, let $f = \sum_{i=1}^{k} \lambda_i p_i$. Since the ground field is $\mathbb{R}$, every λ_i is real and so, taking adjoints and using Theorem 20.8, we obtain $f = f^\star$.

$\Leftarrow$: Suppose conversely that f is self-adjoint and let A be the ($n \times n$ say) matrix of f relative to some ordered orthonormal basis of V. Then A is symmetric. Now let f' be the linear transformation on the complex inner product space $\mathbb{C}^n$ whose matrix relative to the natural ordered orthonormal basis of $\mathbb{C}^n$ is A. Then f' is self-adjoint. By Theorem 20.13, the eigenvalues of f' are all real and, since f' is diagonalisable, the minimum polynomial of f' is a product of distinct linear polynomials over $\mathbb{R}$. Since this is then the minimum polynomial of A, it is also the minimum polynomial of f. Thus we see that f is diagonalisable. That f is ortho-diagonalisable is shown precisely as in Theorem 20.12. $\Diamond$

Corollary *If A is a square matrix over $\mathbb{R}$ then A is orthogonally similar to a diagonal matrix if and only if A is symmetric.* $\Diamond$

We now turn our attention to a useful alternative characterisation of a self-adjoint transformation on a complex inner product space. For this purpose, we observe the following result.

Theorem 20.15 *Let V be a complex inner product space. If $f : V \to V$ is linear and such that $\langle f(x) \mid x \rangle = 0$ for all $x \in V$ then $f = 0$.*

Proof For all $z \in V$ we have

$$0 = \langle f(y+z) \mid y+z \rangle = \langle f(y) \mid z \rangle + \langle f(z) \mid y \rangle;$$
$$0 = \langle f(iy+z) \mid iy+z \rangle = i\langle f(y) \mid z \rangle - i\langle f(z) \mid y \rangle,$$

from which it follows that $\langle f(y) \mid z \rangle = 0$. Then $f(y) = 0$ for all $y \in V$ and so $f = 0$. $\Diamond$

Theorem 20.16 *Let V be a finite-dimensional complex inner product space. If $f : V \to V$ is linear then the following are equivalent :*

(1) *f is self-adjoint;*

Diagonalisation; normal transformations 333

(2) $(\forall x \in V)\ \langle f(x)\,|\,x\rangle \in \mathbb{R}$.

Proof (1) $\Rightarrow$ (2) : If f is self-adjoint then, for every $x \in V$,

$$\overline{\langle f(x)\,|\,x\rangle} = \overline{\langle f^\star(x)\,|\,x\rangle} = \langle x\,|\,f^\star(x)\rangle = \langle f(x)\,|\,x\rangle,$$

from which (2) follows.

(2) $\Rightarrow$ (1) : If (2) holds then

$$\langle f^\star(x)\,|\,x\rangle = \langle x\,|\,f(x)\rangle = \overline{\langle f(x)\,|\,x\rangle} = \langle f(x)\,|\,x\rangle$$

and consequently

$$\langle (f^\star - f)(x)\,|\,x\rangle = \langle f^\star(x)\,|\,x\rangle - \langle f(x)\,|\,x\rangle = 0.$$

Since this holds for all $x \in V$, it follows by Theorem 20.15 that $f^\star = f$.
◊

These results lead to the following notion.

Definition If V is an inner product space then a linear mapping $f : V \to V$ is said to be *positive* (or *semi-definite*) if it is self-adjoint and such that $\langle f(x)\,|\,x\rangle \geq 0$ for every $x \in V$; and *positive definite* if it is self-adjoint and $\langle f(x)\,|\,x\rangle > 0$ for every non-zero $x \in V$.

Theorem 20.17 *Let V be a non-zero finite-dimensional inner product space and $f : V \to V$ a linear transformation. Then the following are equivalent* :

(1) *f is positive*;
(2) *f is self-adjoint and every eigenvalue is real and ≥ 0*;
(3) *there is a self-adjoint $g : V \to V$ such that $g^2 = f$*;
(4) *there is a linear map $h : V \to V$ such that $h^\star \circ h = f$.*

Proof (1) $\Rightarrow$ (2) : Let λ be an eigenvalue of f. By Theorem 20.13, λ is real. Then

$$0 \leq \langle f(x)\,|\,x\rangle = \langle \lambda x\,|\,x\rangle = \lambda\langle x\,|\,x\rangle$$

gives $\lambda \geq 0$ since $\langle x\,|\,x\rangle > 0$.

(2) $\Rightarrow$ (3) : Since f is self-adjoint it is normal and therefore is ortho-diagonalisable. Let its spectral resolution be $f = \sum_{i=1}^{k} \lambda_i p_i$ and, using (2), define $g : V \to V$ by

$$g = \sum_{i=1}^{k} \sqrt{\lambda_i}\, p_i.$$

Since the p_i are ortho-projections and hence self-adjoint, we have that g is self-adjoint. Also, since $p_i \circ p_j = 0$ for $i \neq j$, it follows readily that $g^2 = f$.

(3) $\Rightarrow$ (4) : Take $h = g$.
(4) $\Rightarrow$ (1) : Observe that

$$(h^* \circ h)^* = h^* \circ h^{**} = h^* \circ h$$

and, for all $x \in V$,

$$\langle h^*[h(x)] \mid x \rangle = \langle h(x) \mid h(x) \rangle \geq 0.$$

Thus we see that $h^* \circ h$ is positive. $\diamond$

It is immediate from Theorem 20.17 that every positive linear transformation has a square root. That this square root is unique is shown as follows.

Theorem 20.18 *Let f be a positive linear transformation on a non-zero finite-dimensional inner product space V. Then there is a unique positive linear transformation $g : V \to V$ such that $g^2 = f$. Moreover, there is a polynomial q such that $g = q(f)$.*

Proof Let $f = \sum_{i=1}^{k} \lambda_i p_i$ be the spectral resolution of f and define g by $g = \sum_{i=1}^{k} \sqrt{\lambda_i} p_i$. Since this must be the spectral resolution of g, it follows that the eigenvalues of g are $\sqrt{\lambda_i}$ for $i = 1, \ldots, k$ and so, by Theorem 20.17, g is positive.

Suppose now that $h : V \to V$ is also positive and such that $h^2 = f$. If the spectral resolution of h is $\sum_{j=1}^{m} \mu_j q_j$ where the q_j are orthogonal projections then we have

$$\sum_{i=1}^{k} \lambda_i p_i = f = h^2 = \sum_{j=1}^{m} \mu_j^2 q_j.$$

Now the eigenspaces of f are $\text{Im } p_i$ for $i = 1, \ldots, k$ and also $\text{Im } q_j$ for $j = 1, \ldots, m$. It follows that $m = k$ and that there is a permutation σ on $\{1, \ldots, k\}$ such that $q_{\sigma(i)} = p_i$ whence $\mu_{\sigma(i)}^2 = \lambda_i$. Thus $\mu_{\sigma(i)} = \sqrt{\lambda_i}$ and we deduce that $h = g$.

The final assertion follows by considering the Lagrange polynomials

$$P_i = \prod_{j \neq i} \frac{X - \lambda_j}{\lambda_i - \lambda_j}.$$

Since $P_i(\lambda_t) = \delta_{it}$, the polynomial $q = \sum_{i=1}^{k} \sqrt{\lambda_i} P_i$ is then such that $q(f) = g$. ◊

Corollary 1 *The following are equivalent* :
(1) *f is positive definite*;
(2) *f is self-adjoint and all eigenvalues of f are real and > 0*;
(3) *there is an invertible self-adjoint g such that $g^2 = f$*;
(4) *there is an invertible h such that $h^* \circ h = f$.*

Proof This is clear from the fact that g is invertible if and only if 0 is not one of its eigenvalues. ◊

Corollary 2 *If f is positive definite then f is invertible.* ◊

Of course, the above results have matrix analogues. A square matrix that represents a positive transformation is called a *Gram matrix*. The following characterisation of such matrices is immediate from the above.

Theorem 20.19 *A square matrix is a Gram matrix if and only if it is self-adjoint and all its eigenvalues are real and greater than or equal to 0.* ◊

By Theorem 20.12, the ortho-diagonalisable transformations on a complex inner product space are precisely the normal transformations; and by Theorem 20.14 those on a real inner product space are precisely the self-adjoint transformations. It is natural at this point to ask about the normal transformations on a *real* inner product space; equivalently, to ask about *real* square matrices that commute with their transposes. In particular, can we determine canonical forms for such matrices under orthogonal similarity? The following sequence of results achieves this goal. As a particular case, we shall be able to determine a canonical form for orthogonal matrices.

Definition If V is a finite-dimensional real inner product space and if $f : V \to V$ is a linear transformation then we shall say that f is *skew-adjoint* if $f^* = -f$. The corresponding terminology for real square matrices is *skew-symmetric*; for, the entries of A being real, $A^* = -A$ becomes $A^t = -A$.

Theorem 20.20 *If V is a non-zero finite-dimensional real inner product space and if $f : V \to V$ is a linear transformation then there is a unique self-adjoint transformation $g : V \to V$ and a unique skew-adjoint transformation $h : V \to V$ such that $f = g + h$. Moreover, f is normal if and only if g and h commute.*

Proof Clearly, we have

$$f = \tfrac{1}{2}(f + f^*) + \tfrac{1}{2}(f - f^*)$$

where $\tfrac{1}{2}(f + f^*)$ is self-adjoint and $\tfrac{1}{2}(f - f^*)$ is skew-adjoint. Suppose then that $f = g + h$ where g is self-adjoint and h is skew-adjoint. Then $f^* = g^* + h^* = g - h$ and consequently we see that $g = \tfrac{1}{2}(f + f^*)$ and $h = \tfrac{1}{2}(f - f^*)$.

If now f is normal then $f \circ f^* = f^* \circ f$ gives

$$(g + h) \circ (g - h) = (g - h) \circ (g + h),$$

which reduces to $g \circ h = h \circ g$. Conversely, if $g \circ h = h \circ g$ then

$$f \circ f^* = (g + h) \circ (g - h) = g^2 + h \circ g - g \circ h - h^2 = g^2 - h^2$$

and likewise $f^* \circ f = g^2 - h^2$. $\Diamond$

- In the above expression for f we call g the *self-adjoint part* and h the *skew-adjoint part* of f.

A useful characterisation of skew-adjoint transformations is the following.

Theorem 20.21 *If V is a non-zero finite-dimensional real inner product space and if $f : V \to V$ is a linear transformation then the following are equivalent :*
  (1) *f is skew-adjoint;*
  (2) *$(\forall x \in V) \quad \langle f(x) \,|\, x \rangle = 0$.*

Proof (1) $\Rightarrow$ (2) : If f is skew-adjoint then, since we are dealing with a real inner product space, given $x \in V$ we have

$$\langle f(x) \,|\, x \rangle = \langle x \,|\, f^*(x) \rangle = \langle x \,|\, -f(x) \rangle = -\langle x \,|\, f(x) \rangle = -\langle f(x) \,|\, x \rangle.$$

It follows that $\langle f(x) \,|\, x \rangle = 0$.

(2) $\Rightarrow$ (1) : If (2) holds then for all $x, y \in V$ we have

$$\begin{aligned}0 = \langle f(x + y) \,|\, x + y \rangle &= \langle f(x) \,|\, x \rangle + \langle f(x) \,|\, y \rangle + \langle f(y) \,|\, x \rangle + \langle f(y) \,|\, y \rangle \\ &= \langle f(x) \,|\, y \rangle + \langle f(y) \,|\, x \rangle\end{aligned}$$

whence we obtain

$$\langle f(x) \,|\, y \rangle = -\langle f(y) \,|\, x \rangle = -\langle x \,|\, f(y) \rangle = \langle x \,|\, -f(y) \rangle.$$

It now follows by the uniqueness of adjoints that $f^* = -f$. $\Diamond$

Diagonalisation; normal transformations

Since the main results of our immediate discussion to follow stem from applications of the primary decomposition theorem (Theorem 19.6), the notion of minimum polynomial will play an important role. Now as the ground field in question is $\mathbb{R}$ and since $\mathbb{R}[X]$ is a unique factorisation domain, every non-zero polynomial with real coefficients can be expressed as a product of powers of distinct irreducible polynomials. For our future work, we note that *a monic polynomial over $\mathbb{R}$ is irreducible if and only if it is of the form $X - a$ or $X^2 - (z+\bar{z})X + z\bar{z}$ for some $z \in \mathbb{C}\setminus\mathbb{R}$*. In fact, if $z \in \mathbb{C}\setminus\mathbb{R}$, say $z = a + ib$ with $b \neq 0$, then the polynomial

$$X^2 - (z+\bar{z})X + z\bar{z} = X^2 - 2aX + (a^2 + b^2)$$

is readily seen to be irreducible over $\mathbb{R}$. Conversely, suppose that the monic polynomial p is irreducible over $\mathbb{R}$ and that $p \neq X - a$. Since $\mathbb{C}$ is algebraically closed, we can find $z \in \mathbb{C}$ that is a zero of p. Then since $\overline{p(z)} = p(\bar{z})$ we see that $\bar{z}$ is also a zero of p. It follows that the polynomial

$$X^2 - (z+\bar{z})X + z\bar{z} = (X - Z)(X - \bar{z})$$

is a divisor of p; moreover, we cannot have $z \in \mathbb{R}$ since otherwise $X - z \in \mathbb{R}[X]$ would be a divisor of p in $\mathbb{R}[X]$, contradicting the fact that p is irreducible.

Theorem 20.22 *If V is a non-zero finite-dimensional real inner product space and if $f : V \to V$ is a normal transformation then the minimum polynomial of f is of the form $\prod_{i=1}^{k} p_i$ where $p_1, \ldots, p_k$ are distinct irreducible polynomials.*

Proof Since $\mathbb{R}[X]$ is a unique factorisation domain, we can express the minimum polynomial of f in the form

$$m_f = \prod_{i=1}^{k} p_i^{m_i}$$

where $p_1, \ldots, p_k$ are distinct irreducible polynomials. We have to show that every $m_i = 1$. Suppose, by way of obtaining a contradiction, that for some i we have $m_1 \geq 2$. Let $M_i = \operatorname{Ker}[p_i(f)]^{m_i}$, so that $[p_i(f)]^{m_i}(x) = 0$ for every $x_i \in M_i$. Then we have

$$(\forall x \in M_i) \quad [p_i(f)]^{m_i-1}(x) \in \operatorname{Im} p_i(f) \cap \operatorname{Ker} p_i(f).$$

But since f is normal so is $p_i(f)$ by Theorem 20.10(2). It now follows by Theorem 20.10(3) that the restriction of $[p_i(f)]^{m_i}$ to M_i is the zero transformation. But, by the primary decomposition theorem, $[p_i(f)]^{m_i}$

is the minimum polynomial of the transformation induced on M_i by f. From this contradiction we therefore deduce that each m_i must be 1. ◊

Concerning the minimum polynomial of a skew-adjoint transformation, we have the following result.

Theorem 20.23 *Let V be a finite-dimensional real inner product space and let $f : V \to V$ be a skew-adjoint transformation. If p is an irreducible factor of the minimum polynomial of f then either $p = X$ or $p = X^2 + c^2$ for some $c \neq 0$.*

Proof As skew-adjoint transformations are normal, it follows by Theorem 20.22 that the minimum polynomial of f is of the form $\prod_{i=1}^{k} p_i$ where $p_1, \ldots, p_k$ are distinct irreducible polynomials. We also know that either p_i is linear or p_i is of the form

$$X^2 - 2a_i X + (a_i^2 + b_i^2)$$

where $b_i \neq 0$.

Suppose that p_i is not linear and let $M_i = \operatorname{Ker} p_i(f)$ be the corresponding primary component of f. If f_i denotes the restriction of f to M_i then, by the primary decomposition theorem, the minimum polynomial of f_i is p_i and so

$$0 = p_i(f) = f_i^2 - 2a_i f + (a_i^2 + b_i^2) \operatorname{id}_{M_i}.$$

Since f is skew-symmetric, so also is f_i and consequently we have

$$0 = f_i^2 - 2a_i f_i^* + (a_i^2 + b_i^2)\operatorname{id} = f_i^2 + 2a_i f + (a_i^2 + b_i^2)\operatorname{id}.$$

These equalities give $4a_i f_i = 0$ whence $a_i = 0$; for otherwise $f_i = 0$ whence the minimum polynomial of f_i is $p_i = X$, contradicting the hypothesis. Thus we see that p_i reduces to

$$p_i = X^2 + b_i^2$$

where $b_i^2 > 0$ since $b_i \neq 0$.

Suppose now that p_i is linear, say $p_i = X - a_i$. Then we have $f_i = a_i \operatorname{id}$ and consequently $f_i^* = a_i \operatorname{id} = f_i$. But f_i is skew-adjoint, so $f_i^* = -f_i$. It follows that $0 = f_i = a_i \operatorname{id}$ whence $a_i = 0$ and so $p_i = X$. ◊

Corollary *If f is skew-adjoint then the minimum polynomial of f is given as follows :*

(1) *if $f = 0$ then $m_f = X$;*

(2) *if f is invertible then*
$$m_f = (X^2 + c_1^2)(X^2 + c_2^2) \cdots (X^2 + c_k^2)$$
for distinct real numbers $c_1, \ldots, c_k$;
(3) *if f is neither zero nor invertible then*
$$m_f = (X^2 + c_1^2)(X^2 + c_2^2) \cdots (X^2 + c_k^2)X$$
for distince real numbers $c_1, \ldots, c_k$.

Proof This is immediate from Theorems 20.22, 20.23, and Corollary 3 of Theorem 19.17 on noting that, since the characteristic and minimum polynomials have the same zeros, the constant term in the characteristic polynomial is non-zero if and only if the constant term in the minimum polynomial is non-zero. ◊

Concerning the primary components of a skew-adjoint transformation, we shall require the following result.

Theorem 20.24 *If V is a non-zero finite-dimensional real inner product space and if $f : V \to V$ is a skew-adjoint transformation then the primary components of f are pairwise orthogonal.*

Proof Let M_i, M_j be primary components of f with $i \neq j$. If f_i, f_j are respectively the transformations induced on M_i, M_j by the restrictions of f to M_i, M_j suppose first that the minimum polynomials of f_i, f_j are $X^2 + c_i^2, X^2 + c_j^2$ where $c_i, c_j \neq 0$ and $c_i^2 \neq c_j^2$. Then for all $x_i \in M_i$ and $x_j \in M_j$ we have

$$\begin{aligned}
0 &= \langle (f_i^2 + c_i^2 \mathrm{id})(x_i) \mid x_j \rangle \\
&= \langle f^2(x_i) \mid x_j \rangle + c_i^2 \langle x_i \mid x_j \rangle \\
&= \langle f(x_i) \mid -f(x_j) \rangle + c_i^2 \langle x_i \mid x_j \rangle \\
&= \langle x_i \mid f^2(x_j) \rangle + c_i^2 \langle x_i \mid x_j \rangle \\
&= \langle x_i \mid f_j^2(x_j) \rangle + c_i^2 \langle x_i \mid x_j \rangle \\
&= \langle x_i \mid -c_j^2 x_j \rangle + c_i^2 \langle x_i \mid x_j \rangle \\
&= (c_i^2 - c_j^2)\langle x_i \mid x_j \rangle.
\end{aligned}$$

Since $c_i^2 \neq c_j^2$ we deduce that $\langle x_i \mid x_j \rangle = 0$.

Suppose now that the minimum polynomial of f_i is $X^2 + c_i^2$ with $c_i \neq 0$ and that of f_j is X. Replacing $f_j^2(x_j)$ by 0 in the above string of equalities, we obtain $0 = c_i^2 \langle x_i \mid x_j \rangle$ whence again $\langle x_i \mid x_j \rangle = 0$. ◊

In order to establish our main result on skew-adjoint transformations, we require the following general result.

Theorem 20.25 *Let W be a subspace of a non-zero finite-dimensional inner product space V and let $f : V \to V$ be a linear transformation. Then W is f-stable if and only if $W^\perp$ is f^*-stable.*

Proof By Theorem 11.12 we have $V = W \oplus W^\perp$. If W is f-stable then

$$(\forall x \in W)(\forall y \in W^\perp) \qquad \langle x \mid f^*(y) \rangle = \langle f(x) \mid y \rangle = 0$$

whence we see that $f^*(y) \in W^\perp$ for all $y \in W^\perp$, so that $W^\perp$ is f^*-stable. Applying this observation again, we obtain the converse; for if $W^\perp$ is f^*-stable then $W = W^{\perp\perp}$ is $f^{**} = f$-stable. $\Diamond$

Theorem 20.26 *Let V be a finite-dimensional real inner product space and let $f : V \to V$ be a skew-adjoint transformation whose minimum polynomial is $X^2 + b^2$ where $b \neq 0$. Then $\dim V$ is even and V is an ortho-direct sum of f-cyclic subspaces each of dimension 2. Moreover, there is an ordered orthonormal basis of V with respect to which the matrix of f is*

$$M[b] = \begin{bmatrix} 0 & -b & & & & & \\ b & 0 & & & & & \\ & & 0 & -b & & & \\ & & b & 0 & & & \\ & & & & \ddots & & \\ & & & & & 0 & -b \\ & & & & & b & 0 \end{bmatrix}.$$

Proof Let y be a non-zero element of V. Observe first that $f(y) \neq \lambda y$ for any λ; for otherwise, since $f^2(y) = -b^2 y$, we would have $\lambda^2 = -b^2$ and hence the contradiction $b = 0$. Let W_1 be the smallest f-stable subspace containing y. Since $f^2(y) = -b^2 y$, we see that W_1 is f-cyclic of dimension 2, a cyclic basis for W_1 being $\{y, f(y)\}$. Consider now the decomposition $V = W_1 \oplus W_1^\perp$. This direct sum is orthogonal. By Theorem 20.25, $W_1^\perp$ is f^*-stable and so, since $f^* = -f$, we see that $W_1^\perp$ is also f-stable, of dimension $\dim V - 2$. Now let $V_1 = W_1^\perp$ and repeat the argument to obtain an orthogonal direct sum $V_1 = W_2 \oplus W_2^\perp$ of f-stable subspaces with W_2 f-cyclic of dimension 2. Continuing in this manner, we note that it is not possible in the final such decomposition to have $\dim W_k^\perp = 1$. For, if this were so then $W_k^\perp$ would have a singleton basis $\{z\}$ whence $f(z) \notin W_k^\perp$, a contradiction. Thus $W_k^\perp$ is also of dimension 2. It follows that $\dim V$ is even.

Diagonalisation; normal transformations

We now construct an orthonormal basis for each of the f-cyclic subspaces W_i in the ortho-direct sum representation $V = \bigoplus_{i=1}^{k} W_i$. Consider the basis $\{y_i, f(y_i)\}$ of W_i. Since $\langle y_i \mid f(y_i)\rangle = 0$ we obtain, via the Gram-Schmidt orthonormalisation process, an ordered orthonormal basis

$$B_i = \left\{ \frac{y_i}{\|y_i\|}, \frac{f(y_i)}{\|f(y_i)\|} \right\}$$

for W_i. Now

$$\|f(y_i)\|^2 = \langle f(y_i) \mid -f^*(y_i)\rangle = -\langle f^2(y_i) \mid y_i\rangle = b^2 \|y_i\|^2$$

and so this orthonormal basis is

$$B_i = \left\{ \frac{y_i}{\|y_i\|}, \frac{f(y_i)}{b\|y_i\|} \right\}.$$

Since now

$$f\left(\frac{y_i}{\|y_i\|}\right) = 0 \frac{y_i}{\|y_i\|} + b \frac{f(y_i)}{b\|y_i\|};$$

$$f\left(\frac{f(y_i)}{b\|y_i\|}\right) = -b \frac{y_i}{\|y_i\|} + 0 \frac{f(y_i)}{b\|y_i\|},$$

it follows that the matrix of f relative to B_i is

$$\begin{bmatrix} 0 & -b \\ b & 0 \end{bmatrix}.$$

It is now clear that $\bigcup_{i=1}^{k} B_i$ is an ordered orthonormal basis of V with respect to which the matrix of f is of the form stated. ◊

Corollary 1 *If V is a non-zero finite-dimensional real inner product space an if $f : V \to V$ is a skew-adjoint transformation then there is an ordered orthonormal basis of V with respect to which the matrix of f is of the form*

$$\begin{bmatrix} M[c_1] & & & \\ & M[c_2] & & \\ & & \ddots & \\ & & & M[c_n] \end{bmatrix}$$

where $c_1, \ldots, c_n$ are distinct positive real numbers and $M[c_i]$ is either a zero matrix or a matrix as illustrated in Theorem 20.26.

Proof It suffices to combine the Corollary to Theorem 20.23 with Theorems 20.24 and 20.26. ◊

Corollary 2 *A real square matrix is skew-symmetric if and only if it is orthogonally similar to a matrix of the form given in Corollary 1.* ◊

Let us now turn to the general problem of a normal transformation on a real inner product space. Recall by Theorem 20.20 that such a transformation f can be expressed uniquely as $f = g + h$ where g is self-adjoint and h is skew-adjoint. Moreover, by Theorem 20.14, g is ortho-diagonalisable.

Theorem 20.27 *Let V be a finite-dimensional real inner product space and let $f : V \to V$ be a normal transformation whose minimum polynomial is the irreducible quadratic*

$$m_f = X^2 - 2aX + (a^2 + b^2) \qquad (b \neq 0).$$

If g, h are respectively the self-adjoint and skew-adjoint parts of f then
  (1) *h is invertible;*
  (2) *$m_g = X - a$;*
  (3) *$m_h = X^2 + b^2$.*

Proof (1) Suppose, by way of obtaining a contradiction, that $\operatorname{Ker} h \neq \{0\}$. Since f is normal, we have $g \circ h = h \circ g$ by Theorem 20.20. It follows from this that $\operatorname{Ker} h$ is g-stable. Since $f = g + h$, the restriction of f to $\operatorname{Ker} h$ coincides with that of g. As $\operatorname{Ker} h$ is g-stable, we can therefore define a linear transformation $f' : \operatorname{Ker} h \to \operatorname{Ker} h$ by the prescription

$$f'(x) = f(x) = g(x).$$

Since g is self-adjoint so is f'. By Theorem 20.14, f' is then ortho-diagonalisable and so its minimum polynomial is a product of distinct linear factors. But $m_{f'}$ must divide m_f which, by the hypothesis, is irreducible. This contradiction therefore gives $\operatorname{Ker} h = \{0\}$ whence h is invertible.

(2) Since $f = g + h$ with $g^\star = g$ and $h^\star = -h$ we have $f^\star = g - h$, whence $f^2 - 2af + (a^2 + b^2)\operatorname{id}_V = 0$ and $(f^\star)^2 - 2af^\star + (a^2 + b^2)\operatorname{id}_V = 0$ and consequently $f^2 - (f^\star)^2 = 2a(f - f^\star) = 4ah$. Thus, since f commutes with $f^\star$, we see that

$$g \circ h = \tfrac{1}{2}(f + f^\star) \circ \tfrac{1}{2}(f - f^\star) = \tfrac{1}{4}[f^2 - (f^\star)^2] = ah$$

and so $(g - a\operatorname{id}_V) \circ h = 0$. Since h is invertible by (1), we then have that $g - a\operatorname{id}_V = 0$ whence $m_g = X - a$.

(3) Since $f - h = g = a\operatorname{id}_V$ we have $f = h + a\operatorname{id}_V$ and so

$$\begin{aligned} 0 &= f^2 - 2af + (a^2 + b^2)\operatorname{id}_V \\ &= (h + a\operatorname{id}_V)^2 - 2a(h + a\operatorname{id}_V) + (a^2 + b^2)\operatorname{id}_V \\ &= h^2 + b^2\operatorname{id}_V. \end{aligned}$$

Diagonalisation; normal transformations

Now h is skew-adjoint and by (1) is invertible. It therefore follows by the Corollary to Theorem 20.23 that $m_h = X^2 + b^2$. ◊

We now extend Theorem 20.24 to normal transformations.

Theorem 20.28 *If V is a non-zero finite-dimensional real inner product space and if $f : V \to V$ is a normal transformation then the primary components of f are pairwise orthogonal.*

Proof By Theorem 20.22, the minimum polynomial of f has the general form

$$m_f = (X - a_0) \prod_{i=1}^{k}(X^2 - 2a_i X + a_i^2 + b_i^2)$$

where each $b_i \neq 0$. By Theorem 19.6, the primary components of f are

$$M_0 = \text{Ker}(f - a_0 \text{id}_V),$$
$$(i = 1,\ldots,k) \quad M_i = \text{Ker}[f^2 - 2a_i f + (a_i^2 + b_i^2)\text{id}_V];$$

moreover, the restriction of f to M_i for $i = 0,\ldots,k$ induces a normal transformation $f_i : M_i \to M_i$ the minimum polynomial of which is $X - a_0$ if $i = 0$ and $X^2 - 2a_i X + a_i^2 + b_i^2$ otherwise. Note that each f_i is normal so that $f_i = g_i + h_i$ where g_i is self-adjoint and h_i is skew-adjoint. Now g_i and h_i coincide with the mappings induced on M_i by g and h where g is the self-adjoint part of f and h is the skew-adjoint part of f. To see this, let these induced mappings be g', h' respectively. Then for every $x \in M_i$ we have

$$g_i(x) + h_i(x) = f_i(x) = f(x) = g(x) + h(x)$$

and so $g_i - g' = h' - h_i$. Since the left-hand side is self-adjoint and the right-hand side is skew-adjoint, we deduce that $g_i = g'$ and $h_i = h'$.

Suppose now that $i, j > 0$ with $i \neq j$. Then $m_{f_i} = X^2 - 2a_i X + a_i^2 + b_i^2$ and $m_{f_j} = X^2 - 2a_j X + a_j^2 + b_j^2$, where either $a_i \neq a_j$ or $b_i^2 \neq b_j^2$. By Theorem 20.22, we have

$$m_{g_i} = X - a_i, \quad m_{g_j} = X - a_j, \quad m_{h_i} = X^2 + b_i^2, \quad m_{h_j} = X^2 + b_j^2.$$

Given $x_i \in M_i$ and $x_j \in M_j$, we therefore have

$$\begin{aligned}
0 = \langle (h_j^2 + b_j^2 \text{id}_{V_i})(x_i) \mid x_j \rangle &= \langle h^2(x_i) \mid x_j \rangle + b_j^2 \langle x_i \mid x_j \rangle \\
&= \langle x_i \mid h^2(x_j) \rangle + b_j^2 \langle x_i \mid x_j \rangle \\
&= \langle x_i \mid h_j^2(x_j) \rangle + b_j^2 \langle x_i \mid x_j \rangle \\
&= -b_j^2 \langle x_i \mid x_j \rangle + b_j^2 \langle x_i \mid x_j \rangle \\
&= (b_i^2 - b_j^2)\langle x_i \mid x_j \rangle,
\end{aligned}$$

so that in the case where $b_i^2 \neq b_j^2$ we have $\langle x_i | x_j \rangle = 0$. Likewise,

$$\begin{aligned}
0 = \langle (g_i - a_i \mathrm{id}_V)(x_i) | x_j \rangle &= \langle g(x_i) | x_j \rangle + a_i \langle x_i | x_j \rangle \\
&= \langle x_i | g(x_j) \rangle + a_i \langle x_i | x_j \rangle \\
&= \langle x_i | g_j(x_j) \rangle + a_i \langle x_i | x_j \rangle \\
&= a_j \langle x_i | x_j \rangle + a_i \langle x_i | x_j \rangle \\
&= (a_j - a_i) \langle x_i | x_j \rangle,
\end{aligned}$$

so that in the case where $a_i \neq a_j$ we have $\langle a_i | a_j \rangle = 0$. We thus see that $M_1, \ldots, M_k$ are pairwise orthogonal. That M_0 is orthogonal to each M_i with $i \geq 1$ follows from the above strings of equalities on taking $j = 0$ and using the fact that $f_0 = a_0 \mathrm{id}_V$ is self-adjoint and therefore $g_0 = f_0$ and $h_0 = 0$. $\diamond$

We can now establish a canonical form for real normal matrices.

Theorem 20.29 *If V is a non-zero finite-dimensional real inner product space and $f : V \to V$ is a normal transformation then there is an ordered orthonormal basis of V relative to which the matrix of f is of the form*

$$\begin{bmatrix} A_1 & & & \\ & A_2 & & \\ & & \ddots & \\ & & & A_k \end{bmatrix}$$

where each A_i is either a 1×1 matrix, or a 2×2 matrix of the form

$$\begin{bmatrix} \alpha & -\beta \\ \beta & \alpha \end{bmatrix}$$

in which $\beta \neq 0$.

Proof With the same notation as used above, let

$$m_f = (X - a_0) \prod_{i=1}^{k} (X^2 - 2a_i X + a_i^2 + b_i^2)$$

and let the primary components of f be M_i for $i = 0, \ldots, k$. Then

$$m_{f_i} = \begin{cases} X - a_0 & \text{if } i = 0; \\ X^2 - 2a_i X + a_i^2 + b_i^2 & \text{otherwise.} \end{cases}$$

Given any M_i with $i \neq 0$, we have $f_i = g_i + h_i$ where the self-adjoint part g_i is such that $m_{g_i} = X - a_i$, and the skew-adjoint part h_i is such that

$m_{h_i} = X^2 + b_i^2$. By Theorem 20.26, there is an ordered orthonormal basis B_i of M_i with respect to which the matrix of h_i is

$$M[b_i] = \begin{bmatrix} 0 & -b_i & & & & & \\ b_i & 0 & & & & & \\ & & 0 & -b_i & & & \\ & & b_i & 0 & & & \\ & & & & \ddots & & \\ & & & & & 0 & -b_i \\ & & & & & b_i & 0 \end{bmatrix}.$$

Since the minimum polynomial of g_i is $X - a_i$, we have $g_i(x) = a_i x$ for every $x \in B_i$ and so the matrix of g_i relative to B_i is the diagonal matrix all of whose entries are a_i. It now follows that the matrix of $f_i = g_i + h_i$ relative to B_i is

$$M[a_i, b_i] = \begin{bmatrix} a_i & -b_i & & & & & \\ b_i & a_i & & & & & \\ & & a_i & -b_i & & & \\ & & b_i & a_i & & & \\ & & & & \ddots & & \\ & & & & & a_i & -b_i \\ & & & & & b_i & a_i \end{bmatrix}.$$

In the case where $i = 0$, we have $f_0 = a_0 \text{id}_V$ so f is self-adjoint, hence ortho-diagonalisable. There is therefore an ordered orthonormal basis with repect to which the matrix of f_0 is diagonal.

Now by Theorem 20.28 the primary components of f are pairwise orthogonal. Stringing together the above ordered orthonormal bases for $M_0, \ldots, M_k$ we therefore obtain an ordered orthonormal basis for V with respect to which the matrix of f is of the form stated. ◊

Corollary *A real square matrix is normal if and only if it is orthogonally similar to a matrix of the form described in Theorem* 20.29. ◊

Let us now turn our attention to orthogonal transformations on real inner product spaces. Recall that f is orthogonal if f^{-1} exists and is f^*. An orthogonal transformation is therefore in particular a normal transformation. So our labours produce a bonus : we can use the above result to determine a canonical form for orthogonal matrices.

Theorem 20.30 *If V is a non-zero finite-dimensional real inner product space and $f : V \to V$ is an orthogonal transformation then there is an* ordered orthonormal basis of V with respect to which the matrix of f is of the form

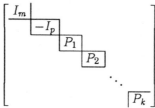

in which each P_i is a 2×2 matrix of the form

$$\begin{bmatrix} \alpha & -\beta \\ \beta & \alpha \end{bmatrix}$$

where $\beta \neq 0$ and $\alpha^2 + \beta^2 = 1$.

Proof With the same notation as in the above, the matrix $M(a_i, b_i)$ that represents f_i relative to the ordered orthonormal basis B_i is an orthogonal matrix (since f_i is orthogonal). Multiplying this matrix by its transpose, we obtain an identity matrix and, equating the entries, we see that $a_i^2 + b_i^2 = 1$. As for the primary component M_0, the matrix of f_0 is diagonal. Since the square of this diagonal matrix must then be an identity matrix, its entries have to be ± 1. We can now rearrange the basis to obtain a matrix of the required form. ◊

The results of this chapter have applications to a variety of problems that occur in other areas of mathematics. By way of illustration, we mention that orthogonal transformations have applications in the study of finite symmetry groups, and the Jordan decomposition theorem is useful in obtaining solutions of systems of simultaneous first order linear differential equations with constant coefficients. Whilst these and other applications are largely another story, we shall end this chapter with a brief description of an application to the study of *quadratic forms*. These in turn have important applications in number theory and in cartesian geometry. In the course of our discussion, we shall shed light on another equivalence relation on square matrices relative to which we shall seek a useful canonical form.

Suppose then that V is a vector space of dimension n over a field F and let $f : V \times V \to F$ be a bilinear form on V. If $(e_i)_n$ is an ordered basis of V then by the *matrix of the bilinear form* f relative to $(e_i)_n$ we shall mean the matrix $A = [a_{ij}]_{n \times n}$ given by

$$a_{ij} = f(e_i, e_j).$$

Diagonalisation; normal transformations

If $x = \sum_{i=1}^{n} x_i e_i$ and $y = \sum_{i=1}^{n} y_i e_i$ then, by the bilinearity of f, we have

$$f(x,y) = \sum_{i=1}^{n}\sum_{j=1}^{n} x_i y_j f(e_I, e_j) = \sum_{i,j=1}^{n} x_i y_j a_{ij}. \qquad (2)$$

Conversely, given any $n \times n$ matrix $A = [a_{ij}]$ over F, it is clear that (2) defines a bilinear form f on V whose matrix relative to $(e_i)_n$ is A.

- In what follows we shall commit the usual abuse of identifying a scalar λ with the 1×1 matrix $[\lambda]$. In this way, we can write (1) as

$$f(x,y) = [x_1 \ \ldots \ x_n] A \begin{bmatrix} y_1 \\ \vdots \\ y_n \end{bmatrix} = \mathbf{x}^t A \mathbf{y}.$$

Example 20.1 Let $f : F^n \times F^n \to F$ be given as follows : for $x = (x_1, \ldots, x_n)$ and $y = (y_1, \ldots, y_n)$ let $f(x,y) = \sum_{i=1}^{n} x_i y_i$. It is readily seen that f is bilinear; in fact,

$$f(x,y) = [x_1 \ \ldots \ x_n] \begin{bmatrix} y_1 \\ \vdots \\ y_n \end{bmatrix}.$$

Let $(e_i)_n$ be the natural ordered basis of F^n. Then the matrix of f relative to $(e_i)_n$ is simply I_n.

Example 20.2 The matrix

$$A = \begin{bmatrix} 0 & 1 & 1 \\ 0 & 0 & 1 \\ 0 & 0 & 0 \end{bmatrix}$$

gives rise to the bilinear form $f : \mathbb{R}^3 \times \mathbb{R}^3 \to \mathbb{R}$ described by

$$f(\mathbf{x}, \mathbf{y}) = x_1(y_2 + y_3) + x_2 y_3.$$

It is natural to ask how the matrix of a bilinear form is affected when we change reference to another ordered basis.

Theorem 20.31 Let V be a vector space of dimension n over a field F. Let $(e_i)_n$ and $(e'_i)_n$ be ordered bases of V. If $f : V \times V \to F$ is a bilinear form on V and if $A = [a_{ij}]_{n \times n}$ is the matrix of f relative to $(e_i)_n$ then the matrix of f relative to $(e'_i)_n$ is $P^t A P$ where

$$P = \text{Mat}[\text{id}_V, (e'_i)_n, (e_i)_n]$$

is the matrix that represents the change of basis from $(e_i)_n$ to $(e'_i)_n$.

Proof We have $e'_j = \sum_{i=1}^{n} p_{ij} e_i$ for $j = 1, \ldots, n$ and so

$$\begin{aligned}
f(e'_i, e'_j) &= f\left(\sum_{t=1}^{n} p_{ti} e_t, \sum_{k=1}^{n} p_{kj} e_k\right) \\
&= \sum_{t=1}^{n} \sum_{k=1}^{n} p_{ti} p_{kj} f(e_t, e_k) \\
&= \sum_{t=1}^{n} \sum_{k=1}^{n} p_{ti} p_{kj} a_{tj} \\
&= \sum_{t=1}^{n} p_{ti} \left(\sum_{k=1}^{n} a_{tk} p_{kj}\right) \\
&= [P^t A P]_{ij},
\end{aligned}$$

from which the result follows. ◇

Definition If A, B are $n \times n$ matrices over a field F then we shall say that B is *congruent* to A if there is an invertible matrix P over F such that $B = P^t A P$.

It is clear that the relation of being congruent is an equivalence relation on $\text{Mat}_{n \times n}(F)$.

Definition A bilinear form $f : V \times V \to F$ is said to be *symmetric* if

$$(\forall x, y \in V) \quad f(x, y) = f(y, x).$$

It is clear that the matrix of a symmetric bilinear form is itself symmetric, and conversely that every symmetric matrix yields a symmetric bilinear form.

Definition Let V be a vector space over a field F. Then a mapping $Q : V \to F$ is called a *quadratic form* on V if there exists a symmetric bilinear form $f : V \times V \to F$ such that

$$(\forall x \in V) \quad Q(x) = f(x, x).$$

In what follows, we shall restrict our attention to the real field $\mathbb{R}$. Our reason for so doing is, quite apart from the fact that most applications involve $\mathbb{R}$, that certain difficulties have to be avoided when the

Diagonalisation; normal transformations

ground field is of characteristic 2. Indeed, instead of working with IR we could work with any field that is not of characteristic 2; but in what we shall discuss no great advantage would be gained in so doing.

Given a symmetric bilinear form $f : V \times V \to$ IR, we shall denote by $Q_f : V \to$ IR the quadratic form given by

$$(\forall x \in V) \quad Q_f(x) = f(x,x).$$

Theorem 20.32 *Let V be a vector space over IR. If $f : V \times V \to$ IR is a symmetric bilinear form then*

(1) $(\forall \lambda \in$ IR$)(\forall x \in V)$ $\quad Q_f(\lambda x) = \lambda^2 Q_f(x);$
(2) $(\forall x, y \in V)$ $\quad f(x,y) = \frac{1}{2}[Q_f(x+y) - Q_f(x) - Q_f(y)];$
(3) $(\forall x, y \in V)$ $\quad f(x,y) = \frac{1}{4}[Q_f(x+y) - Q_f(x-y)].$

Proof (1) : $Q_f(\lambda x) = f(\lambda x, \lambda x) = \lambda^2 f(x,x) = \lambda^2 Q_f(x)$.
(2) : Since f is symmetric we have

$$Q_f(x+y) = f(x+y, x+y) = f(x,x) + f(x,y) + f(y,x) + f(y,y)$$
$$= Q_f(x) + 2f(x,y) + Q_f(y),$$

whence (2) follows.
(3) : By (1) we have $Q_f(-x) = Q_f(x)$ so that, by (2),

$$Q_f(x-y) = Q_f(x) - 2f(x,y) + Q_f(y)$$

and consequently

$$Q_f(x+y) - Q_f(x-y) = 4f(x,y). \quad \Diamond$$

Corollary *A real quadratic form on V is associated with a uniquely determined symmetric bilinear form.*

Proof Suppose that $Q : V \to$ IR is a quadratic form that is associated with both the symmetric bilinear forms $f, g : V \times V \to$ IR. Then clearly $Q = Q_f = Q_g$ and so, by Theorem 20.32(2), we deduce that $f = g$. $\Diamond$

Example 20.3 The bilinear form of Example 20.1 is clearly symmetric. In the case where $F =$ IR the associated quadratic form is given by

$$(\forall x \in V) \quad Q_f(x) = f(x,x) = \sum_{i=1}^{n} x_i^2.$$

Theorem 20.33 *Let V be a vector space of dimension n over IR and let $(e_i)_n$ be an ordered basis of V. If $f : V \times V \to$ IR is a symmetric bilinear* form on V, if Q_f is the associated quadratic form, and if $A = [a_{ij}]_{n \times n}$ is the (symmetric) matrix of f relative to $(e_i)_n$, then for all $x = \sum_{i=1}^{n} x_i e_i \in V$ we have

$$Q_f(x) = \sum_{i,j=1}^{n} x_i x_j a_{ij}.$$

Conversely, if $A = [a_{ij}]_{n \times n}$ is a real symmetric matrix then the above prescription defines a real quadratic form Q_f on V such that A is the matrix of the associated symmetric bilinear form relative to the ordered basis $(e_i)_n$.

Proof The first part is clear since $Q_f(x) = f(x, x)$. As for the converse, we note that the mapping $f : V \times V \to \mathbb{R}$ given by

$$f(x, y) = \sum_{i,j=1}^{n} x_i y_j a_{ij}$$

is (as is readily verified) symmetric and bilinear with $f(e_i, e_j) = a_{ij}$. The associated quadratic form is precisely Q_f. ◊

- By the *matrix of a real quadratic form* we shall mean, by an abuse of language, the matrix of the associated symmetric bilinear form.

Example 20.4 The mapping $Q : \mathbb{R}^2 \to \mathbb{R}$ given by

$$Q(x, y) = 4x^2 + 6xy + 9y^2 = [x \ y] \begin{bmatrix} 4 & 3 \\ 3 & 9 \end{bmatrix} \begin{bmatrix} x \\ y \end{bmatrix}$$

is a quadratic form on $\mathbb{R}^2$. The associated symmetric bilinear form is the mapping $f : \mathbb{R}^2 \times \mathbb{R}^2 \to \mathbb{R}$ given by

$$\begin{aligned} f((x, y), (x', y')) &= \tfrac{1}{2}[Q(x + x', y + y') - Q(x, y) - Q(x', y')] \\ &= 4xx' + 3(xy' + x'y) + 9yy'. \end{aligned}$$

As we have seen in Theorem 20.33, for every quadratic form $Q : V \to \mathbb{R}$ with associated matrix A relative to an ordered basis $(e_i)_n$ of V,

$$Q(x) = \sum_{i,j=1}^{n} x_i x_j a_{ij} = [x_1 \ \ldots \ x_n] A \begin{bmatrix} x_1 \\ \vdots \\ x_n \end{bmatrix}.$$

Our aim now is to determine a canonical form for real symmetric matrices under congruence. This will allow us to obtain a simpler formula for

Diagonalisation; normal transformations 351

$Q(x)$ (relative, of course, to a specific ordered basis). This we achieve by means of the following result, at the heart of whose proof lie the facts that a real symmetric matrix is ortho-diagonalisable (Corollary to Theorem 20.14) and that its eigenvalues are all real (Corollary to Theorem 20.13).

Theorem 20.34 *If A is a real $n \times n$ symmetric matrix then A is congruent to a unique matrix of the form*

$$\begin{bmatrix} I_r & & \\ & -I_s & \\ & & 0 \end{bmatrix}.$$

Proof Since A is real symmetric it is ortho-diagonalisable and its eigenvalues are all real. Let the positive eigenvalues be $\lambda_1, \ldots, \lambda_r$ and let the negative eigenvalues be $-\lambda_{s+1}, \ldots, -\lambda_{r+s}$. Then there is a real orthogonal matrix P such that

$$P^t A P = P^{-1} A P = \mathrm{diag}\{\lambda_1, \ldots, \lambda_r, -\lambda_{r+1}, \ldots, -\lambda_{r+s}, 0, \ldots, 0\}.$$

Let N be the $n \times n$ diagonal matrix whose diagonal entries are

$$n_{ii} = \begin{cases} \dfrac{1}{\sqrt{\lambda_i}} & \text{if } i = 1, \ldots, r+s; \\ 1 & \text{otherwise.} \end{cases}$$

Then it is readily seen that

$$N^t P^t A P N = \begin{bmatrix} I_r & & \\ & -I_s & \\ & & 0 \end{bmatrix}.$$

Since P and N are invertible, so also is PN; and since $N^t P^t = (PN)^t$, it follows that A is congruent to a matrix of the stated form.

As for uniqueness, it suffices to suppose that

$$L = \begin{bmatrix} I_r & & \\ & -I_s & \\ & & 0_{n-(r+s)} \end{bmatrix}, \quad M = \begin{bmatrix} I_{r'} & & \\ & -I_{s'} & \\ & & 0_{n-(r'+s')} \end{bmatrix}$$

are congruent and show that $r = r'$ and $s = s'$. Now if L and M are congruent then they are certainly equivalent; and since equivalent matrices have the same rank, we deduce that

$$r + s = \mathrm{rank}\, L = \mathrm{rank}\, M = r' + s'.$$

Suppose, by way of obtaining a contradiction, that $r < r'$ (so that, by the above, $s > s'$). Let W be the real vector space $\mathrm{Mat}_{n \times 1}(\mathbb{R})$. It is clear that W is an inner product space relative to the mapping described by

$$(\mathbf{x}, \mathbf{y}) \mapsto \langle \mathbf{x} | \mathbf{y} \rangle = \mathbf{x}^t \mathbf{y}.$$

Consider the mapping $f_L : W \to W$ given by

$$f_L(\mathbf{x}) = L\mathbf{x}.$$

If $\mathbf{x} = [x_1 \ \ldots \ x_n]^t$ and $\mathbf{y} = [y_1 \ \ldots \ y_n]^t$, we have

$$\langle \mathbf{x} | f_L(\mathbf{y}) \rangle = \mathbf{x}^t L \mathbf{y}$$
$$= x_1 y_1 + \ldots + x_r y_r - x_{r+1} y_{r+1} - \ldots - x_{r+s} y_{r+s}.$$

Now

$$\langle f_L(\mathbf{x}) | \mathbf{y} \rangle = \langle L\mathbf{x} | \mathbf{y} \rangle = (L\mathbf{x})^t \mathbf{y} = \mathbf{x}^t L^t \mathbf{y} = \mathbf{x} L \mathbf{y}$$

and consequently we see that f_L is self-adjoint. Similarly, so is $f_M : W \to W$ given by $f_M(\mathbf{x}) = M\mathbf{x}$. Consider now the subspaces

$$X = \{ \mathbf{x} \in W \ ; \ x_1 = \ldots = x_r = 0, \ x_{r+s+1} = \ldots = x_n = 0 \};$$
$$Y = \{ \mathbf{x} \in W \ ; \ x_{r'+1} = \ldots = x_{r'+s'} = 0 \}.$$

Clearly, X is of dimension s and for every non-zero $\mathbf{x} \in X$ we have

$$\langle f_L(\mathbf{x}) | \mathbf{x} \rangle = \mathbf{x}^t L \mathbf{x} = -x_{r+1}^2 - \ldots - x_{r+s}^2 < 0. \tag{3}$$

Also, Y is of dimension $n - s'$ and for $\mathbf{x} \in Y$ we have

$$\langle f_M(\mathbf{x}) | \mathbf{x} \rangle = \mathbf{x}^t M \mathbf{x} = x_1^2 + \ldots + x_r^2 \geq 0.$$

Now since L and M are congruent there is an invertible matrix P such that $M = P^t L P$. Since, for all $\mathbf{x} \in Y$,

$$0 \leq \langle f_M(\mathbf{x}) | \mathbf{x} \rangle = \langle M\mathbf{x} | \mathbf{x} \rangle$$
$$= \langle P^t L P \mathbf{x} | \mathbf{x} \rangle$$
$$= \langle (f_{P^t} \circ f_L \circ f_P)(\mathbf{x}) | \mathbf{x} \rangle$$
$$= \langle (f_L \circ f_P)(\mathbf{x}) | f_P(\mathbf{x}) \rangle,$$

we see that

$$(\forall \mathbf{y} \in f_P^{\rightarrow}(Y)) \qquad \langle f_L(\mathbf{y}) | \mathbf{y} \rangle \geq 0. \tag{4}$$

Now since f_P is an isomorphism we have

$$\dim f_P^{\rightarrow}(Y) = \dim Y = n - s'$$

Diagonalisation; normal transformations

and so

$$\dim f_{\vec{P}}(Y) + \dim X = n - s' + s > n \geq \dim[f_{\vec{P}}(Y) + X]$$

and consequently, by Corollary 1 of Theorem 8.10,

$$\dim[f_{\vec{P}}(Y) \cap X] > 0.$$

Suppose then that $\mathbf{z}$ is a non-zero element of $f_{\vec{P}}(Y) \cap X$. Then from (3) we see that $\langle f_L(\mathbf{z}|\mathbf{z}\rangle$ is negative; whereas from (4) we see that $\langle f_L(\mathbf{z}|\mathbf{z}\rangle$ is non-negative. This contradiction shows that we cannot have $r < r'$. In a similar way we cannot have $r' < r$. We conclude therefore that $r = r'$ whence also $s = s'$. ◊

Corollary [Sylvester's law of inertia] *Let V be a vector space of dimension n over $\mathbb{R}$ and let $Q : V \to \mathbb{R}$ be a quadratic form on V. Then there is an ordered basis $(e_i)_n$ of V such that if $x = \sum_{i=1}^{n} x_i e_i$ then*

$$Q(x) = x_1^2 + \ldots + x_r^2 - x_{r+1}^2 - \ldots - x_{r+s}^2.$$

Moreover, the integers r and s are independent of such a basis. ◊

- The integers $r + s$ and $r - s$ are called the *rank* and *signature* of the quadratic form Q.

Example 20.5 Consider the quadratic form $Q : \mathbb{R}^3 \to \mathbb{R}$ given by

$$Q(x, y, z) = x^2 - 2xy + 4yz - 2y^2 + 4z^2.$$

By the process of completing the squares it is readily seen that

$$Q(x, y, z) = (x - y)^2 - 4y^2 + (y + 2z)^2$$

which is in canonical form, of rank 3 and signature 1. Alternatively, we can use matrices. The matrix of Q is

$$A = \begin{bmatrix} 1 & -1 & 0 \\ -1 & -2 & 2 \\ 0 & 2 & 4 \end{bmatrix}.$$

Let P be the orthogonal matrix such that $P^t A P$ is the diagonal matrix D. If $\mathbf{y} = P^t \mathbf{x}$ (so that $\mathbf{x} = P\mathbf{y}$) then

$$\mathbf{x}^t A \mathbf{x} = (P\mathbf{y})^t A P \mathbf{y} = \mathbf{y}^t P^t A P \mathbf{y} = \mathbf{y}^t D \mathbf{y},$$

where the right-hand side is of the form $X^2 - 4Y^2 + Z^2$.

Example 20.6 The quadratic form given by
$$Q(x,y,z) = 2xy + 2yz$$
can be reduced to canonical form either by the method of completing squares or by a matrix reduction. The former is not so easy in this case, but can be achieved as follows. Define
$$\sqrt{2}x = X+Y, \quad \sqrt{2}y = X-Y, \quad \sqrt{2}z = Z.$$
Then the form becomes
$$X^2 - Y^2 + (X-Y)Z = (X + \tfrac{1}{2}Z)^2 - (Y + \tfrac{1}{2}Z)^2$$
$$= \tfrac{1}{2}(x+y+z)^2 - \tfrac{1}{2}(x-y+z)^2,$$
which is of rank 2 and signature 0.

Definition A quadratic form Q is said to be *positive definite* if $Q(x) > 0$ for all non-zero x.

By taking the inner product space V to be $\text{Mat}_{n\times 1}(\mathbb{R})$ under $\langle \mathbf{x}|\mathbf{y}\rangle = \mathbf{x}^t\mathbf{y}$, we see that a quadratic form Q on V is positive definite if and only if, for all non-zero $\mathbf{x} \in V$,
$$0 < Q(\mathbf{x}) = \mathbf{x}^t A\mathbf{x} = \langle A\mathbf{x}|\mathbf{x}\rangle,$$
which is the case if and only if A is positive definite. It is clear that this situation obtains when there are no negative terms in the canonical form, i.e. when the rank and signature are the same.

Example 20.7 Let $f : \mathbb{R} \times \mathbb{R} \to \mathbb{R}$ be a function whose partial derivatives f_x, f_y are zero at (x_0, y_0). Then the Taylor series at $(x_0 + h, y_0 + h)$ is
$$f(x_0, y_0) + \tfrac{1}{2}[h^2 f_{xx} + 2hk f_{xy} + k^2 f_{yy}](x_0, y_0) + \ldots.$$
For small values of h, k the significant term is this quadratic form in h, k. If it has rank 2 then its normal form is $\pm H^2 \pm K^2$. If both signs are positive (i.e. the form is positive definite) then f has a relative minimum at (x_0, y_0), and if both signs are negative then f has a relative maximum at (x_0, y_0). If one is positive and the other is negative then f has a saddle point at (x_0, y_0). Thus the geometry is distinguished by the signature of the quadratic form.

Example 20.8 Consider the quadratic form
$$4x^2 + 4y^2 + 4z^2 - 2xy - 2yz + 2xz.$$

Its matrix is
$$A = \begin{bmatrix} 4 & -1 & 1 \\ -1 & 4 & -1 \\ 1 & -1 & 4 \end{bmatrix}.$$

The eigenvalues of A are $3, 3, 6$. If P is an orthogonal matrix such that $P^t A P$ is diagonal then, changing coordinates by $\mathbf{X} = P^t \mathbf{x}$, we transform the quadratic form to
$$3X^2 + 3Y^2 + 6Z^2$$
which is positive definite.

EXERCISES

20.1 Let $\mathrm{SL}(2, \mathbb{C})$ be the multiplicative subgroup of $\mathrm{Mat}_{2 \times 2}(\mathbb{C})$ consisting of those 2×2 matrices of determinant 1. If $M \in \mathrm{SL}(2, \mathbb{C})$ and $\mathrm{tr}\, M \notin \{2, -2\}$, prove that there exists $P \in \mathrm{SL}(2, \mathbb{C})$ and $t \in \mathbb{C} \setminus \{0, 1, -1\}$ such that
$$P^{-1} M P = \begin{bmatrix} t & 0 \\ 0 & t^{-1} \end{bmatrix}.$$

If $M \in \mathrm{SL}(2, \mathbb{C})$ and $M \neq I_2$, $\mathrm{tr}\, M = 2$ prove that there exists $P \in \mathrm{SL}(2, \mathbb{C})$ such that
$$P^{-1} M P = \begin{bmatrix} 1 & 1 \\ 0 & 1 \end{bmatrix}.$$

If $M \in \mathrm{SL}(2, \mathbb{C})$ and $M \neq -I_2$, $\mathrm{tr}\, M = -2$ prove that there exists $P \in \mathrm{SL}(2, \mathbb{C})$ such that
$$P^{-1} M P = \begin{bmatrix} -1 & 1 \\ 0 & -1 \end{bmatrix}.$$

20.2 If $A, P \in \mathrm{Mat}_{n \times n}(\mathbb{C})$ with P invertible and if $f \in \mathbb{C}[X]$, prove that
$$f(P^{-1} A P) = P^{-1} f(A) P.$$

If $B \in \mathrm{Mat}_{n \times n}(\mathbb{C})$ is triangular with diagonal entries $t_1, \ldots, t_n$, show that $f(B)$ is triangular with diagonal entries $f(t_1), \ldots, f(t_n)$.

Suppose now that $\lambda_1, \ldots, \lambda_n$ are the eigenvalues of A. Deduce from the above that the eigenvalues of $f(A)$ are $f(\lambda_1), \ldots, f(\lambda_n)$ and that
$$\det f(A) = \prod_{i=1}^n f(\lambda_i), \qquad \mathrm{tr}\, A = \sum_{i=1}^n f(\lambda_i).$$

[*Hint.* Let B be the Jordan form of A.]

20.3 By a *circulant matrix* we mean a square matrix over $\mathbb{C}$ of the form

$$M = \begin{bmatrix} \alpha_1 & \alpha_2 & \alpha_3 & \cdots & \alpha_n \\ \alpha_n & \alpha_1 & \alpha_2 & \cdots & \alpha_{n-1} \\ \alpha_{n-1} & \alpha_n & \alpha_1 & \cdots & \alpha_{n-2} \\ \vdots & \vdots & \vdots & & \vdots \\ \alpha_2 & \alpha_3 & \alpha_4 & \cdots & \alpha_1 \end{bmatrix}.$$

If $f = \alpha_1 + \alpha_2 X + \ldots + \alpha_n X^n$ and M is an $n \times n$ circulant matrix, prove that

$$\det M = \prod_{i=1}^{n} f(\omega_i)$$

where $\omega_1, \ldots, \omega_n$ are the n-th roots of unity.

[*Hint.* Observe that $M = F(A)$ where

$$A = \left[\begin{array}{c|c} & I_{n-1} \\ \hline 1 & \end{array}\right]$$

and use Exercise 20.2.]

20.4 Let V be a non-zero finite-dimensional vector space over a field F and let $f : V \to V$ be a linear transformation. Let $V = \bigoplus_{i=1}^{n} V_i$ be the primary decomposition of V as a direct sum of f-stable subspaces. Prove that for each projection $\mathrm{pr}_i : V \to V_i$ there exists $g_i \in F[X]$ such that $\mathrm{pr}'_i = g_i(f)$ where $\mathrm{pr}'_i : V \to V$ is induced by pr_i.

Deduce that if W is an f-stable subspace of V then

$$W = \bigoplus_{i=1}^{n} (W \cap V_i).$$

20.5 Let V be a non-zero finite-dimensional inner product space. Prove that if ϑ is a mapping from $V \times V$ to the ground field of V then ϑ is an inner product on V if and only if there is a positive transformation $f : V \to V$ such that

$$(\forall x, y \in V) \qquad \vartheta(x, y) = \langle f(x) \mid y \rangle.$$

20.6 Let $a, b \in \mathbb{R}$ be such that $a < b$. If V is the real vector space of continuous functions $f : [a, b] \to \mathbb{R}$, prove that the mapping $Q : V \to \mathbb{R}$ described by

$$Q(f) = \int_a^b [f(x)]^2 \, dx$$

is a quadratic form on V.

20.7 Determine the rank and signature of the quadratic form $Q : \mathbb{R}^3 \to \mathbb{R}$ given by

$$Q(x,y,z) = 2x^2 - 4xy + 2xz + 3y^2 - 2yz + 4z^2.$$

20.8 For each of the following quadratic forms write down the symmetric matrix A for which the form is expressible as $\mathbf{x}^t A \mathbf{x}$. Diagonalise each of the forms and in each case find an invertible matrix P such that $P^t A P$ is diagonal with diagonal entries in $\{-1, 0, 1\}$:

(1) $x^2 + 2y^2 = 9z^2 - 2xy + 4xz - 6yz$;
(2) $4xy + 2yz$;
(3) $yz + xz + z^2 - 4t^2 + 2xy - 2xt + 6yz - 8yt - 14zt$.

Index

abelian group, 2
adjoint, 156
adjugate, 251
algebra, 4
alternating, 238
  subgroup, 236
alternator, 258
annihilating submodule, 120
annihilator, 264
artinian, 46

balanced, 194
basis, 80
Bessel's inequality, 150
bicentraliser, 179
bidual, 115
butterfly of Zassenhaus, 39

canonical
  basis, 81
  decomposition, 36
  form, 130
  injection, 61
  projection, 58
cartesian product module, 58
Cauchy-Schwartz inequality, 148
Cayley-Hamilton theorem, 313
centraliser, 179
chain conditions, 45
character module, 166
characteristic polynomial, 312
classical canonical matrix, 306
coefficient matrix, 133
cofactor, 251
cokernel, 42
column
  equivalent, 141
  operation, 134
  rank, 130
commutative, 2
  diagram, 24
companion matrix, 297

composition series, 50
conjugate
  isomorphism, 154
  transformation, 154
  transpose, 158
congruent, 348
coordinate form, 116
coproduct, 60
correspondence theorem, 34
cyclic, 41, 295

derivation, 260
determinant, 247
diagonalisable, 318
diagram chasing, 26
dimension, 91
direct
  sum (external), 61
  sum (internal), 66
  summand, 67
divisible abelian group, 165
division algebra, 5
dual basis, 116
dual module, 115

eigenspace, 319
eigenvalue, 318
eigenvector, 319
elementary
  divisor, 301
  divisor ideal, 281
  Jordan matrix, 306
  matrix, 134
endomorphism, 15
epimorphism, 15
equivalent matrices, 130
exact sequence, 25
exterior
  algebra, 241
  power, 239
  product, 240
external direct sum, 61

INDEX

field, 2
finite-dimensional, 92
finite height, 52
first isomorphism theorem, 36
five lemma, 27
flat, 206
formal power series, 5
four lemma, 26
Fourier coefficients, 152
free, 80

generator, 10
Gram-Schmidt orthonormalisation, 152
group, 2

height, 52
Hermite matrix, 137
hermitian, 159
homogeneous equation, 133
homomorphism, 15
homothety, 247

ideal, 40
idempotent, 70
image, 17
indecomposable, 280
injective, 163
inner product 147
   isomorphism, 153
   space, 147
internal direct sum, 66
invariant factor, 299
   ideal, 284
isomorphism, 15

Jacobson radical, 185
Jordan-Hölder tower, 50
Jordan
   canonical matrix, 307
   morphism, 307

kernel, 17
Krull's theorem, 90

Lagrange polynomial, 123
Laplace expansion, 250
lattice of submodules, 12
left R-module, 2
linear
   combination, 10
   equation, 139
   form (functional), 115
linearly independent, 80

matrix, 126
maximum condition, 45
minimum condition, 46
minimum polynomial, 294
modular law, 12
monomorphism, 15
morphism, 15
multilinear, 218

nilpotent, 190
noetherian, 45
norm, 148
normal, 328
   sequence, 284
nullity, 105

ortho-diagonalisable, 325
ortho-direct sum, 327
orthogonal, 149
   complement, 157
orthonormal basis, 151

Parseval's identity, 152
pivotal condensation, 261
p-module, 270
positive (definite), 354
primary component, 295
principal ideal domain, 263
product, 56
projection, 70
projective, 101
   lifting, 102

quadratic form, 348
quotient module, 33

rank, 105, 269
rational canonical matrix, 299
refinement, 49
regular ring, 212
relatively prime, 270
right R-module, 2
ring, 2
row-echelon matrix, 137
row-equivalent, 136
row
   matrix, 130
   operation, 134
   rank, 131

Schreier's refinement theorem, 49
Schur's lemma, 48
second isomorphism theorem, 37

seff-adjoint, 158
semi-exact sequence, 25
semigroup, 2
semisimple, 174
signature, 353
signum, 235
similar, 140
simple, 48, 172
skew-adjoint, 335
skew-symmetric, 335
snake diagram, 43
spectral resolution, 321
spectrum, 318
split exact sequence, 68
splitting morphism, 68
stairstep matrix, 137
standard inner product, 147
subalgebra, 8
subspace, 8
submodule, 8
   generated by subset, 10
substitution morphism, 291
supplementary, 67

sum, 11
symmetriser, 259
Sylvester's law of inertia, 353

tensor,
   algebra, 229
   product, 194
third isomorphism theorem, 39
torsion free, 211
tower of submodules, 49
trace, 228
triangle inequality, 148

unique factorisation domain, 270
unitarily similar, 159
unitary
   ring, 2
   transformation (matrix), 159

vector space, 2
von Neumann ring, 212

Wedderburn-Artin theorem, 181

Zassenhaus butterfly, 39